V. I. Kukulin, V. G. Neudatchin, I. T. Obukhovski, Yu. F. Smirnow

Clusters as Subsystems in Light Nuclei

D. F. Jackson

Direct Cluster Reactions – Progress Towards a Unified Theory

Clustering Phenomena in Nuclei

Edited by
K. Wildermuth
P. Kramer

Volume 3

Volume 1
Wildermuth/Tang
A Unified Theory of the Nucleus

Volume 2
Kramer/John/Schenzle
Group Theory and the Interaction
of Composite Nucleon Systems

Volume 3
Kukulin/Neudatchin/Obukhovski/Smirnov
Clusters as Subsystems in Light Nuclei
Jackson
Direct Cluster Reactions – Progress Towards a Unified Theory

V. I. Kukulin/V. G. Neudatchin/I. T. Obukhovski/
Yu. F. Smirnov

Clusters as Subsystems in Light Nuclei

D. F. Jackson

Direct Cluster Reactions – Progress Towards a Unified Theory

With 52 Figures

Friedr. Vieweg & Sohn Braunschweig / Wiesbaden

CIP-Kurztitelaufnahme der Deutschen Bibliothek

Clusters as subsystems in light nuclei/V. I. Kukulin ... Direct cluster reactions – progress towards a unified theory/D. F. Jackson. – Braunschweig; Wiesbaden: Vieweg, 1983.
(Clutering phenomena in nuclei; Vol. 3)
ISBN 3-528-08493-6

NE: Kukulin, V. I. [Mitverf.]; Jackson, Daphne F.: Direct cluster reactions – progress towards a unified theory; GT

Set by Vieweg, Braunschweig
Produced by W. Langelüddecke, Braunschweig
Cover design: Peter Morys, Salzhemmendorf
Printed in Germany-West

ISBN 3-528-08493-6

Preface

Volume 3 of the series deals with nuclear reactions on the basis of subsystem or cluster dynamics.

In the first contribution, the authors formulate this dynamics within the rigorous framework of many-body scattering theory due to Faddeev. The implications of the Pauli principle on cluster dynamics, the role of virtual cluster excitations, and the synthesis of this approach with the shell model are studied. New Approximation schemes are developed, compared to known computational methods, and applied to various types of nuclear reactions.

The theory of subsystem dynamics is then transfered to the subnuclear level by considering nucleons as being made out of quarks. The algebraic tools for systems of quarks are developed. Basic properties of the two-nucleon system are derived from an analysis of the system of six quarks.

The second contribution of this volume presents a review of direct cluster reactions. The specific features of these reactions are exposed and the corresponding reaction theory is discussed. The relation of these reactions to nuclear structure theory and corresponding models is analyzed, with special attention devoted to the spectroscopic factors. Applications include alpha-particle knock-out, many-body trasnfer, and heavy-ion peripheral reactions.

Karl Wildermuth
Peter Kramer

Tübingen, November 1982

Contents

Clusters as Subsystems in Light Nuclei

V. I. Kukulin, V. G. Neudatchin, I. T. Obukhovski, Yu. F. Smirnow

Direct Cluster Reactions – Progress Towards a Unified Theory

Daphne F. Jackson

V. I. Kukulin, V. G. Neudatchin, I. T. Obukhovski, Yu. F. Smirnov
Nuclear Research Institute, Moscow State University

Clusters as Subsystems in Light Nuclei

Contents

1 Introduction

The problem of clusters in atomic nuclei involves many different aspects. Our task here is to supplement substantially the already available good reviews [Sa77, Ho78, Wi77] by the material which, we hope, may turn out to be of a rather general importance. In the present review, we shall be concerned with two aspects.

The first of them is the eternal "naive" question of to what extent simple cluster potential model may be valid, in which, for certain specific situations, a system of n clusters ($n \ll A$) with the two-body potential interactions ($\alpha + \alpha$ as a two-body system, the ^{9}Be or ^{12}C nuclei, as three-body systems, etc.) is considered rather than a system of A nucleons. The central point here is the question of how far these models are compatible with the Pauli exclusion principle, which, as is well known, is of critical significance in substantiating and constructing a nuclear shell model.

The second aspect dealt with is the extraction of clusters, as subsystems in the nucleus, on the basis of a fully microscopic shell-model description.

Both the aspects indicated above are intimately related with each other. Indeed, the modern theory of many-body scattering enables one to reduce, quite rigorously, the problem of A particles to that of several clusters [Kr66, Be71], the interaction between which implicitly includes all the integrated degrees of freedom of the whole A-particle system. Such a formal approach including the coupling between many internal cluster excitations leads however to an unfeasible computational scheme. Therefore, for a further advance in the field under discussion, it is necessary to combine the rigorous N-particle equations and the heuristically valuable physical conceptions based on the nuclear shell model and deeply rooted in modern nuclear physics. Accordingly, the first part of the present review is an attempt to give an outline of such a synthesis. More specifically, we discuss the following problem: how, proceeding from the microscopic shell-model considerations, one may construct an interaction between clusters, so that, at the next stage, this cluster-cluster interaction might be used in a more complicated system of several clusters to be treated either on the basis of the rigorous equations of scattering theory, i.e., the Faddeev-Yakubovsky approach, or on the basis of the variational principle involving the use of a complete basis. Such a combined approach may serve as a broad foundation for solving a vast number of problems in modern nuclear physics, in particular, those of the theory of direct nuclear reactions with the deuterons and lithium nuclei (^{6}Li, ^{7}Li), which comprises elastic scattering, breakup, stripping, pickup, etc., the study of the structure of the ground and lowest excited states of practically all light nuclei [1]) and also many other problems. Obviously, in order to develop such a combined approach, one has to solve as the integral parts of the general assignment, the following problems:

i) how to take into account the Pauli exclusion principle at the levels of two-cluster and many-cluster systems, i.e., the modification of the Lippmann-Schwinger and

---

[1]) A simple analysis shows that, in the region of excitation energies not higher than 10 to 15 MeV, almost all p-shell nuclei are presentable as a systems of not more than four composite particles (for example, $^{11}B \rightarrow \alpha + \alpha + t$, $^{13}C \rightarrow 3\alpha + n$, $^{15}N \rightarrow 3\alpha + t$, etc.).

Faddeev-Yakubovsky equations enabling at least the main features of the Pauli principle to be taken into account;

ii) how to take into account the virtual cluster excitations within the framework of the integral equation formalism and, what is closely related to the two preceding problems,

iii) how to take into account the fact that the reduced witdths of clusters are different from unit.

These questions are discussed in the first three chapters of the present review. It will be noted that the formalism for taking account of excited cluster configurations within the framework of the shell model and the possibilities of experimental study of such states in reactions with high-energy particles have already been discussed in detail (see, a review [Ne79]). The above problem is in turn contiguous with the important field of cluster spectroscopy, which, during the last twenty years, has been developed significantly, starting with the first investigations [Ba59, Sm61], both in theory and in experiment [Ne69, Al75, Ho77, Ne79], and includes, in particular, the data on direct nuclear reactions with heavy ions, quasi-elastic cluster knockout, and cluster absorption of pions.

Chapter 2 of the present review has a formal character and contains the main facts of scattering theory necessary for understanding the subsequent material and the formalism of the recently proposed method of orthogonal projection in many-body scattering theory [Ku76a, 76b, 78], while Chapter 3 deals with a more specific material concerning a system made up of two nucleons in the field of the nucleus, with a allowance for the effects listed above (see, (i)–(iii)).

Chapter 4 describes two simplified models for cluster interaction (the model of orthogonal conditions OCM and the micriscopically-substantiated optical potential (MSOP) with forbidden states), which make it possible to take into account the main features of Pauli's exclusion principle and the potential character of cluster interaction. Although the resonating group method (RGM) is the best developed present-day model for a system composed of two or several clusters (see the detailed and comprehensive presentation of RGM theory in the monographs by Wildermuth and Tang [Wi77], by Kramer et al. [Kr79] and in the review by Horiuchi [Ho77]), it obviously is a rather tedious approach making difficult the separation of the principal features of cluster interaction from the less important ones. However, a remarkable progress is expected here, as far as we may except that out of the large number of exchange terms in such systems, as say, $\alpha + {}^{16}O$ or $\alpha + {}^{40}Ca$ only a certain limited number of them is essential for us, whereas the contribution of the remaining exchange terms can be treated using perturbation theory [Ku 77]. On the other hand, the RGM approach fails to take into account the sometimes important effects due to cluster breakup in a nuclear reaction. The simplified approach based on RGM, i.e. OCM has the same disadvantage. Therefore it will be meaningful to trace in detail the effects due to the Pauli exclusion principle together with effects of cluster break-up, taking as an example the 3N and 4N systems, which can be studied rigorously on the basis of the exact integral equations of Faddeev and Yakubovsky. In analyzing the structure of 3N- and 4N-integral equations from the point of view of the Pauli principle it was suprising to see that the expressions which are obtained in the orthogonal rearrangement of three- and four-body integral equations and corresponding iterative series are very similar to OCM-terms, but contrary to the latters have exact *dynamical sense* and take into account the break-up of the reaction fragments involved. A subsequent investigation [Ku 78c] has made it possible to elucidate the exact meaning of the forbidden states in the RGM and OCM and develop a generalized model of orthogonality conditions, which take into account (at least in part) the virtual cluster break-up in the process of scattering. This material is dealt with in Chapter 5. Chapter 6 discusses the problems of calculating three-cluster systems using the complete-basis variational principle. It also describes the procedure for construction of the complete variational basis in the orthogonal subspace of allowed states [Sm 74], as applied to the $^{12}C \rightarrow 3\alpha$ system. In § 2 of that chapter, an alternative approach toward the same problem is discussed on the example of a calculation of the structure of low-lying states in nuclei with A = 6 ($^6He - {}^6Li - {}^6Be$) in the $\alpha + 2N$ model. Here the complete basis is used instead of truncated one and the space is truncated by going over from the true Hamiltonian to the pseudo-Hamiltonian [Kr 74] with very large orthogonalizing coupling constant. Methods developed in the field of cluster physics have recently been elaborated to a significant extent and have found application in other fields of physics, which are distantly related or quite unrelated to cluster physics. One should mention the application of the method of orthogonal projecting to the problem of rearrangement of the Born series in scattering theory in order to make them convergent also at low energies [Ku 76, 78a], as well as the application of a very similar approach to calculate electron scattering from atoms and molecules [Lip 61, Bur 70, 72, etc.] and to calculate electron band structure in solids [Ju 72, Ku 78a]. One of the most interesting applications of cluster physics is its application to study the quark structure of the lightest nuclei. In fact, if quarks are fermions, as has been rather firmly established, and the wave function for a many-quark system should satisfy the generalized Pauli principle, the N–N system should be in some respects similar to the $^3H + {}^3He$ system well studied in cluster physics. However, the N–N system will undoubtedly display the fundamentally new effects as well, which are due to the colour degrees of freedom, specific nature of quark interaction, relativism, etc. Accordingly, in Chapter 7 we discuss the structure of a six-quark system and its relation to the N–N interactions, both strong and weak.

Recently, there have been made known interesting tentative data on resonances in the NN-system at excitation energies of about 400 MeV (for an analysis of these data, see [Ho 78]). This obviously supports the familiar theoretical works [Ja 78, Cr 75] which treated the 6 q system on the basis of the ideas of the shell model in the MIT bag (true, it was only the $(Os)^6$ configuration). A stimulating analog here is furnished by the $\alpha + \alpha$ system, the resonance levels of which are well reproduced by the shell model.

In one of the works by the present authors [Ob 79a] it has been demonstrated that, in contrast to what was supposed previously, the symmetry of the colour magnetic interaction between quarks admits of the colour magnetic attraction of nucleons (Jaffe [Ja 77] has earlier mentioned the colour magnetic attraction in the $\Lambda + \Lambda$ system). It arises because the orbital permutation symmetry with respect to quarks in the N–N system depends on the nature of the wave function for the relative motion of these nucleons. This fact is well known in cluster physics [Ku 71, Ku 75] and is illuminated in the first part of the present review. Indeed, in the oscillator representation one can readily obtain, for example, two nodes of the wave function in the $\alpha + \alpha$ freedom degree in the ground-state ^8Be with the shell-model structure $(Os)^4(1p)^4$, which corresponds to the symmetry [f] = [44], whereas the wave function without nodes would correspond to the $(Os)^8$ configuration with the symmetry [f] = [8]. Analogously in the 6q system, if the wave function in the N–N channel has a node, then we have the $(Os)^4(1p)^2$ configuration, which permits the orbital symmetry [f] = [42] and, as a consequence, a strong colour magnetic attraction between nucleons in this channel (it is natural to associate the resonances with this fact). The appearance of the s^4p^2-type configurations calls for the development of a many-body shell-model technique for many-quark systems [Ob 79b], which is what we discuss in this review. This technique is rather different from that used in the nuclear shell model, because now we have a longer chain of symmetry indices in various spaces.

We illustrate this technique by calculating the energy of attraction in going over from the configuration $(Os)^6$, [f] = [6] with the colour magnetic repulsion between nucleons to the configuration $(Os)^4(1p)^2$, [f] = [42] and also by calculating the matrix elements for a weak interaction [Du 79]. The last example is interesting, because this weak interaction (the parity nonconservation effects) can serve as a good "analyzer" just for the region of small distances in the N–N system [Du 78, Ne 78] and may be helpful in answering the question whether or not the nucleons "penetrate" into each other.

Another good illustration of the extent to which cluster effects are basically present in many-quark systems is the interesting problem of "hidden colour" in the NN-system, which has been pointed out by Matveev [Ma 78]. This is a complete analogy of virtual excited clusters in atomic nuclei [Go 76, Ne 79]. All this is the inavoidable consequence of the total antisymmetrization "in the constituent particles" in a system composed of several unexcited clusters in those cases when the Young schemes for the system in certain spaces are different from the completely (anti)symmetric ones. In a similar fashion, in the NN system (in particular, for the $(Os)^6$ quark configuration as well) there appears an "automatic" admixture of the $\Delta\Delta$-component [Sm 78].

A relativistic quark RGM calculation for the NN or $\pi\pi$ system is of course of current importance (for some representative estimates, see [Ri 78] and [Br 74]), by a shell-model treatment it yields a satisfactory microscopic picture in the most interesting region of overlapping of nucleons.

The present review is primarily based on studies carried out over the past five years by the sector of theoretical nuclear physics at Moscow State University, Institute of Nuclear Physics. The authors wish to thank their colleagues Dr. V. M. Dubovik, Dr. V. M. Krasnopol'sky, Dr. Yu. M. Tchuvil'sky and Dr. V. N. Pomerantsev for many illuminating discussions, helpful suggestions, and also for constructive criticism.

To sum up, we have illuminated the overall purpose and structure of the review and now we turn to the exposition of its physical content.

2 Quantum Theory of Scattering in the System of Few Composite Particles

2.1 On- and Off-Shell Scattering, Equation for Three-Body Scattering

Presented in this Section will some basic facts from the quantum theory of scattering in two- and three-body systems, which we are called upon to use below. Emphasis will be made to the facts concerning the systems of composite particles. The basic relations for two-body scattering will be presented first of all.

The equation for the resolvent $g(z) = (z - h)^{-1}$ can easily be obtained through simple algebraic operations [Sc74a]

$$g(z) = g_0(z) + g_0(z) v g(z) \tag{2.1a}$$

$$= g_0(z) + g(z) v g_0(z), \tag{2.1b}$$

where $g_0(z) \equiv (z - h_0)^{-1}$, h_0 is the free Hamiltonian; $h = h_0 + v$ is the total two-body Hamiltonian. If the scattering functions are determined [Go64] as usually:

$$\psi^{(\pm)} = \lim_{\epsilon \to 0} \pm i\epsilon g(E \pm i\epsilon)\psi_0 ,$$

where ψ_0 is the eigenfunction (EF) of the free Hamiltonian h_0 and eq. (2.1) is used, we can readily obtain the Lippmann-Schwinger equation (LSE) for the scattering wave functions $\psi^{(\pm)}$

$$\psi^{(\pm)} = \lim_{\epsilon \to 0} \pm i\epsilon[g_0(E \pm i\epsilon) + g_0(E \pm i\epsilon) v g(E \pm i\epsilon)]\psi_0 =$$

$$= \lim_{\epsilon \to 0} \pm i\epsilon g_0(E \pm i\epsilon)\psi_0 + \lim \pm i\epsilon g_0(E \pm i\epsilon) v g(E \pm i\epsilon)\psi_0 = \psi_0 + g_0 v \psi^{(\pm)}. \tag{2.2a}$$

In the momentum representation, we get:

$$\psi_k^{(\pm)}(p) = \psi_{0k}(p) + \left(E \pm i\epsilon - \frac{p^2}{2\mu}\right)^{-1} \int dp' v(p, p') \psi_k^{(\pm)}(p') , \tag{2.2b}$$

where $E = k^2/2\mu$ and $\psi_{0k}(p) = \delta(k - p)$.

After that, the t-operator (or t-matrix) will be introduced through the relation

$$S_{pp'} = \delta(p - p') - 2\pi i \delta\left(\frac{p'^2}{2\mu} - \frac{p^2}{2\mu}\right) \langle p' | t(E + i0) | p \rangle .$$

All information about the scattering process can be found in the t-operator which gives the differential cross section through the relation

$$\frac{d\sigma}{d\Omega} = (2\pi)^4 \mu^2 |\langle p' | t(E + i0) | p \rangle|^2 ,$$

where **p** and **p**′ are the momenta of the incident and ejected particles respectively. After substituting (2.1b) in (2.1a), we get the expression

$$t(z) = v + vg(z)v \, , \tag{2.3}$$

which relates the total resolvent $g(z)$ to the t-matrix. Eq. (2.3) can readily be used to find the off-shell t-matrix at $p'^2/2\mu \neq p^2/2\mu \neq E$. By combining again (2.1) and (2.3), we shall obtain the integral equation for the t-matrix:

$$t(z) = v + vg_0(z)t(z) \tag{2.4a}$$

$$= v + t(z)g_0(z)v \, , \tag{2.4b}$$

which are called LSE for t-matrix. This equation also permits the t-matrix to be off-shell continued. In particular, (2.4a) transformed to the momentum respresentation gives

$$t(\mathbf{p}, \mathbf{p}'; E) = v(\mathbf{p}, \mathbf{p}') + \int \frac{v(\mathbf{p}, \mathbf{p}'')\, t(\mathbf{p}'', \mathbf{p}'; E)}{\frac{p''^2}{2\mu} - E - i\epsilon}\, d\mathbf{p}'' \, , \tag{2.5a}$$

and we get after expanding in partial waves:

$$t_l(p, p'; E) = v_l(p, p') + \frac{1}{2\pi^2} \int \frac{v_l(p, p'')\, t_l(p'', p'; E)\, p''^2\, dp''}{\frac{p''^2}{2\mu} - E - i\epsilon} \, , \tag{2.5b}$$

whereupon we may write for the on-shell t-matrix

$$t_l(k, k; k^2/2\mu) \equiv t_l(k) = -\frac{1}{k}\, e^{i\delta_l(k)} \sin \delta_l(k) \, , \tag{2.6}$$

where $\delta_l(k)$ is the phase shift of scattering in the l-the partial wave.

Now, the question arises what is the physical meaning of the off-shell t-matrix? If a pure two-body scattering is experimentally studied, then $p^2/2\mu = p'^2/2\mu = E$ and none of the off-shell effects are observable. In other words, all information about the scattering is contained in the phase shifts $\delta_l(k)$ through which the on-shell t-matrix may be parametrized according to (2.6). After that, if but one of the momenta (say p' in eq. (2.5)) proves to be on the energy shell, while another off shell, then the value $t_l(p, k; k^2/2\mu)$ called the half-shell t-matrix may in principle be also observable experimentally [Ba71a], for example in case of bremsstrahlung. If LSEq. (2.5a) is written for the on-shell t-matrix, the kernel of the equation will comprise the half-shell t-matrix. As to the fully off-shell t-matrix, it will be contained in the equations for three-body scattering. This is explained by the fact that two particles involved in the processes of three-body scattering interact in the force field of the third particle and that it is such processes that give rise to the off-shell effects since, in this case, the energy conservation law is valid only for final and initial states and may be violated for the intermediate (or virtual) states thereby resulting in the escape from the energy shell.

An attempt will be made below to demonstrate that the inner part of the wave function determines the t-matrix off-shell continuation just to the extent to which the scattering function asymptotic behaviour determines the on-shell t-matrix.

The Schrödinger equation for the l-th partial wave in the coordinate representation may be written (using the units with $\hbar = 1$) as

$$\left[\frac{1}{r^2}\frac{d}{dr}\left(r^2\frac{d}{dr}\right)+k^2-\frac{2\mu l(l+1)}{r^2}\right]\psi_l^{(+)}(k,r)=2\mu\int_0^\infty v_l(r,r')\psi_l^{(+)}(k,r')r'^2\,dr' \tag{2.7}$$

where the potential $v_l(r,r')$ is nonlocal in the general case.

The solution $\psi_l^{(+)}(k,r)$ is constrained at the origin and has the asymptotic form[1])

$$\psi_l^{(+)}(k,r)\underset{r\to\infty}{\sim}\frac{e^{i\delta_l}}{kr}\sin\left(kr+\delta_l(k)-\frac{\pi l}{2}\right).$$

Since the off-shell t-matrix (2.6) may be written in the conventional form [Ne66]

$$t_l(k,k;k^2/2\mu)=2\mu\int_0^\infty dr\,r^2\,j_l(kr)\int_0^\infty v_l(r,r')\psi_l^{(+)}(k,r')r'^2\,dr'$$

then the half-shell t-matrix may be written as

$$t_l(p,k;k^2/2\mu)=2\mu\int_0^\infty dr\,r^2\,j_l(pr)\int_0^\infty v_l(r,r')\psi_l^{(+)}(k,r')r'^2\,dr'=$$

$$=\int_0^\infty dr\,r^2 j_l(pr)\left[\frac{1}{r^2}\frac{d}{dr}\left(r^2\frac{d}{dr}\right)+k^2-\frac{2\mu l(l+1)}{r^2}\right]\psi_l^{(+)}(k,r), \tag{2.7a}$$

where use was made of eq. (2.7).

Introduce now the phase-shifted free solution

$$\chi_l(kr)=e^{i\delta_l}[\cos\delta_l\,j_l(kr)+\sin\delta_l\,n_l(kr)].$$

Since, when acting on $\chi_l(k,r)$, the operator in square brackets of eq. (2.7a) gives zero, the eq. (2.7a) may be rewritten as

$$t_l(p,k;k^2/2\mu)=\int_0^\infty dr\,r^2\,j_l(pr)\left[\frac{1}{r^2}\frac{d}{dr}\left(r^2\frac{d}{dr}\right)+k^2-\frac{l(l+1)\cdot 2\mu}{r^2}\right]\times$$

$$\times\left(\psi_l^{(+)}(k,r)-\chi_l(k,r)\right). \tag{2.7b}$$

Introduce now the difference function

$$\Delta_l(k,r)=kr\,e^{-i\delta_l(k)}\left[\psi_l^{(+)}(k,r)-\chi_l(kr)\right] \tag{2.8}$$

which is vanishing at large r.

[1]) It will be noted that the normalization of this solution is different from the normalization given in numerous standard manuals. For example, the asymptotic normalization in [Wu62] differs from our normalization by a factor of $e^{i\delta}/r$.

After substituting eq. (2.8) in eq. (2.7b) and integrating twice in parts, we get the sought result [Pi 71].

$$t_l(p,k;k^2/2\mu) = \left(\frac{p}{k}\right)^l t_l(k) + (k^2-p^2)\frac{e^{i\delta_l}}{k}\int_0^\infty dr\, r\, j_l(pr)\Delta_l(k,r) \quad (2.9)$$

The result (2.9) is of importance from several viewpoints. First, it indicates that the half-shell (and eventually the fully off-shell) t-matrix (see below) is determined by the behaviour of the scattering wave function in the inner region, i.e. in the interaction region where the exact solution $\psi_l^{(+)}(k,r)$ differs from the asymptotic one $\chi_l(kr)$. In this case, the on-shell t-matrix $t_l(k)$ (and the phase shifts $\delta_l(k)$) are assumed to be *preset*. Second, the result (2.9) permits the half-shell t-matrix to be determined through only the difference function $\Delta_l(k,r)$ thus *avoiding a potential*. This fact makes it possible to construct the half-shell t-matrix by directly parametrizing the difference function $\Delta_l(k,r)$ [Pi 71]. Yet it is highly important to the cluster physics that the two-body off-shell t-matrix comprised in the kernels of three-body integral equations (see below) at a *preset on-shell behaviour* and, thereby, the solution for the three-body problem is determined by the structure of the scattering wave function $\psi_l^{(+)}(k,r)$ *in the cluster overlapping region.* This means that, among the vast diversity of the phenomenological potentials fitting the phase shifts of cluster-cluster scattering (say, in such systems as $\alpha+\tau$, $\alpha+\alpha$, etc.), only those should be selected which will satisfactorily reproduce the "true" structure of the scattering wave function $\psi_l^{(+)}(k,r)$ in the cluster overlapping region (such "true" structure may be obtained from microscopic calculations). The same concerns probably with the N–N interaction where quarks may play the role of constituents (see Chapter 7).

Convenient relationships between the half-shell and fully off-shell t-matrices has been established in [No65, Ko65]. Introduce the half-shell function

$$f_l(p,k) = \frac{t_l(p,k,k^2/2\mu)}{t_l(k,k;k^2/2\mu)} .$$

After that, the representation (2.9) was used in [No65, Ko65] to obtain the following useful representation for the fully off-shell t-matrix:

$$t_l(p,q;k^2/2\mu) = f_l(p,k)\, t_l(k)\, f_l(k,q) + R_l(p,q;k^2/2\mu)\,, \quad (2.10)$$

where the first term is separable, while the second term contributes only fully off-shell, i.e. R_l vanishes when *any* of the momenta (p or q) sets on the energy shell.

Presented below in brief (for detailed and comprehensive presentation, see [Sc 74a]) will be the formalism for three-body scattering which is necessary to understanding the material of the subsequent Chapters.

First, we shall obtain the Faddeev equations (FE) for the Green functions (GF). Similarly to the two-body problem (see eq. (2.1)), we may write for three-body GF:

$$G = G_0 + G_0VG \quad (2.11a)$$

$$= G_0 + GVG_0\,, \quad (2.11b)$$

where $V = v_1 + v_2 + v_3$, $G_0 = (z-H_0)^{-1}$; H_0 is the free three-body Hamiltonian. The equation (2.11a) may be rewritten as

$$G = G_0 + G^{(1)} + G^{(2)} + G^{(3)}, \quad (2.12)$$

where

$$G^{(i)} \equiv G_0 v_i G, \qquad i = 1, 2, 3 . \tag{2.13}$$

After substituting eq. (2.12) in eq. (2.13), transposing the diagonal part to the left, and inverting it (using the determination of the t-matrix $t_i = (1 - v_i G_0)^{-1} v_i$ and the partial GF $G_i(z) = (z - H_0 - v_i)^{-1}$), we find

$$G^{(i)}(z) = G_i(z) - G_0(z) + \sum_{j=1}^{3} G_0(z) F_{ij}(z) G^{(j)}(z) , \tag{2.14}$$

where the Faddeev kernel

$$F_{ij}(z) = (1 - \delta_{ij}) t_j ; \quad i, j = 1, 2, 3$$

contains zeroes in diagonal. After that, the relationship between GF and t-matrix can readily be used to find FE for the components of the total three-body t-matrix:

$$T = T^{(1)} + T^{(2)} + T^{(3)}$$

and

$$T^{(i)}(z) = t_i(z) + \sum_{j=1}^{3} F_{ij}(z) G_0(z) T^{(j)}(z) . \tag{2.15}$$

The operator FE written above for the components of the three-body t-matrix may be visually illustrated by the following diagrams:

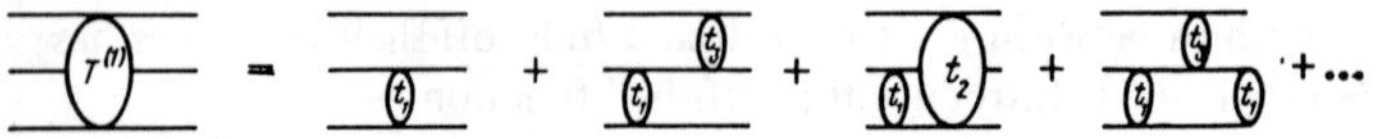

Fig. 1 The graphical representation of the first terms in iterative series for FE's.

Owing to the specific form of the Faddeev kernel, only the t-matrices of *different* two-body sybsystems will follow each other in succession, the fact that will eventually give a connected (i.e. Fredholm-type) kernel. It is obvious from the diagrams presented above that all intermediate two-body scatterings are off the energy-shell and, therefore, the problem of the correct off-shell behaviour of two-body t-matrices becomes to be of decisive importance.

The above presented three-body equations have been proved strictly in [Fa63] to be the Fredholm integral equations which may be solved by the known numerical methods. We shall use them below as the starting point for deriving the equations which describe the system of three composite particles including the Pauli principle and the internal virtual excitations of the particles.

2.2 The Method of Orthogonal Projecting in Two-Body Scattering; Construction of the Off-Shell t-Matrix and Green Function in Orthogonal Subspaces

We shall slightly digress here from the physical essence of the problem and give consideration to the pure formal problem of constructing the scattering operators in an orthogonal subspace H_Q of a certain full Hilbert subspace H. The main difficulty with such problem is that the plane-wave basis which is always used in the scattering theory [Ne66, Go64] in the initial, final, and intermediate states *cannot, despite its being complete, satisfy the property of orthogonality to* H_Γ. If, however, the scattering theory is constructed using so called orthogonalized plane waves (OPW) [Zi64], significant difficulties arise in the theory since such orthogonalized waves *are not orthogonal to each other* [Zi64] and form an overcomplete set in H_Q.

Thus, the essential difficulty is that one has to construct the full (in the subspace H_Q which is the orthogonal addition to H_Γ, so that $H_\Gamma \oplus H_Q = H$) basis of the free solutions for the Schrödinger equation

$$(h_0 - \epsilon)\widetilde{\psi}_0 = 0\,, \quad \Gamma\widetilde{\psi}_0 = 0 \tag{2.16}$$

(where h_0 is the free Hamiltonian; Γ is the projector on the "forbidden" subspace H_Γ) and the functions $\widetilde{\psi}_0$ should be orthogonal to each other (at different energies) and to the subspace H_Γ (see eq. (2.16)).

The required basis $\{\widetilde{\psi}_0\}$ has successfully been constructed [Sc73, 74b] using the Feshbach projection formalism [Fe58, 62]. The prime idea of the construction was very simple. The solutions for eq. (2.16) are assumed to be the solutions for equation

$$(Qh_0Q - \epsilon)\widetilde{\psi}_0 = 0\,, \tag{2.17}$$

where the projection operator $Q = 1 - \Gamma$. In this case, the new Hamiltonian Qh_0Q is hermitian and herefore the solutions for eq. (2.17) form, as always, the complete and orthogonal basis in H_Q. Yet the actual construction of the functions $\widetilde{\psi}_0(\mathbf{k})$ involves fairly tedious, through simple, transformations [Sc73, 74b] and gives

$$\widetilde{\psi}_0(\mathbf{k}) = \psi_0(\mathbf{k}) - g_0(\epsilon)\Gamma(\Gamma g_0(\epsilon)\Gamma)^{-1}\Gamma\psi_0(\mathbf{k})\,, \tag{2.18}$$

where $\epsilon = k^2$ (use is made of the system of units $\hbar = 2m = 1$); $\psi_0(\mathbf{k}) = (2\pi)^{3/2}\exp(i\mathbf{kr})$ is the plane wave. The designation $\Gamma(\Gamma g_0(\epsilon)\Gamma)^{-1}\Gamma$ in case of m-dimensional projector $\Gamma = \sum_{i=1}^{m} |\varphi_i\rangle\langle\varphi_i|$ means

$$\Gamma(\Gamma g_0(\epsilon)\Gamma)^{-1}\Gamma_{-1} = \sum_{ij} |\varphi_i\rangle(\langle\varphi|g_0(\epsilon)|\varphi\rangle)^{-1}_{ij}\langle\varphi_j|\,,$$

where $(\langle\varphi|g_0(\epsilon)|\varphi\rangle)^{-1}_{ij}$ denotes the ij-th element of the matrix inverse to the matrix

$$M = \left\{\begin{matrix} \langle\varphi_1|g_0(\epsilon)|\varphi_1\rangle \ldots \langle\varphi_1|g_0(\epsilon)|\varphi_m\rangle \\ \\ \langle\varphi_m|g_0(\epsilon)|\varphi_1\rangle \ldots \langle\varphi_m|g_0(\epsilon)|\varphi_m\rangle \end{matrix}\right\}. \tag{2.19}$$

The eq. (2.18) gives immediately the required orthogonality

$$\Gamma\widetilde{\psi}_0(\mathbf{k}) = 0 \ \text{ at all } \mathbf{k}\,.$$

In case of one-dimensional projector $\Gamma = |\varphi\rangle\langle\varphi|$, the expression (2.18) can readily be used to find:

$$\widetilde{\psi}_0(\mathbf{k}) = \psi_0(\mathbf{k}) - \frac{g_0(\epsilon)|\varphi\rangle\langle\varphi|\psi_0(\mathbf{k})\rangle}{\langle\varphi|g_0(\epsilon)|\varphi\rangle} \tag{2.18a}$$

whence one can clearly seen the differences of the basis $\widetilde{\psi}_0(\mathbf{k})$ from the basis of orthogonalized plane waves (OPW):

$$\psi_{OPW}(\mathbf{k}) = \psi_0(\mathbf{k}) - |\varphi\rangle\langle\varphi|\psi_0(\mathbf{k})\rangle\,. \tag{2.20}$$

In particular (and contrary to $\psi_{OPW}(\mathbf{k})$), the functions $\widetilde{\psi}_0(\mathbf{k})$ in the representation of partial waves exhibit the asymptotic behaviour $\widetilde{\psi}_{0,l}(\mathrm{k},\mathrm{r}) \sim A_l \sin(\mathrm{kr} - \pi l/2 + \widetilde{\delta}_l)$ where δ_l is the phase shift. In other words, the basis $\{\widetilde{\psi}_0(\mathbf{k})\}$ displays the properties of distorted (owing to the requirement of orthogonality) waves and is therefore called elsewhere [Sc 73, 74b] the orthogonality scattering basis. We shall return below to discussing the basis because of its other numerous useful properties.

We shall turn now to constructing the scattering operators (t-matrix, GF, etc.) in the orthogonal subspace H_Q. Use may be made in this case of the orthogonality scattering basis $\{\widetilde{\psi}_0(\mathbf{k})\}$. However, the method of orthogonalizing pseudopotentials (OPM) proposed in [Kr 74, Ku 76a] is more convenient and universal. In this approach, we are not in need of constructing first a functional basis in H_Q and, after that, writing the scattering operators in the basis constructed. Instead, the construction as a whole is made directly in the operator form, thus avoiding the evident construction of the orthogonality scattering basis which may also be constructed using the above mentioned method, if desired.

In the conventional approach [Fe 58, 62], the main difficulty is to find the procedure of inverting the operator in the orthogonal subspace. For example, the complete GF gives rise to the operator

$$\widetilde{G} = Q\,[Q(E-H)Q]^{-1}\,Q \tag{2.21}$$

with $Q = 1 - \Gamma$ has to be constructed and allowance should be made for the case where Γ is an *infinite-dimensional projector.*

The idea of the orthogonalizing pseudopotential method is quite clear and bears close relation to the concept of penalty functions in the theory of dynamic programming [Glo 76]. The general approach traces its roots in Courant's work of 1943 [Co43] where it was used to simplify the solution for the variational problem of vibrations of a plane with fastened boundary.

The reasoning may approximately develop as following.

Let some part of the space (a part of the Hilbert space in our case and a part of the space of the optimization parameters in the optimization theory), be forbidden or inaccessible by a dynamic system. It proves convenient in this case rather to introduce a penalty function in the examined equation and then solve the equation *in the full space* than to directly solve problem in the orthogonal subspace H_Q of the full Hilbert space H ($H = H_\Gamma \oplus H_Q$). The penalty function should be constructed in such a way that, in case the equation is solved in the full space and the state vector appears in the forbidden subspace H_Γ the function would "displace'' the dynamic system into the allowed subspace H_Q.

In case of the Schrödinger equation

$$(H - E)\,\psi = 0$$

with r additional conditions of orthogonality

$$\langle \varphi_1 | \psi \rangle = 0, \;\; \langle \varphi_2 | \psi \rangle = 0, \ldots \langle \varphi_r | \psi \rangle = 0$$

the separable potential (finite-rank operator)

$$V_{OP} = \lambda\Gamma$$

with

$$\Gamma = \sum_{i=1}^{r} |\varphi_i\rangle\langle\varphi_i|$$

can conveniently be used as a penalty function, with the coupling constant $\lambda \to \infty$. Accordingly, the operator $\widetilde{G}$ from eq. (2.21) is given by the equality [Ku 76a, b]

$$\widetilde{G} = \lim_{\lambda \to \infty} (E - H - \lambda\Gamma)^{-1} \tag{2.22}$$

irrespectively of the nature of the Hamiltonian H (for example of the number of particles in the system) and of the dimension of the projector Γ. Using the standard operator identity

$$\widetilde{G} = G + \lambda G\,\Gamma\widetilde{G}$$

and then multipying by Γ, we get

$$(1 - \lambda\Gamma\,G\Gamma)\,\Gamma\widetilde{G} = \Gamma G\,,$$

whence

$$\Gamma\widetilde{G} = (\Gamma(1 - \lambda G)\Gamma)^{-1}\,\Gamma G$$

and, after that,

$$\widetilde{G}(\lambda) = G + G\Gamma\,[\Gamma(\lambda^{-1} - G)\Gamma]^{-1}\,\Gamma G\,. \tag{2.23}$$

Turning to the limit $\lambda \to \infty$, we can readily find the eventual expression

$$\widetilde{G} = G - G\Gamma\,(\Gamma G\Gamma)^{-1}\,\Gamma G \tag{2.24}$$

whence it is clear that GF constructed in H_Q exhibits the required orthogonality properties:

$$\Gamma\widetilde{G} = \widetilde{G}\Gamma = 0\,.$$

Therefore, the constructed GF (2.24) will be called henceforth the projected GF.

The example of the (2.23) → (2.24) transition shows immediately that the OPM is advantageous over the known Feshbach method in that all the intermediate calculations may be made with the finite constant λ (which means the *operations with full basis*), whereas the limiting transition $\lambda \to \infty$ (i.e. the transition to orthogonal subspace) can be made already in the eventual formulas. In other words, the pseudohamiltonian $\widetilde{H}_\lambda = H + \lambda\Gamma$ is introduced instead of the initial Hamiltonian H; after that, all the conventional formulas of the scattering theory are written for $\widetilde{H}_\lambda$ and then the limiting transition is made in the eventual results. In its essence, such approach resembles much the classical method for introducing the correct boundary conditions in the scattering operators through the small imaginary addend to the energy:

$$\frac{1}{\epsilon - h} + \begin{pmatrix}\text{correct}\\ \text{boundary}\\ \text{conditions}\end{pmatrix} \xrightarrow{\quad\quad\quad} \text{corresponds to} \quad \lim_{\eta \to 0} \frac{1}{\epsilon \pm i\eta - h}.$$

In this case, the action of the operator in the intermediate stages is examined with finite η, while the limit $\eta \to 0$ is taken in only the eventual expressions [Go64]. Evidently, the meaning of all the formulas at finite η has been determined exactly.

Quite the same situation holds in OPM. Apart from offering the possibility of strict determination of the operators in an orthogonal subspace, OPM proves to be handy in practice. Consider as an example the solution for the Schrödinger equation in the subspace H_Q (the Saito problem [Sa68, 69]),

$$(E - H)\psi_E = 0; \qquad \langle \varphi | \psi_E \rangle = 0, \tag{2.25}$$

where $\Gamma = |\varphi\rangle\langle\varphi|$. Following OPM we shall introduce the pseudo-Hamiltonian $H_\lambda = H + \lambda|\varphi\rangle\langle\varphi|$ and solve the modified Schrödinger equation

$$(E - \widetilde{H}_\lambda)\widetilde{\psi}_E = 0 \tag{2.26}$$

yet *without additional constraints.* It follow from eq. (2.26) that

$$(E - H)\,|\widetilde{\psi}_E\rangle = \lambda\,|\varphi\rangle\langle\varphi|\widetilde{\psi}_E\rangle$$

whence

$$|\widetilde{\psi}_E\rangle = |\psi_E\rangle + C(\lambda)G(E)|\varphi\rangle,$$

where $G(E) = (E - H)^{-1}$ and the constant

$$C(\lambda) = \frac{\lambda\langle\varphi|\psi_E\rangle}{1 - \lambda\langle\varphi|G(E)|\varphi\rangle}$$

Turning now to the extreme $\lambda \to \infty$, we find

$$\lim_{\lambda \to \infty} C(\lambda) = -\frac{\langle\varphi|\psi_E\rangle}{\langle\varphi|G(E)|\varphi\rangle}$$

and, eventually, the complete solution

$$|\widetilde{\psi}_E\rangle = \lim_{\lambda \to \infty} |\widetilde{\psi}_E(\lambda)\rangle = |\psi_E\rangle - \frac{G(E)\,|\varphi\rangle\langle\varphi|\psi_E\rangle}{\langle\varphi|G(E)|\varphi\rangle} \tag{2.27}$$

which immediately gives the necessary requirement of orthogonality

$$\langle\varphi|\widetilde{\psi}_E\rangle = 0, \qquad \lambda \to \infty.$$

The orthogonality scattering functions $\widetilde{\psi}_0$ (k) (2.18a) are immediately obtainable by using the free Hamiltonian h_0 from eq. (2.27) instead of the total Hamiltonian H. The possibility of directly constructing the operators in orthogonal subspace is an important advantage of the described approach, in particular for off-shell amplitudes, since the off-shell behaviour of the scattering amplitudes in two-body subsystems is essential to applications, especially to the systems of few composite particles. There exist numerous methods of orthogonal projections which are on-shell equivalent to each other but yield *different off-shell results* (i.e. scattering operators). In particular, one of the widely used methods for solving the problem (2.25) is the transition to QHQ operator, i.e. *solution of the equation*

$$(E - QHQ)\,\widetilde{\psi}_E = 0 \tag{2.28}$$

where $Q = 1 - \Gamma$. Just such an equation was used by Saito [Sa68, 69] to describe the two-cluster interaction.

GF of eq. (2.28) $G^s(E) = (E - QHQ)^{-1}$

i.e. the resolvent of QHQ operator, can be found from the identity

$$G^s = G - G(\Gamma H + H\Gamma - \Gamma H\Gamma)G^s$$

whence

$$G^s = G - G\Gamma(\Gamma G\Gamma)^{-1}\Gamma G + \Gamma/E \tag{2.29}$$

In other terms, $G^s = \widetilde{G} + \Gamma/E$ and differs from our projected GF $\widetilde{G}$ by the term Γ/E, i.e. by N-multiple pole at $E = 0$ (N-dimension of the projector Γ). This extra pole will result in a strong difference of G^s from $\widetilde{G}$ off the energy shell. Such singularity of G^s results actually from the fact that any vector $\varphi \in H_\Gamma$ may be added to the solution $\widetilde{\psi}_E$ in eq. (2.28) at $E = 0$, i.e. an ambiguity appears in the on-shell solution for eq. (2.28) (though the asymptotic behaviour of $\widetilde{\psi}_E$ proves to be invariant), which results in the off-shell singularity (see section 1 above).

The above remarks are to prelude the construction of the projected off-shell t-matrix. Let the modified t-operator be defined by the conventional relation

$$\widetilde{t} = -G_0^{-1} + G_0^{-1}\widetilde{G}G_0^{-1} \tag{2.30}$$

whence

$$G_0\widetilde{t} = \widetilde{G}\widetilde{V} = \lim_{\lambda\to\infty} \widetilde{G}_\lambda(V + \lambda\Gamma)$$

$$\widetilde{t}\,G_0 = \widetilde{V}\widetilde{G} = \lim_{\lambda\to\infty} (V + \lambda\Gamma)\widetilde{G}_\lambda\,. \tag{2.31}$$

Such products are the kernels in FE (see in Section 3) and, therefore, explicit expressions of them will be obtained. We shall find first the λ-extreme in eq. (2.31). With this purpose, $\widetilde{G}(\lambda)$ should be expanded in powers of $1/\lambda$ using eq. (2.23):

$$\widetilde{G}(\lambda) = \widetilde{G} - \lambda^{-1}\,G\Gamma(\Gamma G\Gamma)^{-2}\,\Gamma G + \dots$$

whence

$$\lim_{\lambda\to\infty} \widetilde{G}\lambda\Gamma = -G\Gamma(\Gamma G\Gamma)^{-1}\,\Gamma$$

and, by substituting in (2.31), we get eventually

$$\lim_{\lambda\to\infty} \widetilde{G}_\lambda\widetilde{V}_\lambda = \widetilde{G}V - G\Gamma(\Gamma G\Gamma)^{-1}\,\Gamma\,. \tag{2.32}$$

It will be noted that, apart from the projected t-matrix (2.30), it is possible to introduce some additional similar expressions [Po78] which are coincident with $\widetilde{t}$ on the energy shell but different in the off-shell behaviour. The more natural definition (2.30) will be used henceforth. It is the actually projected operators (2.24) and (2.30) and the operator (2.32) that should be used in the problem of few composite particles whose set of equations will be derived in the next Section.

2.3 Faddeev Equations (FE) in the Orthogonality Condition Model; Construction of Many-Body Projection Operators

2.3.1 Derivation of orthogonalized FE

Consideration will be given to a system of three clusters the interaction between which can be described in terms of the orthogonality condition model (OCM). In the general case, this procedure is, generally speaking, insufficient because the successive reduction of A-body problem to the problem of three or more clusters gives rise to both the two-cluster and three-cluster, etc. exchange interactions. It seems to be reasonable that the major contribution from such many-body exchange forces be included through the orthogonality projection but now to the three-cluster forbidden states.

For the sake of simplicity we confine ourselfes here the inclusion of two-cluster exchanges, whereas the many-body exchange interactions, through the orthogonality projection has been considered in [Ku76b, 77a; Po78].

Thus, our model will be formulated as follows. The system Hamiltonian is

$$H_{123} = H_0 + V_1 + V_2 + V_3 \,, \tag{2.33}$$

where H_0 is the operator of kinetic energy; $V_k \equiv V_{ij}$ (ijk = 123, 231, 312) denotes the two-body interactions. Taken as the latter may be either the two-cluster folding-potentials V_D or the microscopically substantiated cluster-cluster optical potentials proposed in [Ne71, 72; Ku75b] (see Chapter 5).

We have to solve the three-body Schrödinger equation

$$(H_{123} - E)\,\psi_{123} = 0 \tag{2.34}$$

including the requirement of orthogonality

$$\Gamma_1 \psi_{123} = 0, \qquad \Gamma_2 \psi_{123} = 0, \qquad \Gamma_3 \psi_{123} = 0, \tag{2.35}$$

where the two-body projectors Γ_k are constructed on the forbidden states in the k-th two-body subsystem (k = 1, 2, 3).

It is also necessary to include the boundary conditions relevant to a specific problem (elastic or inelastic scattering of a particle by two-cluster system, break-up in final state, etc.). Now, the projectors Γ_k are infinite dimensional and can be written for example, in the coordinate representation as

$$\Gamma_k = \sum_n \varphi_k^{(n)}(\mathbf{r}_k)\,\varphi_k^{(n)*}(\mathbf{r}'_k)\,\delta(\rho_k - \rho'_k), \tag{2.36}$$

where $(\mathbf{r}_k, \rho_k)$ is the Jacobi coordinates in the k-th channel; the sum in n is taken for the forbidden states of the k-th pair. Bearing in mind the most general formulation of the three-body problem admitting any possible boundary conditions, the Schrödinger equation

(2.34) has to be converted to the FE set retaining the orthogonality conditions (2.35). With this purpose, we shall follow the common rule and replace the two-body potentials V_k with the pseudopotentials

$$\widetilde{V}_k = V_k + \lambda_k \Gamma_k\,, \quad (\lambda_k \text{ are finite})\,. \tag{2.37}$$

This procedure permits us to proceed to the three-body problem without constraints, i.e. to the complete basis in intermediate states. Thus, the standard scheme of FE derivation proves to be also applicable here so that the correct solutions exhibiting the required orthogonality properties (2.35) can be obtained in the eventual results when the extremes $\lambda_k \to \infty$ (k = 1, 2, 3) are taken. For the sake of completeness, we shall reproduce here the scheme for deriving FE (see Section 1 above where the true potentials are to be replaced by the pseudopotentials). First, we shall introduce the pseudohamiltonian $\widetilde{H}_{123} = H_0 + \widetilde{V}_1 + \widetilde{V}_2 + \widetilde{V}_3$ and the appropriate GF, namely the total GF $\widetilde{G} = (E - \widetilde{H})^{-1}$ and channel GF $\widetilde{G}_i = (E - H_0 - \widetilde{V}_i)$ (i = 1, 2, 3). After that, the standard identities

$$\widetilde{G} = G_0 + G_0 \widetilde{V}\widetilde{G}\,, \tag{2.38}$$

where

$$\widetilde{V} = \widetilde{V}_1 + \widetilde{V}_2 + \widetilde{V}_3 \tag{2.39}$$

and

$$\widetilde{G}_i = G_0 + G_0 \widetilde{V}_i \widetilde{G}_i \tag{2.40}$$

will be used, whence

$$\widetilde{G}_i = (1 - G_0 \widetilde{V}_i)^{-1} G_0\,. \tag{2.41}$$

Substituting (2.39) in (2.38), we get the Faddeev partition

$$\widetilde{G} = G_0 + \widetilde{G}^{(1)} + \widetilde{G}^{(2)} + \widetilde{G}^{(3)} \tag{2.42}$$

where

$$\widetilde{G}^{(i)} \equiv G_0 \widetilde{V}_i \widetilde{G}\,. \tag{2.43}$$

After that, by substituting again (2.42) in (2.43), transposing the diagonal terms to the left, and converting the corresponding diagonal parts with due account of (2.41), we find:

$$\widetilde{G}^{(i)}(\lambda) = \widetilde{G}_i(\lambda)\, \widetilde{V}_i(\lambda)\, (G_0 + \widetilde{G}^{(j)}(\lambda) + \widetilde{G}^{(k)}(\lambda))\,, \qquad (ijk = 123, 232, 312). \tag{2.44}$$

The argument λ has been introduced here to emphasize that we still examine the case of finite λ.

Turn now to the limiting transition where all $\lambda_i \to \infty$. Using the ready result (2.32) we find eventually

$$\widetilde{G}^{(i)} = [\widetilde{G}_i V_i - G_i \Gamma_i (\Gamma_i G_i \Gamma_i)^{-1} \Gamma_i]\, (G_0 + \widetilde{G}^{(j)} + \widetilde{G}^{(k)})\,. \tag{2.45}$$

The equation (2.45) just constitutes the modified set of FE satisfying the necessary requirements (2.35). In fact, on acting in (2.45) from the left by the projector Γ_i (using $\Gamma_i \widetilde{G}_i = 0$): $\Gamma_i \widetilde{G}^{(i)} = -\Gamma_i (G_0 + \widetilde{G}^{(j)} + \widetilde{G}^{(k)})$ or $\Gamma_i (G_0 + \widetilde{G}^{(i)} + \widetilde{G}^{(j)} + \widetilde{G}^{(k)}) = 0$ and considering (2.42), we find

$$\Gamma_i \widetilde{G} = 0; \quad (i = 1, 2, 3)\,. \tag{2.46}$$

In other words, the total GF found from the solution of the modified FE (2.45) proves to be orthogonal to all forbidden states in all three two-body subsystems $i = 1, 2, 3$.

The coupling constants λ have been labelled with subscript i to include the particular cases (see below) where the orthogonalization has to be made not in all the two-body interactions, for example if we have but a single composite particle (nucleus), while the other two particles are simple nucleons. In this case, the corresponding λ_i may be assumed to be zero. The conventional procedure [Sc74a] can be used to obtain FE for the wave functions from (2.45), and the complete wave function

$$\widetilde{\Psi} = \lim_{\epsilon \to 0} i\epsilon \widetilde{G}(E + i\epsilon)\Phi$$

proves naturally to be orthogonal to all the forbidden two-body states $\varphi_i^{(n)}$. For example, in case of scattering of particle 1 by the bound state of clusters (2,3), we get

$$\widetilde{\Psi}_1 = \widetilde{\psi}_1^{(1)} + \widetilde{\psi}_1^{(2)} + \widetilde{\psi}_1^{(3)}$$

and

$$\widetilde{\psi}_1^{(i)} = \widetilde{\phi}_1 \delta_{i1} + (\widetilde{G}_i V_i - G_i \Gamma_i (\Gamma_i G_i \Gamma_i)^{-1} \Gamma_i)(\widetilde{\psi}_1^{(j)} + \widetilde{\psi}_1^{(k)}) \tag{2.46a}$$

where the initial function $\widetilde{\phi}_1 = \widetilde{\phi}_1(\mathbf{k}_1)\,\delta(\mathbf{p}_1 - \mathbf{p}_1^0)$ ($\mathbf{p}_1^0$ is the momentum of the projectile) should be taken including the orthogonality conditions in subsystem 1, i.e. the bound state function $\widetilde{\phi}_1(\mathbf{k}_1)$ should be found considering the requirement of orthogonality to the forbidden states in subsystem 1.

In case the cluster interaction can be described by a microscopically substantiated optical potential (MSOP) [Ne71, 72; Ku75b], the functions of both the forbidden states $\varphi_i^{(n)}$ and the allowed states $\widetilde{\phi}_i$ are the eigenfunctions of two-body Hamiltonians h_i $(i = 1, 2, 3)$. In this case, the operators Γ_i and G_i commute, and it can readily be shown that

$$\widetilde{G}_i = (1 - \Gamma_i)\, G_i\,,$$

i.e. the terms corresponding to the forbidden states are excluded from the spectral representation of two-body GF, whereupon the FE renormalization is especially simple, namely the addend to the kernel is

$$\Delta R_i = \widetilde{G}_i \widetilde{V}_i - G_i V_i = \sum_n |\varphi_i^{(n)}\rangle\langle\varphi_i^{(n)}| \left(\frac{V_i}{E - E_i^{(n)} - p_i^2/2\mu_i} + 1 \right) \delta(\mathbf{p}_i - \mathbf{p}_i')\,,$$

where $E_i^{(n)}$ are the energies of the forbidden states.

2.3.2 Construction of many-body projectors on the basis of FE

It will be shown in this brief Subsection how the formulas for three-body projectors can be derived from the modified FE. The basic problem when constructing the total many-body projectors Γ from the two-body projectors Γ_i is the noncommutativity of different Γ_i, i.e. $[\Gamma_i \Gamma_j]_- \neq 0$. Therefore, the general formula for Γ has to be derived using particular technique. The total projector Γ is determined by the conditions

$$\Gamma_i \Gamma = \Gamma \Gamma_i = \Gamma, \qquad \Gamma^2 = \Gamma \tag{2.47}$$

It will be shown that, if Γ_i and Γ satisfy eq. (2.47), then

$$\lim_{\lambda\to\infty} (K-\lambda\Gamma)^{-1} = \lim_{\lambda\to\infty} \left(K-\lambda \sum_{i=1}^{3} \Gamma_i\right)^{-1} \tag{2.48}$$

for *any* operator K. Use will be made of the standard identity

$$\widetilde{K}_\lambda^{-1} \equiv \left(K-\lambda \sum_{i=1}^{3} \Gamma_i\right)^{-1} = (K-\lambda\Gamma)^{-1} - (K-\lambda\Gamma)^{-1}\,\lambda\Delta\Gamma \times \widetilde{K}_\lambda^{-1}\,, \tag{2.49}$$

where $\Delta\Gamma = \Gamma - \sum_{i=1}^{3} \Gamma_i$. Turning to the limit $\lambda\to\infty$ in eq. (2.49) and considering that $\lim_{\lambda\to\infty} (K-\lambda\Gamma)^{-1}\,\lambda\Delta\Gamma$ is finite and the $\lim_{\lambda\to\infty} \Delta\Gamma\,\widetilde{K}_\lambda^{-1} = 0$ (this relation follows from eq. (2.45) if one puts $V_i = 0$ and $E - H_0 = K$), we get the relation (2.48).

The formula (2.48) shows visually that the orthogonality projecting method described here makes it possible to do with only two-body projectors Γ_i *without constructing the total projector* Γ; in this case, the eventual result will be the same as that *obtained with the total projector.*

If $K = 1$ is taken in eq. (2.48) and the known relation for the projector resolvent

$$(1-\lambda\Gamma)^{-1} = (1-\Gamma) + \Gamma/(1-\lambda)$$

is used, we get the following handy formula for the total projector:

$$1-\Gamma = \lim_{\lambda\to\infty} (1-\lambda\Gamma)^{-1} = \lim_{\lambda\to\infty} \left(1-\lambda \sum_{i=1}^{3} \Gamma_i\right)^{-1}. \tag{2.50}$$

If now the Faddeev reduction

$$R(\lambda) = 1 + R_1 + R_2 + R_3\,,$$

where

$$R_i \equiv \lambda\Gamma_i R(\lambda) = \lambda\Gamma_i (1 + R_1 + R_2 + R_3)$$

is introduced for the resolvent $R(\lambda) = (1-\lambda \sum_{i=1}^{3} \Gamma_i)^{-1}$ then the simple transformations similar to those used when deriving eq. (2.45) will give in the limit $\lambda\to\infty$

$$R_i = -\Gamma_i (1 + R_j + R_k)\,. \tag{2.51}$$

Since, however, it follows from eq. (2.50) that the total projector

$$\Gamma = 1 - R(\lambda)|_{\lambda\to\infty} = -R_1 - R_2 - R_3 \tag{2.52}$$

the iteration of eq. (2.52) will give the desired formulas for Γ. In particular, the first iteration of eq. (2.51) gives

$$\Gamma^{(1)} = \Gamma_1 + \Gamma_2 + \Gamma_3\,,$$

the second iteration gives

$$\Gamma^{(2)} = \Gamma_1 - \Gamma_1\Gamma_2 - \Gamma_1\Gamma_3 + \Gamma_2 - \Gamma_2\Gamma_1 - \Gamma_2\Gamma_3 + \Gamma_3 - \Gamma_3\Gamma_1 - \Gamma_3\Gamma_2$$

etc.

In such a way, a complete series can be obtained for Γ.

Returning now to eq. (2.46), we may write

$$\Gamma\widetilde{G} = \widetilde{G}\,\Gamma = 0 \tag{2.53}$$

and, similarly, for the complete three-body wave function $\widetilde{\Psi} = \lim_{\epsilon \to 0} i\epsilon \lim \widetilde{G}(E + i\epsilon)\Phi$ which is simultaneousely a solution for the initial three-body Schrödinger equation (2.34) to (2.35).

2.4 Faddeev Equations in Three-Body Orthogonal Subspace; Rearrangement and Improvement of the Born Series Convergence in the Scattering Theory

2.4.1 General considerations

Presented above was the formalism of the three-body scattering problem which makes it possible to include the additional requirements of orthogonality of the wave functions and scattering operators to *two-body* forbidden states in two-cluster subsystems. Developed in this Section will be the appropriate formalism permitting FE to be derived in *three-body orthogonal subspace*, i.e. in the three-body subspace H_Q orthogonal to all the three-body states included in the three-body projector $\Gamma = \sum_{i=1}^{N} |\Phi_i\rangle\langle\Phi_i|$, where the states $|\Phi_i\rangle$ are assumed to be quadratically integrable. In other words, we seek for the solution of the Schrödinger equation

$$(H_{123} - E)\Psi_{123} = 0 \tag{2.54}$$

with the additional conditions of orthogonality

$$\langle\Phi_i|\Psi_{123}\rangle = 0, \qquad i = 1, \ldots N \tag{2.55}$$

to some of the three-body states $|\Phi_i\rangle$. In practice, this problem is of importance to the numerical solution of the three-body problem (see below) and for the inclusion of many-body exchange forces. Thus, in order to find the total projected GF $\widetilde{G}$ we shall again follow the common procedure and turn from the total three-body Hamiltonian H_{123} to the pseudohamiltonian

$$\widetilde{H}_{123} = H_{123} + \lambda\Gamma = H_0 + \lambda\Gamma + V,$$

where

$$V = V_1 + V_2 + V_3 .$$

After that, introduce the projected channel GF

$$\widetilde{G}_i = (E - H_0 - \lambda\Gamma - V_i)^{-1}, \qquad (i = 0, 1, 2, 3)$$

related to the conventional G_i through the standard expressions of the type of eq. (2.24). Besides that, we shall use another standard identity for GF:

$$\widetilde{G}(\lambda) = \widetilde{G}_0(\lambda) + \widetilde{G}_0(\lambda)\,V\,\widetilde{G}(\lambda), \tag{2.56}$$

where

$$\widetilde{G}(\lambda) = (E - \widetilde{H}_\lambda)^{-1} \quad \text{and} \quad \widetilde{G}_0(\lambda) = (E - H_0 - \lambda\Gamma)^{-1}.$$

By the analogy with eq. (2.24), we can easily find that

$$\widetilde{G}_0 = \lim_{\lambda \to \infty} (E - H_0 - \lambda \Gamma)^{-1} = G_0 - G_0 \Gamma (\Gamma G_0 \Gamma)^{-1} \Gamma G_0 .$$

After that, using the definition $V = V_1 + V_2 + V_3$ in eq. (2.56) and making the standard Faddeev reduction, we shall find the set of projected FE

$$\widetilde{G}^{(i)} = \widetilde{G}_i - \widetilde{G}_0 + \widetilde{G}_i V_i (\widetilde{G}^{(j)} + \widetilde{G}^{(k)}) \tag{2.57}$$

(ijk = 1, 2, 3 and cyclic transpositions), which differs from the ordinary FE by the tilde sign over *all* GF. The total projected GF is

$$\widetilde{G} = \widetilde{G}_0 + \widetilde{G}^{(1)} + \widetilde{G}^{(2)} + \widetilde{G}^{(3)} , \tag{2.58}$$

where

$$\widetilde{G}^{(i)} \equiv \widetilde{G}_0 V_i \widetilde{G} .$$

It will be noted that the equations (2.57) differ from (2.44) in that the three-body projector in (2.57) is included not in the potential but the three-body free GF $\widetilde{G}_0$.

Now, the scattering wave function $\widetilde{\psi}_i$ will be determined by the relation

$$\widetilde{\psi}_i = \lim_{\epsilon \to 0} i\epsilon \widetilde{G} (E + i\epsilon) \Phi_i$$

and then the FE set for the wave function of three-body scattering will be obtained from (2.57). For example, in case of scattering of particle 1 by the bound state of particles (2,3), we get

$$\widetilde{\psi}_1 = \widetilde{\psi}_1^{(1)} + \widetilde{\psi}_1^{(2)} + \widetilde{\psi}_1^{(3)}$$

$$\widetilde{\psi}_1^{(i)} = \widetilde{\Phi}_1 \delta_{i1} + \widetilde{G}_i V_i (\widetilde{\psi}_1^{(j)} + \widetilde{\psi}_1^{(k)}) \tag{2.59}$$

(i, j, k = 1, 2, 3 and cyclic transpositions), where

$$\widetilde{\Phi}_1 = \Phi_1 - G_i \Gamma (\Gamma G_i \Gamma)^{-1} \Gamma \Phi_i . \tag{2.60}$$

In the matrix designation, the kernel of the equation sets (2.57) and (2.59) is

$$\widetilde{K} = \widetilde{G} V \mathbb{A} = (1 - \mathbb{I}\Gamma)\, G V \mathbb{A} = (1 - \mathbb{I}\Gamma) G_0\, T \mathbb{A} = (1 - \mathbb{I}\Gamma) K \tag{2.61}$$

Here, the three-row matrices G, V, T, $\widetilde{G}$ and $\mathbb{I}\Gamma$ are diagonal and their elements are G_i, V_i, T_i, $\widetilde{G}_i$ and Γ_i respectively, whereas

$$\Gamma_i = G_i \Gamma (\Gamma G_i \Gamma)^{-1} \Gamma$$

is a nonhermitian projector. The numerical matrix $\mathbb{A}$ is of the form

$$\mathbb{A} = \begin{pmatrix} 0 & 1 & 1 \\ 1 & 0 & 1 \\ 1 & 1 & 0 \end{pmatrix}$$

and the matrix operator $K = G_0 T \mathbb{A}$ is the kernel of the conventional FE set.

Now, the developed formalism of orthogonal projecting will be applied to the important aspect of the scattering theory, namely to the problem of rearrangement of the Born series to enhance its convergence if it is convergent or else make it convergent if it is divergent. The solution of this problem, being interesting per se, is necessary for the practical calculations with the three-body equations derived. The basic idea of such an applications stems from the well known fact that the continuum functions are orthogonal at all energies to the functions of the discrete spectrum, i.e. to the functions of the bound states of the system. On the other hand, it well known [Ne66] that the singularities due to bound states just result in the Born series divergence at low energies. Therefore, if the scattering problem is solved from the very beginning in the orthogonal subspace H_Q from which the bound-state vectors are excluded, one may expect to get a convergent Born series. When treating the problems outlined above, we shall follow the works [Ku76, 78; Po76].

2.4.2 Two-body problem; potential scattering

Following the common rule, the interaction potential will be replaced by the pseudo-potential $\widetilde{V} = V + \lambda\Gamma$ where the projector Γ includes all the bound states of the initial Hamiltonian. After that, the standard algebraic transformations will give the orthogonalized LSE for GF:

$$\widetilde{g} = \widetilde{g}_0 + \widetilde{g}_0 V \widetilde{g} \tag{2.62}$$

and for the scattering wave function

$$\widetilde{\psi} = \widetilde{\psi}_0 + \widetilde{g}_0 V \widetilde{\psi}\,, \tag{2.63}$$

where $\widetilde{g}_0$ is given by a relation of the type of (2.24); $\widetilde{\psi}_0$ is the wave function of orthogonality scattering (2.12). It will be emphasized that, in case of so called eigenstate projecting (i.e. when the exact eigenstates (ES) of the Hamiltonian are included in the projector Γ) $\widetilde{\psi} = \psi$ and therefore the solution for eq. (2.63) will *directly* give the required wave function. The iterations of eq. (2.63) give

$$\widetilde{\psi} = \widetilde{\psi}_0 + \widetilde{g}_0 V \widetilde{\psi}_0 + \widetilde{g}_0 V \widetilde{g}_0 V \widetilde{\psi}_0 + \ldots \tag{2.64}$$

whence it can easily be seen that the *orthogonal scattering basis only appears* in *all* the intermediate states, the fact that has become the key point in constructing the theory of scattering in an orthogonal subspace [Ku76a, 78b; Po76]. It may strictly be proved [Ku78b] that the kernel of orthogonalized LSE (2.63), i.e. the operator $\widetilde{g}_0 V$, is smaller in norm than the initial kernel $g_0 V$; in other words, the convergence of the series (2.64) is better than the convergence of the initial series. Moreover, it may be proved for a broad class of potentials (for details, see [Po78, Ku78b] that the Born series (2.64) is convergent from the threshold, i.e. *at all energies*. This conclusion can qualitatively be understood from the following. If the ordinary LSE is iterated on the plane wave basis, i.e.

$$\psi = \psi_0 + g_0 V \psi_0 + g_0 V g_0 V \psi_0 + \ldots \tag{2.65}$$

then the exact wave function ψ in the presence of bound states in the potential V must be orthogonal to the functions of such bound states and, therefore, have nodes in the inner region. Moreover, the theorem has been known [Al65] that the number of the nodes of the scattering wave function at zero energy equals the number of bound states in the system. On the other hand, the plane waves ψ_0 at low energies have no nodes in the inner region (because $kR \ll 1$). Hence, the functions of the initial iterations of the series (2.65) i.e. $\psi_0, g_0 V \psi_0$ etc., are quite dissimilar to the exact ψ. As far as the series (2.64) is concerned, *each of its terms* is orthogonal to all the functions of the discrete spectrum and has the required number of nodes in the inner region [Ku78c]. In other words, contrary to (2.65), each terms of the rearranged Born series (2.64) already exhibits the properties of the exact solution, the fact which gives rise to a more rapid convergence of (2.64) as compared with (2.65).

A more rigorous proof may be outlined as follows [Ku76a, 78b]. The rate of the Born series convergence is determined by the greatest-modulus EV of kernel $K = G_0 V$:

$$K \,|\chi_n\rangle = \alpha_n \,|\chi_n\rangle, \qquad n = 1, \dots .$$

The hermitian operators will be obtained by symmetrizing the kernel $G_0(E)V$ at $E < 0$ in such a way that

$$K = (-G_0)^{1/2} K_s (-G_0)^{1/2}$$

where

$$K_s = K_s^* = -(-G_0)^{1/2} V (-G_0)^{1/2}$$

is the hermitian operator. This is admissible since at $E < 0$ the operator $(-G_0)$ is positively determined. EV's of the operators K and K_s coincide, and hence EV's of the operator K are real (at $E < 0$). The projector $P = G_0\Gamma(\Gamma G_0 \Gamma)^{-1}\Gamma G_0$ will also be symmetrized using $(-G_0)^{1/2}$:

$$P = (-G_0)^{1/2} P_s (-G_0)^{-1/2}$$

where

$$P_s = P_s^* = -(-G_0)^{1/2}\, \Gamma(\Gamma G_0 \Gamma)^{-1}\, \Gamma\, (-G_0)^{1/2}$$

is the hermitian projector. Therefore, the projected kernel $\widetilde{K} = \widetilde{G}_0 V$ may be rewritten as

$$\widetilde{K} = (1 - P) K = (-G_0)^{1/2}\, (1 - P_s) K_s (-G_0)^{1/2}$$

In this case, however, EV's $\widetilde{\alpha}_n$ of kernel $\widetilde{K}$ coincide with EV's of the operator $(1 - P_s)K_s$ which are determined by the equation

$$(1 - P_s)\, K_s \,|\hat{\chi}\rangle = \hat{\alpha}\, |\hat{\chi}\rangle .$$

The action of the projector P_s on this equation gives the condition $\alpha P_s |\hat{\chi}\rangle = 0$ which means that the operator $(1 - P_s)K_s$ and, therefore, $\widetilde{K}$ also has an N-multiple zero EV while all the rest EV's and EF's are determined by the equation

$$(1 - P_s)\, K_s\, (1 - P_s) |\hat{\chi}\rangle = \alpha \,|\hat{\chi}\rangle$$

i.e. are EV's and EF's for the hermitian operator $\widetilde{K}_s = (1 - P_s)K_s(1 - P_s)$ too.

Thus, EV's $\widetilde{\alpha}_n$ of the operators $\widetilde{K}$, $(1 - P_s)K_s$ and $\widetilde{K}_s$ coincide and, therefore, are real and may be found from the minimax principle, for $\widetilde{K}_s$ is the hermitian and quite continuous operator. In particular, we get in case of positive EV's:

$$\widetilde{\alpha}_1^+ \equiv \widetilde{\alpha}_{max} = \max_{\|\varphi\| = 1} \langle \varphi | \widetilde{K}_s | \varphi \rangle \leqslant \max_{\|\varphi\| = 1} \langle \varphi | K_s | \varphi \rangle = \alpha_{max} .$$

For negative EV's we get correspondingly:

$$\widetilde{\alpha}_1^- \equiv \widetilde{\alpha}_{min} = \min_{\|\varphi\| = 1} \langle \varphi | \widetilde{K}_s | \varphi \rangle \geqslant \min_{\|\varphi\| = 1} \langle \varphi | K_s | \varphi \rangle = \alpha_{min} .$$

This means that the conditions

$$\alpha_{max} \geqslant \widetilde{\alpha}_n \geqslant \alpha_{min}$$

are satisfied for any projector Γ.

Further, in case of the total ES projecting, $\widetilde{G} = (1 - \Gamma)G$ i.e. all bound state poles are excluded from projected GF. In this case, however, the corresponding poles are also absent in the resolvent $(1 - \widetilde{G}_0 V)^{-1}$ of the kernel $\widetilde{K}$ for

$$\widetilde{G} = (1 - \widetilde{G}_0 V)^{-1} \widetilde{G}_0 = (1 - \widetilde{G}_0 V) G_0 - \Gamma G_0 .$$

But, because the condition $\widetilde{\alpha}_n(E_p) = 1$ is satisfied by the poles of $\widetilde{G}$ (or the poles of $(1 - \widetilde{G}_0 V)^{-1}$), we get the condition $\widetilde{\alpha}_n(E) \neq 1$. On the other hand, it is obvious that $\| G_0(E) \widetilde{V} \| \underset{E \to -\infty}{\to 0}$ and hence $\alpha_n \underset{E \to -\infty}{\to 0}$. In combination with the found boundaries of $\widetilde{\alpha}_n$, this means that $\widetilde{\alpha}_n(E) \underset{E \to -\infty}{\to 0}$. Then, considering the monotonic function $\alpha_n(E)$, we find that

$$1 > \widetilde{\alpha}_n(E) \geqslant \alpha_{min}(E) \geqslant \alpha_{min}(0)$$

for all n, the fact that demonstrates the convergence of the orthogonalized Born series (2.64) at $E \leqslant 0$. Since, however, EV's $\widetilde{\alpha}_n(E)$ are continuous at $E = 0$, the convergence holds also for some band at $E > 0$, though it is difficult in general to find a rigorous proof for the *entire* range of positive energies. In case of scattering in the S-wave, the convergence of the series (2.64) probably takes place at all energies. Considered as an example will be the results of the calculations of EV's $\widetilde{\alpha}_n$ at $E > 0$ for the Hulthen potential. Fig. 2 shows the convergence of orthogonal iterations for the Hulthen potential:

$$V(r) = \frac{g}{1 - \exp(r/a)}$$

with $g = 2$. The initial Born series (2.65) is strongly divergent ($\sim$as 2^n, where n is the order of iteration) in this case.

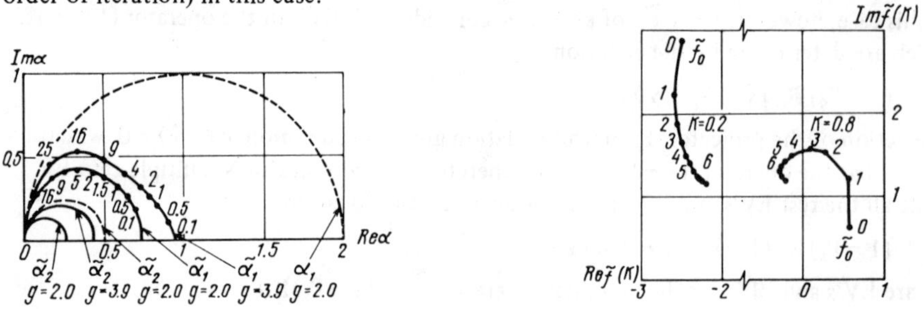

Fig. 2 a) The trajectories of the projected kernel eigenvalues $\widetilde{\alpha}_n(E)$ in α-plane as compared with initial α_n-values (with the two values of coupling constant of the potential, $g = 3.9$ and $g = 2$).
b) The Argand diagram for the S-wave amplitude of scattering by the Hulten potential. The numerals at the curves indicate the order of orthogonal iterations.

2.4.3 Three-body scattering

Examine now the three-body scattering. The iterations of the orthogonalized FE (2.59) will now be of the form

$$\begin{aligned}
\widetilde{\psi}^{(1)} &= \Phi_1 + \widetilde{G}_1 V_1 \widetilde{G}_2 V_2 \widetilde{\Phi}_1 + \widetilde{G}_1 V_1 \widetilde{G}_3 V_3 \widetilde{\Phi}_1 + \dots \\
\widetilde{\psi}^{(2)} &= \widetilde{G}_2 V_2 \widetilde{\Phi}_1 + \widetilde{G}_2 V_2 \widetilde{G}_3 V_3 \widetilde{\Phi}_1 + \dots \\
\widetilde{\psi}^{(3)} &= \widetilde{G}_3 V_3 \widetilde{\Phi}_1 + \widetilde{G}_3 V_3 \widetilde{G}_2 V_2 \widetilde{\Phi}_1 + \dots
\end{aligned} \tag{2.66}$$

and we shall take into account that $\widetilde{G}_i$ and $\widetilde{\Phi}_1$ (see (2.60)) are orthogonal to H_Γ. In its turn, this means that all the intermediate states in the iterations of three-body equations (2.66) are constructed, as in the two-body problem, on the orthogonalized basis, so the convergence of such orthogonal iteration in this case should be better than the convergence of the initial iterations of FE. Though a rigorous proof in this case is much more difficult than in the two-body problem, we can still demonstrate [Ku78b; Po78] that, on imposing definite constraints on the two-body potentials, the iterations (2.66) will be convergent right away from the threshold, i.e. *at all energies.*

As an illustrative example we consider the scattering of particle 1 by the bound state of two other particles 2 and 3 (all particles are assumed to be bosons with the same masses). Only two interations, V_1 and V_2, will be considered, which gives a single bound three-body state at the corresponding values of the parameters. The convergence of the orthogonalized Born series for three-body scattering amplitude is shown by the Argand-diagram in Fig. 3. A fairly rapid convergence can be seen.

There exist a number of cases, however, where the direct application of the technique described here fails to yield the desirable results. We mean the cases where two- or three-body resonances appear. Such resonances, similarly to the bound states, give rise to the divergence of the Born series at low energies. To include them, the orthogonal projecting technique described here should be somewhat generalized [Ku78b; Po78]. Refraining from detailed treatment of the corresponding generalization, we shall present here the basic relations. In this case, the general concept of construction proves to be different from the case of bound states where the equality $\widetilde{\Psi} = \Psi$ holds for all energies and any number of particles. It is more convenient here to use the standard formalism developed by Feshbach [Fe58, 62]

$$\Psi = \Gamma\Psi + Q\Psi \equiv \Psi_\Gamma + \Psi_Q$$

where

$$\Gamma^2 = \Gamma, \; Q^2 = Q, \; \Gamma Q = Q\Gamma = 0 .$$

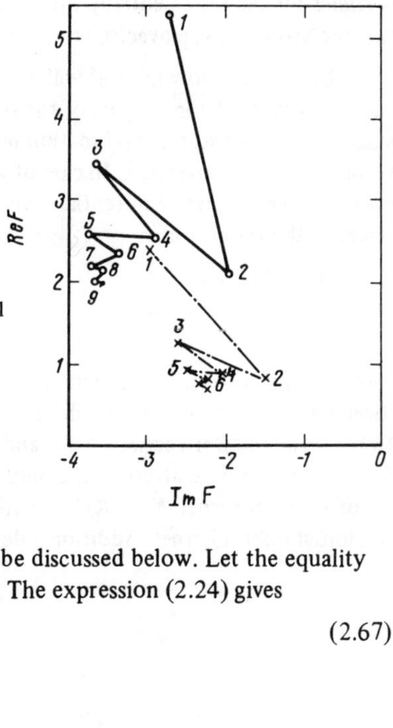

Fig. 3
The Argand diagram for the convergence of the orthogonal iterations for the S-wave amplitude of three-body scattering of a particle by two-body bound state. The lower and upper broken lines correspond to the values of the dimensionless wave vector of "deuteron" $\alpha = 0.46$ and $\alpha = 0.23$ respectively. The numerals at the angles of the brokens indicate to the order of iteration.

The selection of the functions contained in Γ will be discussed below. Let the equality (2.24) be inverted, i.e. G be expressed through $\widetilde{G}$. The expression (2.24) gives

$$G = \widetilde{G} + G\Gamma(\Gamma G \Gamma)^{-1}\Gamma G . \tag{2.67}$$

Multiplying (2.67) by ΓG^{-1} from the left and by $G^{-1}\Gamma$ from the right and using minor algebraic manipulations, we get

$$\Gamma G \Gamma = \Gamma[\Gamma(G^{-1} - G^{-1}\widetilde{G}G^{-1})\Gamma]^{-1}\Gamma .$$

After substituting the above expression in (2.67) and replacing G^{-1} by $(E - H)$, we get

$$G = \widetilde{G} + (1 - \widetilde{G}H)\Gamma[\Gamma(E - H - H\widetilde{G}H)\Gamma]^{-1}\Gamma(1 - H\widetilde{G}) . \tag{2.68}$$

The expression (2.68) permits the total GF G to be restored through the projected GF $\widetilde{G}$ and the total Hamiltonian H. Quite the same partition of the total GF can also be obtained from standard approach [Fe58, 62] where it is of the form (in our designations):

$$G = G_{QQ} + G_{\Gamma Q} + G_{Q\Gamma} + G_{\Gamma\Gamma}$$

where

$$\left.\begin{aligned} G_{QQ} &= \widetilde{G} + \widetilde{G}H\,D(E)\,H\widetilde{G} \\ G_{Q\Gamma} &= -\widetilde{G}H\,D(E) \\ G_{\Gamma Q} &= -D(E)\,H\widetilde{G} \\ G_{\Gamma\Gamma} &= D(E) \end{aligned}\right\} \tag{2.69}$$

and

$$D(E) = \Gamma[\Gamma(E - H - H\widetilde{G}H)\Gamma]^{-1}\Gamma .$$

It is easy now to obtain the expression for the complete wave function:

$$\Psi = \lim_{\epsilon \to 0} i\epsilon\, G\Phi = \widetilde{\Psi} - (1 - \widetilde{G}H)\,D(E)\,H\widetilde{\Psi} . \tag{2.70}$$

It will be emphasized that the relations (2.68), (2.69), (2.70) are valid for *any number of particles* for the are essentially the formal equations relating the various projections of the operators and eigenvectors to the operators and eigenvectors themselves.

Discussed below in brief will be the selection of Γ [Ku78b]. In case of resonances, the inclusion of the inner part of the resonance function in Γ, will eliminate the resonant singularity in the scattering function in the Q-space, whereupon the Born series for $\widetilde{\Psi}$ and $\widetilde{G}$ proves to be convergent. In case of the two-body problem with attractive interaction potential (i.e. in case of potential scattering) and the Hamiltonian $H = H_0 + V$, the bound states of the Hamiltonian

$$H_\beta = H_0 + \beta V ,$$

with

$$\beta > 1$$

should be included in Γ. In such projecting procedure, the EV's $\eta(E) > 1$ resulting in a divergence of the initial series will be excluded from the LSE kernel $\beta G_0 V$ [Ku78b; Po78]. In this case, the Born series for $\widetilde{G}$ and $\widetilde{\Psi}$ will be convergent. Similar situation holds also for the three-body scattering; the only difference is that the solutions for the homogeneous set of FE with kernel $\bar{K} = \beta K(E)$ (with $\beta > 1$) should be included in Γ, where $K(E)$ is the initial matrix kernel. Additional details may be found in [Ku78b; Po78].

3 Two Particles in the Field of a Core (The Model of Direct Nuclear Reactions With Deuterons)

Discussed in this Section will be the problem of two particles in field of a core including the effects of antisymmetrization of valent particles and core particles and the core excitations. Such problem may be used as a very handy model in the theory of direct nuclear reactions and as a substantiation of the widely adopted approximations (DWBA, coupling channel method etc.). It is not surprising, therefore, that the problem has been examined in numerous theoretical works which dealt with both conceptual and calculational aspects. Far from being able to discuss them in details, we shall present here the main features of the formalisms proposed and the results obtained.

3.1 Preliminary Discussion

It should be emphasized right away that all the calculations made may clearly be divided, for the practical purposes, into two trends.

The first trend includes all the calculations based on the formalism of three-body integral equations. Such calculations are aimed at comparison between the 'exact' results found by solving the three body integral equations and the approximate results obtained using the widely adopted approximations of the type of the coupled-channel method [Be70], the distorted-wave method for ordinary stripping [Aa66] and for the stripping to unbound states [Bri78], and at studying the effect of the reduced nucleon widths on the behaviour of the differential cross sections of the stripping or elastic scattering [Be73]. Unfortunately, all the works made in such directions have been based on simple separable potentials used as the nucleon-nucleus and nucleon-nucleon interactions, which are much unlike the optical potentials which in practice are always used in nuclear physics to describe the nucleon-nucleus interactions. More specifically, such separable models suffer from the following drawbacks.

i) The selected interaction potentials are effective in one-or two partial waves, i.e. where the contribution from the bound states and near-threshold resonances dominate. At the same time, apart from the said singularities, the nucleon-nucleus scattering contains a smooth but significant "optical" background (of non-separable nature) which is much responsible for the interference effects notable in the direct nuclear processes.

ii) The separable potential can properly approximate the accurate t-matrix in only some vicinity of a resonant pole; when moving away from the pole, the quality of approximation will be markedly deteriorated (see the discussion of UPA-model in [Le74]). Meanwhile, consideration has to be given mainly to direct nuclear reactions with deuterons at the incident deuteron energies $E_d > 10$ MeV. In this case, the contribution from the "optical" background which cannot be described by the separable interaction may even exceed the contribution from, for example, the bound-state pole.

iii) There are numerous cases where the contribution from the lowest partial waves in nucleon-nucleus interaction (it is here that the separable interaction model is operative) are important. Yet the target nucleus is well known to contain the occupied one-particle states, corresponding to the filled or partially filled shells of the target nucleus, just in the lowest partial waves ($s_{1/2}$, $p_{1/2}$, $p_{3/2}$, etc.). For example, the ^{16}O nucleus includes the $1s_{1/2}$, $2p_{1/2}$ and $2p_{3/2}$ filled shells. Therefore, the nucleon-target interaction potential should take account of the existence of these filled shells. Since the fully filled shells fail explicitly to affect the scattering amplitudes (at low energies), the standard approach proceeds as follows. If, apart from the filled shell, the lowest partial waves do not contain other observable (bound or resonance) states, the interaction in such waves is taken to be of repulsive nature, namely the repulsive separable potential. The S-wave in the $n(p) + {}^4He$ nucleus [Sh69, Ch77] is a good example.

However, as it was mentioned in the Introduction and will be discussed in detail below, the interaction in such channels is rather of *attractive* nature which is indicated by, say, the behaviour of the appropriate phase shifts [Ne71, Sa77, Ku75b]. At the same time, the existence of filled orbitals should be included by orthogonalizing the interaction operators or the scattering functions in the respective channels to the functions of the filled orbitals. It is this model that permits a coordination of the resultant scattering wave functions with the shell model.

It may be concluded on summarizing the factors outlined in i)–iii) that, at the best, the separable interaction model used in all the calculations within the frame of three-body integral equations may give only a qualitative or semiquantitative description of the three-body effects in direct nuclear reactions (at least for the usually treated deuteron energies $E_d > 10$ MeV). At lower energies, however, the applicability of the separable model may prove to be much better. The correspondence between the off-shell behaviour of the optical-model t-matrices for the local and the separable potential that approximates the former was studied in most detail in [Ya75] using the optical potential in the n-^{27}Al system as an example. Used as the approximating interaction in the work were the single-term energy-dependent separable potential and the three-term energy independent separable potential. The two approximating potentials have proved to give a very good fitting of the on-shell t-matrix, i.e. the phase shifts for the initial optical potential in a rather wide-energy range (~ 20 MeV). A good agreement can also observed in case of the half-shell behaviour. At the same time *a discouraging disagreement* is observed between the separable and exact fully off-shell t-matrices. However, the FE kernels are known to include just the fully off-shell t-matrices. Figs. 4 and 5 (borrowed from [Ya75]) are good illustrations of this situation. Fig. 4 presents the above comparison for the $s_{1/2}$ and $p_{1/2}$ partial waves; Fig. 5 shows the comparison for the $p_{3/2}$ and $d_{3/2}$ partial waves. Besides that, it has been found that the separable potential

$$V \Rightarrow V_s^{(N)} = \sum_{i,j=1}^{N} V|\phi_i\rangle \Lambda_{ij} \langle \phi_j^* | V$$

(at N = 3) gives a good approximation of the exact t-matrix both on- and off-shell for all partial waves, except for the $S_{1/2}$ wave for which a four-, or even six-, term separable approximation has to be used.

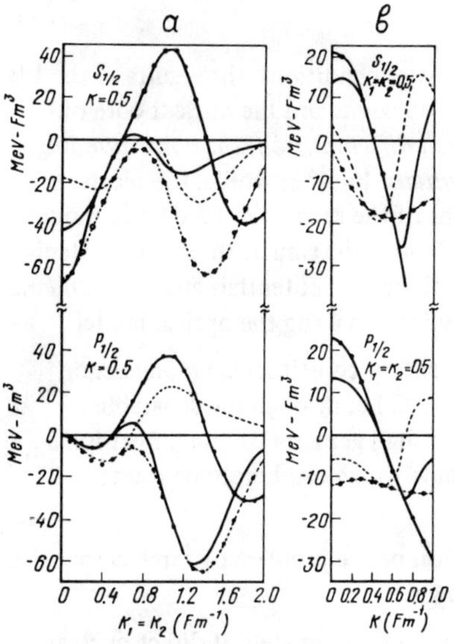

Fig. 4

Comparison between the fully off-shell behaviours of the separable and accurate (i.e. local-potential) t-matrix for the $s_{1/2}$ and $p_{1/2}$ channels of the n-^{27}Al optical potential. The wave numbers k_1, k_2, k relate to $\langle k_1 | t_\nu(k) | k_2 \rangle$. The solid and dashed lines present the real and imaginary components of the separable t-matrix respectively. The dots, or solid lines connecting the dots, give the real part of the local-potential t-matrix. The dots, or dashed lines connecting the dots, give the imaginary part of the local-potential t-matrix.

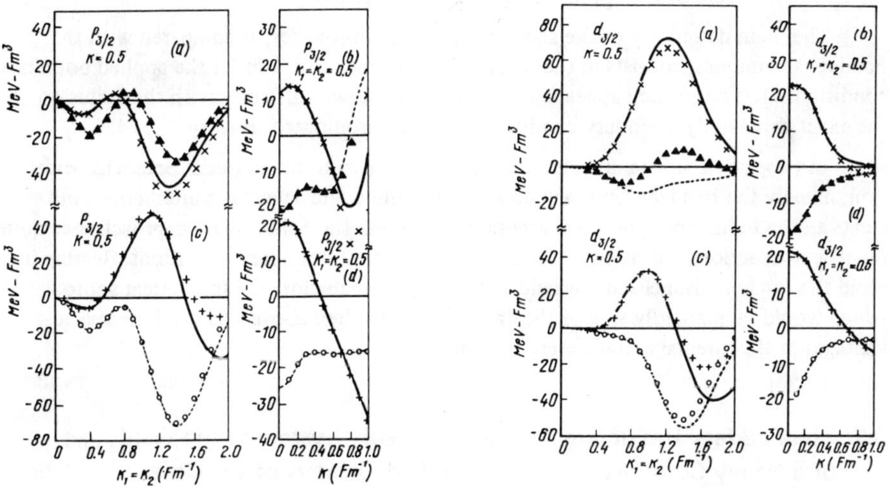

Fig. 5 The same as in Fig. 4 but for the $p_{3/2}$ (side I) and $d_{3/2}$ (side II) channels. The solid and dashed lines in (a) and (b) (for I and II sides) relate to the real and imaginary parts of the separable energy-dependent t-matrix (see in [Ya 75]). The crosses and dark triangles show the real and imaginary parts of the separable term in the Kowalski-Noyes representation of the t-matrix. The pluses and circles show the real and imaginary parts of the separable rank-3 approximation.

Thus, we may summarize as follows.

(1) If one refrains from selecting any special kinematic conditions, the results of the FE solution for the processes involving two valent nucleons and the nuclear core obtained with local optical potentials will be *much different from the corresponding results obtained with one-term separable potentials.* In other words, the results of calculations of direct nuclear reactions in terms of the separable three-body model may be compared *at the best qualitatively* with both the similar three-body calculations (which have not been carried out yet) with optical potentials and the *standard* processings of experimental data (of DWBA type) also using the optical model.

(2) If the nuclear three-body problem whose elementary constituents are, nuclei or clusters is solved by analogy with the 3N problem, i.e. through the separable approximations of the scattering potentials, we shall get a set of one-dimensional equations of a very high dimension, i.e. a scheme which can hardly be treated numerically.;

The above considerations mean that the solution of the problem of three composite particles requires that alternative approaches be sought.

The second trend includes the studies of the three-body models of direct nuclear reactions on the basis of the more habitual Schrödinger approach, i.e. the approximate solution of the three-body Schrödinger equation in the coordinate space [Po75; Au73, 78 etc.]. This approach is much more similar to the conventional approaches used in nuclear physics than the approach with integral equations in both the type of the interactions used and the nature of the approximation made [Au73, 78].

The main deficiency of the above mentioned approaches, as compared with the Faddeev techniques, consists in the schematic and simplified form of the applied boundary conditions. The deficiency appears due to the well-known difficulty with the inclusion of the exact three-body boundary conditions in the Schrödinger formalism [Sc74a].

In view of circumstances discussed above, the two trends suffer a certain imperfection, namely the Faddeev approach uses the simplified and schematic interactions but treats all the boundary conditions accurately, whereas the Schrödinger approach uses quite realistic interactions but, in turn, the simplified form of boundary conditions. Bearing in mind the further advances in the field, we shall describe below the theoretical approach which would be markedly free of the drawbacks described above and combine the advantages of the integral and differential approaches.

3.2 The Faddeev Reduction for the Two-Body System in the Field of a Heavy Core; Account of the Filled Orbitals of the Core

If the recoil of target nucleus is neglected, the Faddeev reduction will conveniently be carried out in the following way. The total interaction will be broken into *two* parts

$$V \equiv V_{12} + (V_{13} + V_{23}) \equiv V_{12} + V_3 , \tag{3.1}$$

where the core is labelled "3", and the nucleons – "1" and "2" respectively. Introduce the partial GF

$$G_3 = (E - H_0 - V_3)^{-1} \tag{3.2a}$$

$$G_{12} = (E - H_0 - V_{12})^{-1}, \tag{3.2b}$$

where the potential V_3 corresponds to the interaction of two nucleons with the core; V_{12} denotes the interaction of the nucleons between each other.

In accordance with eq. (3.1), the reduction of the total GF

$$G(E) = (E - H)^{-1}$$

will be

$$G = G_0 + G^{(12)} + G^{(3)}, \tag{3.3}$$

where

$$G^{(12)} \equiv G_0 V_{12} G, \quad G^{(3)} \equiv G_0 V_3 G.$$

Such reduction is expedient here because GF $G_3(E)$ can be found directly [Ba71b; Fu71]:

$$G_3(E) = g_{13} * g_{23} = \frac{1}{2\pi i} \int g_{13}(\epsilon) g_{23}(E - \epsilon)\, d\epsilon \tag{3.4}$$

where the sign $*$ denotes convolution. The method for calculating the convolution (3.4) including the one-body bound and resonant states is presented in [Fu71]. In such a way, we get the set of *two* FE for the resolvent:

$$G^{(12)} = G_{12} - G_0 + G_{12} V_{12} G^{(3)} \tag{3.5a}$$

$$G^{(3)} = G_3 - G_0 + G_3 V_3 G^{(12)}. \tag{3.5b}$$

The equations for the wave function are obtained, as usual, through the limit

$$\Psi = \lim_{\epsilon \to 0} i\epsilon G(E + i\epsilon)\Phi,$$

where Φ is the function of the initial state and the expressions (3.3) and (3.5) must be substituted for $G(E + i\epsilon)$. Then, in case of, for example, deuteron scattering by nucleus, we get:

$$\Psi = \Psi^{(12)} + \Psi^{(3)} \tag{3.6a}$$

$$\Psi^{(12)} = \Phi_{12} + G_{12} V_{12} \Psi^{(3)} \tag{3.6b}$$

$$\Psi^{(3)} = G_3 V_3 \Psi^{(12)}. \tag{3.6c}$$

The results of the work [Ba66] may be used to reduce the set (3.6) to a *single* equation for the function $X_{12} = V_{12}\Psi$:

$$X_{12} = V_{12}\Phi_{12} + T_{12}\,\Delta G_3 X_{12}, \tag{3.7}$$

where $\Delta G_3 = G_3 - G_0$ and $\Phi_{12} = \varphi_{12}(\mathbf{k}_{12})\delta(\mathbf{p}_3 - \mathbf{p}_3^{(0)})$ is the function of the initial state.

It can easily be shown that, if the complete three-body function X_{12} is used, the amplitudes of the stripping and break-up reactions can be obtained merely in the form of projections on the final channel function [Ba71b]

$$A_{str} = -\langle \varphi_{13}\chi_2 | X_{12} \rangle \tag{3.8a}$$

$$A_{break\text{-}up} = -\langle \chi_1 \chi_2 | X_{12} \rangle, \tag{3.8b}$$

where $\varphi_{13}(\mathbf{k}_{13})$ is the final state wave function of the stripped particle in the core field; χ_i is the scattering wave function of particle i in the core field. It can easily be seen from eq. (3.7) that, in case the t-matrix of two-nucleon interaction T_{12} corresponds to the zero range forces, i.e. if

$$T_{12}(\mathbf{k}_{12}, \mathbf{p}_3, \mathbf{k}'_{12}, \mathbf{p}'_3, E) = \frac{\delta(\mathbf{p}_3 - \mathbf{p}'_3)\,\delta(\mathbf{k}_{12} - \mathbf{k}'_{12})}{4\pi^2(\kappa + \sqrt{2m_{12}(E - p_3^2/2\mu_3)})}$$

then equation (3.7) reduces to a three-dimensional integral equation of Lippmann-Schwinger type [Ba66, 71b]:

$$Y_{\mathbf{p}_0}(\mathbf{p}) = h_{\mathbf{p}_0}(\mathbf{p}) + \int \frac{U_{\mathbf{p}_0}(\mathbf{p}, \mathbf{p}'; E)}{p'^2 - p_0^2 + i\epsilon}\, Y_{\mathbf{p}_0}(\mathbf{p}')d\mathbf{p}' \tag{3.9}$$

for the wave function of relative motion of the deuteron center of mass $Y_{\mathbf{p}_0}(\mathbf{p})$ with the generalized (i.e. nonlocal and energy-dependent) potential $U_{\mathbf{p}_0}(\mathbf{p}, \mathbf{p}'; E)$. Similar equations arise also in the case where T_{12} is separable.

After being expanded in partial waves, the equation (3.9) turns out to be the ordinary *one-dimensional* integral equation whose solution is not, generally speaking, difficult, except for the rather tedious method for calculating its kernel $U_{\mathbf{p}_0}(\mathbf{p}, \mathbf{p}'; E)$ [Ba66, 71b] due to its complex singular structure. It will be emphasized that, when the NN-forces of zero radius (or separable S-wave potential) is used in terms of the Faddeev formalism, we get a *noncoupled* set of one-dimensional equations, contrary to the statement of [Au78]. The entire coupling of various channels is "confined" in the kernel of the obtained equation (3.9) and fails to result in coupling between various equations.

The equations written above, can be applied to the real nuclear problem when two important effects are taken into account, namely, the Pauli principle, i.e. the presence of filled orbitals, and the virtual excitations of the core. The two-effects are conveniently examined on the basis of orthogonal projecting techniques. Considered first will be the Pauli principle. We shall follow our work [Ku76b, 78b]. The general idea is the same as that discussed above, namely the Pauli principle will be included by orthogonalizing the wave functions and scattering operators to the filled orbitals of the core. After introducing the projectors Γ_{13} and Γ_{23} onto the filled orbitals we shall follow the general ansatz (see Chapter 2, Sections 2.2–2.4) and turn to the pseudohamiltonian

$$\widetilde{H} = H_0 + V_{12} + \widetilde{V}_3\,, \tag{3.10}$$

where

$$\widetilde{V}_3 = (V_{13} + \lambda\Gamma_{13}) + (V_{23} + \lambda\Gamma_{23}).$$

In case of an infinitely heavy core, the projectors Γ_{13} and Γ_{23} commute with each other, so the total projector is

$$\Gamma = \Gamma_{13} + \Gamma_{23} - \Gamma_{13}\Gamma_{23}\,. \tag{3.11}$$

On making the Faddeev reduction following the general rules (see in Chapter 2, Sections 2.1 and 2.2) on the basis of the partition (3.10), we obtain the modified FE set

$$\tilde{G} = G_0 + \tilde{G}^{(12)} + \tilde{G}^{(3)} \tag{3.12a}$$

$$\tilde{G}^{(12)} = G_{12} V_{12} (G_0 + \tilde{G}^{(3)}) \tag{3.12b}$$

$$\tilde{G}^{(3)} = \tilde{G}_3 \tilde{V}_3 (G_0 + \tilde{G}^{(12)}), \tag{3.12c}$$

where

$$\tilde{G}^{(12)} = G_0 V_{12} \tilde{G} \text{ and } \tilde{G}^{(3)} = G_0 \tilde{V}_3 \tilde{G}.$$

It follows from the results of Chapter 2, Section 3.2 that GF $\tilde{G}_3 = (E - H_0 - \tilde{V}_3)^{-1}$ in the limit $\lambda \to \infty$ may also be written through the *total* projector Γ:

$$\lim_{\lambda \to \infty} \tilde{G}_3 = \lim_{\lambda \to \infty} (E - H_0 - V_3 - \lambda \Gamma)^{-1} \tag{3.13}$$

so that

$$\lim_{\lambda \to \infty} \tilde{G}_3 \tilde{V}_3 = \tilde{G}_3 V_3 - G_3 \Gamma (\Gamma G_3 \Gamma)^{-1} \Gamma . \tag{3.14}$$

The action of projector Γ on (3.12c) and the application of eq. (3.14) give

$$\lim_{\lambda \to \infty} \Gamma \tilde{G} = 0 \tag{3.15}$$

and, simultaneously,

$$\lim_{\lambda \to \infty} \Gamma_{13} \tilde{G} = 0, \quad \lim_{\lambda \to \infty} \Gamma_{23} \tilde{G} = 0 . \tag{3.16}$$

This means that we have achieved our purpose, for the solutions of the modified set of three-body equations (3.12) prove to be orthogonal to all the wave functions of the filled orbitals of the core.

Further, the Hartree-Fock potential $U(r_n)$ will be used as the nucleon-core interaction. In this case, the filled orbitals of the core will either the same as or very close to EF of the one-particle Hartree-Fock hamiltonian $h(r_n) = h_0 + U(r_n)$. While the operators Γ_α and U_α commutate and the projected GF ($\alpha = 13$ or 23) is

$$\tilde{g}_\alpha(\epsilon) = g_\alpha(\epsilon) - g_\alpha \Gamma_\alpha (\Gamma_\alpha g_\alpha \Gamma_\alpha)^{-1} \Gamma_\alpha g_\alpha = g_\alpha - \sum_i \frac{|\varphi_{\alpha i}\rangle\langle\varphi_{\alpha i}|}{\epsilon - \epsilon_{\alpha i}} , \tag{3.17}$$

where the summation over i includes the filled orbitals in subsystem α; $\epsilon_{\alpha i}$ are their energies.

It can also be shown without difficulties that (considering the Pauli principle) the projected GF $\tilde{G}_3(E)$ corresponding to two noninteracting particles in the core field is as usual (see eq. (3.4)):

$$\tilde{G}_3 = \frac{1}{2\pi i} \int_{-\infty}^{\infty} \tilde{g}_{13}(\epsilon) \tilde{g}_{23}(E - \epsilon) \, d\epsilon . \tag{3.18}$$

Substituting eq. (3.17) in eq. (3.18), we find

$$\tilde{G}_3 = (1 - \Gamma_{13})(1 - \Gamma_{23}) G_3 = (1 - \Gamma) G_3 \tag{3.19}$$

i.e. what is to be obtained according to eq. (3.13). The extended writing is

$$\widetilde{G}_3 = G_3 - \sum_i g_\alpha(E - \epsilon_{\beta i})|\varphi_{\beta i}\rangle\langle\varphi_{\beta i}| - \sum_j g_\beta(E - \epsilon_{\alpha j})|\varphi_{\alpha j}\rangle\langle\varphi_{\alpha j}| + $$

$$+ \sum_{ij} \frac{|\varphi_{\alpha i}, \varphi_{\beta j}\rangle\langle\varphi_{\beta j}, \varphi_{\alpha i}|}{E - \epsilon_{\alpha i} - \epsilon_{\beta j}} \tag{3.20}$$

where $\alpha = 12$; $\beta = 13$; the sums over i and j include the filled orbitals.

The equation for the wave function can be obtained in the same way as eq. (3.6) and is of the form

$$\widetilde{X}_{12} = V_{12}\Phi_{12} + T_{12}\Delta\widetilde{G}_3\widetilde{X}_{12}\,, \tag{3.21}$$

where $\Delta\widetilde{G}_3 = \widetilde{G}_3 - G_0$.

Equation (3.21) differs from the nonprojected eq. (3.7) by only the *separable* addition in its kernel:

$$\Delta K = T_{12}(\widetilde{G}_3 - G_3) = -T_{12}\left(\sum_i |\varphi_{\alpha i}\rangle\langle\varphi_{\alpha i}|\, g_\beta\left(E - \epsilon_{\alpha i}\right) + \right.$$

$$\left. + \sum_j |\varphi_{\beta j}\rangle\langle\varphi_{\beta j}|\, g_\alpha(E - \epsilon_{\alpha j}) + \sum_{ij} \frac{|\varphi_{\alpha i}\,\varphi_{\beta j}\rangle\langle\varphi_{\beta j}\,\varphi_{\alpha i}|}{E - \epsilon_{\alpha i} - \epsilon_{\beta j}}\right). \tag{3.22}$$

In accordance with this, the solutions of the three-body equation modified taking into account the Pauli principle will not be more difficult than the solution of the initial eq. (3.7). Moreover, if eq. (3.21) is solved using the separable expansion of the one-particle GF g_α and g_β in Mittag-Leffler series [Ro78; Ga76], the solution of eq. (3.21) will be even simpler than that of eq. (3.7), since the number of the terms of the expansion of each of the two-body GF g_α ($\alpha = 13, 23$) will be reduced by the number of the filled orbitals.

Described in Chapter 2 was the iteration technique designed for solving the similar three-body integral equations. Since, however, the procedure of Faddeev reduction in the problem of two particles in the field of core is somewhat different from the conventional scheme, some modification will be necessary (for details, see [Po78]).

3.3 Inclusion of the Pauli Principle by the Schrödinger Approach; Deuteron Break-up in the Field of Nucleus

Examine now the Schrödinger equation for two particles in the core field:

$$\{H_0 + V_{np} + V(n) + V(p) - (E - \epsilon_d)\}\,\Psi(p, n) = 0\,, \tag{3.23}$$

where the labels 1, 2, 3 have been replaced by the more visual denominations p, n, d for proton, neutron, and deuteron respectively. As before, V(p) and V(n) are taken to be the Hartree-Fock potentials for the proton-nucleus and neutron-nucleus interactions respectively. The presence of the filled orbitals will be included again through the additional requirements of orthogonality of the wave function $\Psi(p, n)$ of the form

$$\Gamma_p\Psi(p, n) = 0\,,\quad \Gamma_n\Psi(p, n) = 0\,, \tag{3.24}$$

where the projector Γ_p (Γ_n) relates to the filled orbitals for protons (neutrons). As above, the additional orthogonality conditions (3.24) can easily be included by turning to the pseudohamiltonian

$$\widetilde{H} = H_0 + V_{np} + V(n) + V(p) + \lambda\Gamma_n + \lambda\Gamma_p \tag{3.25}$$

and, after that, solving the Schrödinger equation corresponding to eq. (3.23) with pseudohamiltonian (3.25). Since, however, eq. (3.23) is an on-shell equation, the additional conditions (3.24) may well be included otherwise [Sa68, 69a; Fe62] (cf. eq. (2.11) and thereafter):

$$(1-\Gamma)(H_0 + V_{np} + V(n) + V(p) - E + \epsilon_d)(1-\Gamma)\Psi(p,n) = 0, \tag{3.26a}$$

where $\Gamma = \Gamma_n + \Gamma_p - \Gamma_n\Gamma_p$ is the total projector.

It can easily be shown that, if $E - \epsilon_d \neq 0$, any solution for eq. (3.26a) will satisfy the necessary orthogonality conditions (3.24). Another equation was used by Austern et al. [Au73]

$$(1-\Gamma)(H - E + \epsilon_d)\Psi(p,n) = 0. \tag{3.26b}$$

After that, the commutation of Γ_p and V(p), and also Γ_n and V(n), may be used to reduce eq. (3.26) to the Bethe-Goldstone form

$$(H_0 + V(p) + V(n) - E + \epsilon_d)\Psi(p,n) = -(1-\Gamma_p)(1-\Gamma_n)V_{pn}\Psi(p,n). \tag{3.26c}$$

It will be noted, however, that the transition from eq. (3.26b) to eq. (3.26c) is possible only if the requirements (3.24) are assumed to have been satisfied. In the opposite case, i.e. when the *general form* of the solution $\Psi(p,n)$ is sought, the expression (3.26b) is not, generally speaking, equivalent to the eq. (3.26c). Nevertheless, the equivalence may be conserved if the boundary condition of the incident deuteron wave are included [Ga74].

Now, if the projector Γ_p or Γ_n acts from the left on eq. (3.26c), we obtain immediately:

$$\Gamma_p\Psi(p,n) = 0, \quad \text{and} \quad \Gamma_n\Psi(p,n) = 0,$$

i.e. the eq. (3.26c) gives automatically the orthogonalized solution. After that, turning from the one-particle variables to the Jacobi coordinates in the deuteron channel

$$\mathbf{r} = \mathbf{r}_p - \mathbf{r}_n, \quad \mathbf{R} = \frac{1}{2}(\mathbf{r}_p + \mathbf{r}_n)$$

and using the approximation of the zero-range n-p interaction, i.e. $V_{pn} = V(r) \approx -D\,\delta(r)$ (where D is the volume integral of V(r)), we may obtain [Ga74] that

$$\Gamma_p V_{pn}\Psi(\mathbf{r}_p,\mathbf{r}_n) = -D\sum_i \varphi_i(\mathbf{r}_p)\varphi_i^*(\mathbf{r}_n)\Psi(\mathbf{r}_n,\mathbf{r}_n) \tag{3.27a}$$

$$\Gamma_n V_{pn}\Psi(\mathbf{r}_p,\mathbf{r}_n) = -D\sum_j \varphi_j^*(\mathbf{r}_p)\varphi_j(\mathbf{r}_n)\Psi(\mathbf{r}_p,\mathbf{r}_p) \tag{3.27b}$$

$$\Gamma_p\Gamma_n V_{pn}\Psi(\mathbf{r}_p,\mathbf{r}_n) = -D\sum_{ij} \varphi_i(\mathbf{r}_p)\varphi_j(\mathbf{r}_n)\int\varphi_i^*(\mathbf{r}')\varphi_j^*(\mathbf{r}')\,\Psi(\mathbf{r}',\mathbf{r}')\,d\mathbf{r}'. \tag{3.27c}$$

It is physically obvious that the inclusion of the Pauli principle will effectively weaken the neutron and proton correlation in the deuteron and, therefore, will give rise to an additional contribution to deuteron break-up in the nucleus field. The usual reason for the break-up is the large gradient of the nuclear potential at the nuclear boundary. It can clearly be seen from eq. (3.27) and is quite understandable from the general physical considerations that the contribution from the Pauli-terms is most significant, generally speaking, in the inner region of nucleus, i.e. when the nucleon density is maximum. Therefore, two operators which cause the deuteron break-up (i.e. the nuclear and Pauli terms) are localized in different spatial regions. Besides that, the corresponding contributions are differently dependent on energy and therefore, may be separated experimentally. The contribution from the effects of Pauli projecting may be estimated quantitatively on the basis of the difference

$$[V_{np} - QV_{np}]\Psi(n, p) = (\Gamma_p + \Gamma_n - \Gamma_n\Gamma_p)\Psi(n, p), \tag{3.28}$$

where $Q = (1 - \Gamma_p)(1 - \Gamma_n)$ the projection operator. The expression (3.28) may be interpreted to be the difference between the effective n-p interaction with and without the Pauli-effects. It is proposed in [Ga 74] to estimate such a difference through the *dimensionless relation*

$$\mu = \frac{\langle V_{np} - QV_{np}\rangle}{\langle V_{np}\rangle}, \tag{3.29}$$

where the angular brackets denote the diagonal matrix elements. To perceive the general trend, the relation (3.29) may be estimated by applying to the nucleus the Fermi-gas model with the Fermi-momentum k_F related conventionally to the nucleon density (for the normal nuclear density $k_F = 1.36\ \mathrm{fm}^{-1}$). Let the approximation

$$\Psi(\mathbf{r}_p, \mathbf{r}_n) = \varphi_0(r)\exp(i\,\mathbf{K}_d\,\mathbf{R}_d) \tag{3.30}$$

(where $\varphi_0(r)$ is the normalized Hulthen function) be used as the three-body wave function $\Psi(\mathbf{r}_p, \mathbf{r}_n)$. Fig. 6 (borrowed from [Ga 74]) presents the dependence of μ on the incident deuteron energy obtained on the above assumptions for two values (total and half) of the nuclear density.

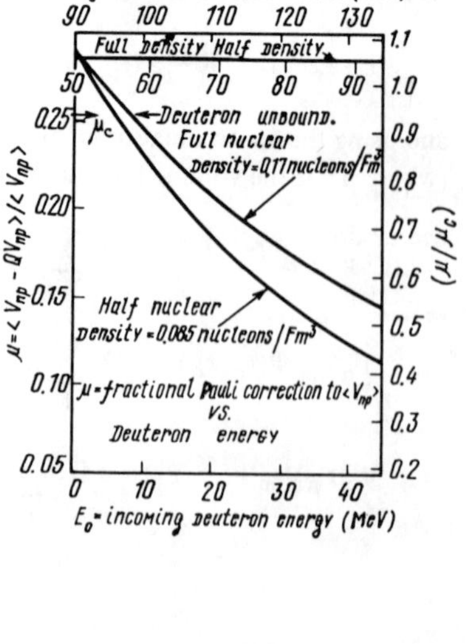

Fig. 6

The dimensionless matrix element $\mu = \dfrac{\langle V_{np} - QV_{np}\rangle}{\langle V_{np}\rangle}$ as a function of deuteron energy inside the nucleus for the normal and halved nuclear densities (i.e. with $\epsilon_F = 38.4$ MeV and 24.2 MeV respectively). The energy scale of the bombarding deuteron below is obtained from the internal deuteron energy by substracting 90 MeV in the case of total density, and 50 MeV in the case of halved density. It will be noted that the critical value of μ ($= \mu_c$) at which the model deuteron becomes unbound in nuclear matter of normal density is 0.27. Shown to the right is the μ-scale in units μ_c. At $\mu \cong \mu_c$, the effects of the Pauli principle are expected to play the major role in the reaction process.

The value of μ is the direct measure of the variation in the neutron-proton interaction due to the Pauli principle. $\mu_c \simeq 0.27$ is the critical value at which the model deuteron becomes to be unbound in the nuclear matter of normal density. It can clearly be seen from Fig. 6 that the contribution from the Pauli-effects at low energies of the incident deuterons is essentially independent of the nuclear region in which the reaction proceeds, namely in the inner region (the total density of nuclear matter) or on the surface (the halved density). Since, however, most of the reactions are localized on the nuclear surface and the wave function of incident deuteron is suppressed in the inner region of nucleus, the accurate quantitative predictions should be made by studying the effects of the Pauli-break-up in terms of a more realistic model including the Coulomb repulsion and the imaginary part of the optical potential. In any case, it is physically clear that the influence of the effects due to the Pauli principle will be significant at low energies of incident deuterons. Unfortunately, as far as we know, such studies in terms of a realistic nuclear model have not been made yet.

As to the pure nuclear break-up of deuteron, the authors of the recent study [Au78], which was also made in terms of the three-body model using the Schrödinger approach, have revealed a very interesting compensation effect of the imaginary potential. It has been found that, as the imaginary part of the nucleon-nucleus optical potential increases (from zero), the deuteron break-up cross section decreases, though the total cross section of the reaction (as determined from the decrease in the incident particle flux) increases significantly with increasing the strength of the absorbing potential. In such a way, although the imaginary potential gives rise to absorption, it simultaneousely causes the break-up cross section to decrease in the low partial waves, so the two effects prove to compensate each other almost completely [Au78]. At the same time, the stripping cross section (studied for the particular reaction ^{40}C(d, p) at $l = 2$, $E_d = 25.5$ MeV, $Q = 4.4$ MeV) decreases markedly when the absorbing potential is introduced in the nucleon-nucleus interaction, although the form of the differential cross section remains almost invariable. It will be noted in summing up that the regularities of the Pauli- and nuclear break-up of deuterons in the nuclear field found in the above described studies will probably permit the two effects to be experimentally separated and studied individually.

3.4 Inclusion of the Core Excitations and the Effect of Nucleon Reduced Widths

In the present section, we shall follow the works [Lo75; Be73] in discussing the formalism which makes it possible to take account of the virtual or real excitations of the core in direct nuclear reactions and the modifications required to include the nucleon reduced widths S_i (more strictly, the modification in case of $S_i < 1$). The logical chain of reasoning here is as follows. Assume first that the core in the initial and final states is not excited, i.e. it is in the ground state. At the same time, our aim is to take account of the virtual excitations of the core in the course of three-body scattering. In this case, the following scheme can conveniently be used. A many-channel generalization of FE is to be written at first (this is an almost trivial task); after that, we shall apply orthogonal projection technique (in the form of the Feshbach projection formalism [Fe58]) in view of replacing the *many-channel* three-body problem by the *one-channel* problem including the contribution from the virtual channels to the effective operators of three-body scattering. Quite the same concept has been the basis of the Feshbach theory of the nucleon-nucleus optical potential where the contribution of virtual excitations of nucleus is transferred to nonlocality and energy dependence of the effective optical potential [Fe58]. If, however, the core is actually excited (say, in the final state) in the cource of three-body reaction, we have to isolate two states, namely the ground state and the observable excited state. As to the contribution from the virtual states, they can be conveniently included, as before, through the effective scattering operators. Such an approach, however, gives rise to two significant difficulties. The first is associated with the tedious (or even unnecessary) calculations of the accurate many-channel off-shell t-matrix of nucleon scattering by the nucleus (these are the t-matrices involved in the kernels of many-channel three-body equations). According to [Lo75] the latter difficulty can be avoided in the same way as in the nuclear optical model, i.e. by averaging the accurate singular t-matrix $t(E)$ over some energy interval Δ, thereby resulting in the replacement of the operator $t(E)$ by the operator $t(E + i\Delta)$ which is much smoother than $t(E)$ and may be taken from the nuclear optical model.

Another difficulty is associated with the highly complicated form of the obtained three-body integral equations and can be eliminated by turning to distorted waves and treating the coupling between the channels with nonexcited and excited cores on the basis of distorted waves. It may be expected that such a procedure will permit the perturbation theory or the first iterations of three-body integral equations to be used instead of the solutions of exact equations to take account of the coupling between the channels.

Thus, the Hamiltonian of the $A + 2$ nucleon system may be written as

$$H = H_0 + V = (t_1 + t_2 + h_3) + V_1 + V_2 + V_3 ,$$

where the particles are labelled as in Section 2 above, t_1 and t_2 denote the kinetic energies of neutron and proton respectively; h_3 is the Hamiltonian of the core. For the nucleon-nucleus channel, the distorting potentials u_α will be introduced:

$$V_\alpha = u_\alpha + v_\alpha \tag{3.31}$$

so that only the residual interactions v_α will give rise to the transitions between the states of the core $|\varphi_c\rangle$:

$$\langle\varphi_c|u_\alpha|\varphi_{c'}\rangle = \delta_{cc'}\langle\varphi_c|u_\alpha|\varphi_c\rangle .$$

Introduce a new Hamiltonian

$$\mathcal{H}_0 = H_0 + u = t_1 + t_2 + h_3 + u_1 + u_2 + V_3 \tag{3.32}$$

and the appropriate resolvent $\mathcal{G}_0 = (z - \mathcal{H}_0)^{-1}$ which are diagonal in the space of the core states

$$\langle\varphi_c|\mathcal{H}_0|\varphi_{c'}\rangle = \delta_{cc'}\mathcal{H}_0^c \; ; \quad \langle\varphi_c|\mathcal{G}_0|\varphi_{c'}\rangle = \delta_{cc'}\mathcal{G}_0^c ,$$

and describe the conventional three-body system with the core in the fixed state $|\varphi_c\rangle$. After that, repeating the ordinary derivation of FE for the transition operators $U_{\beta\alpha}$ [Sc74a] on the basis of distorted waves, we shall readily obtain

$$U_{\beta\alpha} = \bar{\delta}_{\alpha\beta}\,\mathcal{G}_0^{-1} + \sum_\gamma \bar{\delta}_{\beta\gamma}\,t_\gamma\,\mathcal{G}_0\,U_{\gamma\alpha} \tag{3.33}$$

where $\bar{\delta}_{\alpha\beta} \equiv 1 - \delta_{\alpha\beta}$; the operator t_γ is determined through the channel resolvent $G_\gamma = (z - H_0 - v_\gamma)^{-1}$ as

$$t_\gamma = v_\gamma + v_\gamma G_\gamma v_\gamma$$

and satisfies LSE

$$t_\gamma = v_\gamma + v_\gamma G_0 t_\gamma \tag{3.34}$$

Introduce, as usual, the orthogonal subspaces H_P and H_Q, and the corresponding projectors

$$P = \sum_{c=1}^{c_0} |\varphi_c\rangle\langle\varphi_c| \quad \text{and} \quad Q = 1 - P$$

where P comprises the real or important states of the core; in most cases, H_P includes but a single (ground) state of the core. Now, we shall derive the FE set for our model. With this purpose, we shall first project LSE (3.34) on to the subspaces H_P and H_Q:

$$t_\alpha^{PP} = v_\alpha^{PP} + v_\alpha^{PP} G_0 t_\alpha^{PP} + v_\alpha^{PQ} G_0 t_\alpha^{QP} \tag{3.35a}$$

$$t_\alpha^{QP} = v_\alpha^{QP} + v_\alpha^{QP} G_0 t_\alpha^{PP} + v_\alpha^{QQ} G_0 t_\alpha^{QP}\ , \tag{3.35b}$$

where we took into account that G_0 is diagonal in the core states $|\varphi_c\rangle$. After excluding t^{QP} from eq. (3.35a) and eq. (3.35b), we get the following LSE for projection t_α^{PP}:

$$t_\alpha^{PP} = W_\alpha^{PP} + W_\alpha^{PP} G_0 t_\alpha^{PP}\ , \tag{3.36}$$

where the effective potential W_α^{PP} is determined in the usual way [Fe 58]

$$W_\alpha^{PP}(z) = v_\alpha^{PP} + v_\alpha^{PQ} \tilde{G}_\alpha^{QQ} v_\alpha^{QP}\ , \tag{3.37}$$

where (as in the previous Section)

$$\tilde{G}_\alpha^{QQ} = Q[Q(z - H_0 - v_\alpha)Q]^{-1} Q\ .$$

Let us appeal now to the three-body equations which determine the P-projection of the three-body transition operator $U_{\beta\alpha}^{PP}$ corresponding to eq. (3.33). Acting by operators P and Q on eq. (3.33) similarly to the two-body case and excluding the nondiagonal transition operator U^{PQ}, we find the basic equation [Lo 75]:

$$U_{\beta\alpha}^{PP} = \bar{\delta}_{\beta\alpha} (G_0^{-1})^{PP} + \sum_{\nu\mu} \bar{\delta}_{\beta\nu} R_{\nu\mu}^{PP} G_0 U_{\mu\alpha}^{PP}\ , \tag{3.38}$$

where the operator

$$R_{\nu\mu}^{PP} = \delta_{\nu\mu} t_\nu^{PP} + t_\nu^{PQ} G_{\nu\mu}^{QQ} t_\mu^{QP} \tag{3.39}$$

and the generalized propagator

$$G_{\nu\mu}^{QQ} = \sum_\rho \bar{\delta}_{\nu\mu} G_0 (T^{QQ})_{\nu\rho}^{-1} \tag{3.40}$$

(in which $T_{\nu\rho}^{QQ} = \delta_{\nu\rho} - \bar{\delta}_{\nu\rho} t_\rho G_0$) satisfy the equations

$$G_{\nu\mu}^{QQ} = \delta_{\nu\mu} G_0 + \sum_\sigma \bar{\delta}_{\nu\sigma} G_0 t_\sigma^{QQ} G_{\sigma\mu}^{QQ}\ . \tag{3.41}$$

Since the structure of the basic eq. (3.38) is very complex, its kernel is expedient to interpreted physically. In the lowest order for eqs. (3.38) and (3.39), we find (neglecting the distortions)

$$\bar{\delta}_{\beta\mu}\delta_{\nu\mu}t_{\mu}^{PP}G_0 + \bar{\delta}_{\beta\nu}t_{\nu}^{PQ}G_{\nu\mu}^{QQ}t_{\mu}^{QP} \tag{3.42}$$

and t_ν^{PP} is fully determined by the effective potential $W_\nu^{PP}(z)$ (see eq. (3.36)). In the lowest order, the first term of the kernel, $\delta_{\nu\mu}t_\mu^{PP}$ gives $V_\mu^{PP}\delta_{\mu\nu}$, i.e. the pure potential interaction of one of the nucleons with the core. In the second order, the first term gives the terms of the form $V_\mu^{PQ}G_0V_\nu^{QP}$, i.e. the interaction of one of the nucleons with the core through its excitation and subsequent deexcitation (see Fig. 7a). For the second term of eq. (3.42) in the lowest order again, we find

$$V_\nu^{PQ}G_0V_\mu^{QP}\bar{\delta}_{\mu\nu}, \tag{3.43}$$

i.e. this term makes a nonzero contribution at only $\mu \neq \nu$ and, therefore, describes the two-step process where the first step is the nucleon interaction with the core through the potential V_ν^{PQ} resulting in the core excitation to the Q-subspace, and the second stage is the scattering of another nucleon by the core accompanied by core deexcitation. The process may be illustrated by the diagram shown in Fig. 7b. Such a diagram describes the indirect interaction of the valence nucleons, i.e. the interaction through core excitations and corresponds to the effective three-body interaction.

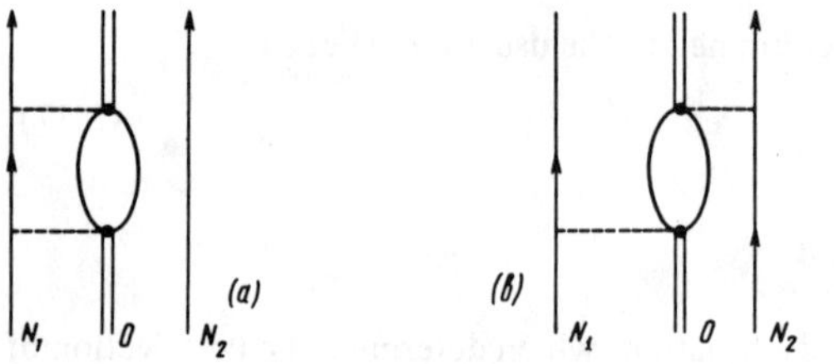

Fig. 7 a) and b) see text.

If the above discussed conclusion is generalized for the general cluster system, it will be seen that additional *three-body* forces appear in the three- and, in general, many-cluster system due to the virtual excitations of the clusters and that the looser the clusters involved, the greater, generally speaking, the contribution from these effects.

Examine now the further simplifications of the modified FE set eq. (3.38). Since the projected t-matrix $t_\nu^{PP}(E)$ is of complex singular analytical structure (such as the poles corresponding to the bound states in various channels, the branching points, the cuts, etc.) and because the averaged optical potential

$$V_{opt}(E) = \langle W^{PP}(E)\rangle = W^{PP}(E + i\Delta) \tag{3.44}$$

(where Δ is the averaging interval) is almost always used in practice instead of the exact optical potential (3.37), it proves also expedient in the three-body problem to carry out the appropriate averagings in energy and to use the corresponding averaged transition operators $\langle U_{\beta\alpha}^{PP}\rangle$ instead of the exact values $U_{\beta\alpha}^{PP}$. Such averaging is difficult to carry out accurately in the eq. (3.38); it is possible, however, to approximately take into account that the many-channel t-matrices t_ν^{PP} are the most sharply variable values in eq. (3.38).

The following method for averaging t_ν^{PP} has been proposed in [Lo75]: the operator t_ν^{PP} should be weighed with a certain Lorentz-factor and then integrated over the interval Δ. As usual, this gives:

$$\langle t(\mathrm{E}) \rangle = t(\mathrm{E} + \mathrm{i}\Delta) \tag{3.45}$$

The physical meaning of both eq. (3.45) and eq. (3.44) is quite understandable, namely the replacement $\mathrm{E} \to \mathrm{E} + \mathrm{i}\Delta$ shifts all the singularities of $\mathrm{V_{opt}(E)}$ and $\langle t(\mathrm{E}) \rangle$ downwards to the complex plane. This means the exclusion of the reaction channels which are in correspondence with the shifted singularities and the replacement of the contribution of such channels by a certain absorption generated by the optical potential and by the averaged t-matrices. This replacement gives rise to nonunitary three-body transition operators. In this case, according to [Lo75], it is the t-matrices and not the potentials that should be averaged, for the singularities of the effective potentials and the t-matrices are not the same (in the given channel, the t-matrix has the poles corresponding to the bound states and resonances in the channel, while the optical potential includes the cuts corresponding to inelastic channels in the given subsystem, so it is the singularities of the t-matrices that should be shifted to the complex plane).

It is clear that the eq. (3.38) is very difficult to solve even with the averaged operators and, therefore, we shall apply the approach with distorted waves. If the distorted waves u_α (see eq. (3.31)) are so selected that the residual interactions v_α prove to be "small", the obtained equations may be solved iteratively.

Let, for example, the equation for scattering of a nucleon by another nucleon coupled with the core be written (in this case, $\beta = \alpha = 1$):

$$\mathrm{U}_{11}^{\mathrm{PP}} = \bar{t}_2^{\mathrm{PP}} + \bar{t}_3^{\mathrm{PP}} + \bar{t}_2^{\mathrm{PQ}}\, G_{22}^{\mathrm{QQ}}\, \bar{t}_2^{\mathrm{QP}}\,, \tag{3.46}$$

where the upper lines denote the t-matrices averaged according to eq. (3.45). However, it follows from eq. (3.41) that the diagonal matrix element $G_{22}^{\mathrm{QQ}} = 0$ in the lowest order. Introduce now the distorted waves. The eigenstates of the channel Hamiltonians $\mathrm{H}_\alpha = \mathrm{H}_0 + \mathrm{V}_\alpha$ will be denoted as $|\Phi_{\alpha\mathrm{m}}\rangle$, so that

$$\mathrm{H}_\alpha |\Phi_{\alpha\mathrm{m}}\rangle = \mathrm{E}_{\alpha\mathrm{m}} |\Phi_{\alpha\mathrm{m}}\rangle .$$

In this case, as usual [Sc 74a], the amplitudes of transitions $\alpha\mathrm{m} \to \beta\mathrm{n}$ are determined by the operator $\mathrm{U}_{\beta\alpha}$ sandwiched between the channel states (n and m are the quantum numbers of bound states in the two-body subsystems)

$$\mathrm{A}_{\beta\mathrm{n},\alpha\mathrm{m}} = \langle \Phi_{\beta\mathrm{n}} | \mathrm{U}_{\beta\alpha} | \Phi_{\alpha\mathrm{m}} \rangle .$$

The distorted waves $|\chi_{\alpha\mathrm{m}}^{(\pm)}\rangle$ are determined as EF of the Hamiltonian $\mathrm{H}_\alpha + (\mathrm{u}_\beta + \mathrm{u}_\gamma)$, i.e. they should be found by solving the three-body problem[1]). They can readily be expressed through the channel states $|\Phi_{\alpha\mathrm{m}}\rangle$ as

$$|\chi_{\alpha\mathrm{m}}^{(\pm)}\rangle = [1 + G_\alpha^{(\pm)}(\mathrm{u}_\beta + \mathrm{u}_\gamma)]\, |\Phi_{\alpha\mathrm{m}}\rangle . \tag{3.47}$$

[1]) A different procedure was used in [Ku 78d, 81b] to introduce the distorted-wave basis in the three-body equations.

After that, the transition amplitude $A_{\beta n,\alpha m}$ may be introduced through distorted waves in the general case as the sum of two terms:

$$A_{\beta n,\alpha m} = \delta_{\alpha\beta} \langle \Phi_{\beta n} | u_\alpha + u_\gamma | \chi^{(+)}_{\alpha m} \rangle + \langle \chi^{(-)}_{\beta n} | U_{\beta\alpha} | \chi^{(+)}_{\alpha m} \rangle . \tag{3.48}$$

Substituting eq. (3.46) in eq. (3.48), we find (bearing in mind the remark made after eq. (3.46)):

$$A_{1n,1m} = \langle \Phi_{1n} | u_2 | \chi^{(+)}_{1m} \rangle + \langle \chi^{(-)}_{1n} | \bar{t}_2^{PP} + \bar{t}_3^{PP} | \chi^{(+)}_{1m} \rangle . \tag{3.49}$$

The first term describes the *potential* scattering of the incident particle by the core and, therefore, this term contributes the elastic scattering. The second term corresponds to interference of two scatterings, namely from the core and from the second nucleon in the core field in the distorted wave representation. The identity of the nucleons is to be taken into account by adding to the amplitude (3.49) the contribution from the exchange scattering [Am67]

$$A_{ex} \equiv A_{2n,1m} = \langle \chi^{(-)}_{2n} | (z - H_0)^{PP} + \bar{t}_3^{PP} + \bar{t}_1^{PQ} G_0 \bar{t}_2^{QP} | \chi^{(+)}_{1m} \rangle . \tag{3.50}$$

On the energy shell, the first term may be replaced by v_2^{PP} since $\chi^{(-)}_{2n}$ is the eigenfunction of the operator $(H_0 + v_2^{PP})$. This gives eventually:

$$A_{2n,1m} = \langle \chi^{(-)}_{2n} | v_2^{PP} + \bar{t}_3^{PP} + \bar{t}_1^{PQ} G_0 \bar{t}_2^{QP} | \chi^{(+)}_{1m} \rangle . \tag{3.50b}$$

Here, the contribution of the term with $\bar{t}^{PQ} G_0 \bar{t}_2^{QP}$ may be graphically interpreted as in Fig. 8 where the double line shows the core. It follows from the figure that the exchange interaction of two nucleons in such a process takes place through the excitation and deexitation of the core.

Fig. 8 See the text.

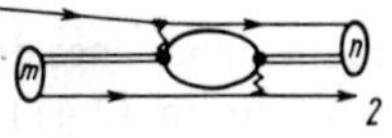

The antisymmetrized amplitude is given by the expression

$$A^{(a)}_{nm} = A_{1n,1m} - A_{2n,1m} . \tag{3.51}$$

Examine now the treatment of stripping and pick-up in terms of our model. The stripping reaction is described by the transition operator U_{13}^{PP} (the ejected particle is labelled $\beta = 1$). Again, we obtain for U_{13}^{PP} in the lowest approximation

$$A_{str} \equiv A_{1n,3m} = \langle \chi^{(-)}_{1n} | (z - H_0)^{PP} + \bar{t}_2^{PP} + \bar{t}_2^{PQ} G_0 \bar{t}_1^{QP} | \chi^{(+)}_{3m} \rangle$$

where $|\chi^{(+)}_{3m}\rangle$ is the eigenfunction of the operator $(H_0 + v_3)^{PP}$.

Therefore, we obtain on the energy shell:

$$A_{1n,3m} = \langle \chi^{(-)}_{1n} | v_3^{PP} | \chi^{(+)}_{3m} \rangle + \langle \chi^{(-)}_{1n} | \bar{t}_2^{PP} + \bar{t}_2^{PQ} G_0 \bar{t}_1^{QP} | \chi^{(+)}_{3m} \rangle . \tag{3.52}$$

In case of the natural assumption $u_3 = 0$, i.e. $v_3 = V_3$, the first term of eq. (3.52) coincides with the standard DWBA amplitude[1]), whereas the second term corresponds to the

[1]) An improved version of the DWBA formalism based on a different basis of distorted waves has been described in [Ku78d, 81b].

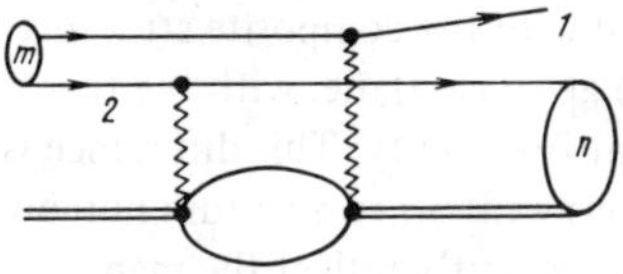

Fig. 9 See the text.

transitions accompanied by excitation of the core [St 74a; Da 74], see Fig. 9. The pick-up amplitude is of quite similar form:

$$A_{3n,1m} = \langle \chi_{3n}^{(-)} | v_3^{PP} | \chi_{1m}^{(+)} \rangle + \langle \chi_{3n}^{(-)} | \bar{t}_2^{PP} + \bar{t}_1^{PQ} G_0 \bar{t}_2^{QP} | \chi_{1m}^{(+)} \rangle . \tag{3.53}$$

Our reasonings may be summarized as follows. The formalism [Lo 75] set forth above makes it possible to calculate the amplitudes of all the direct processes in the core + two particles system including the possible excitiation of the core using the three-body integral equations on the basis of distorted waves in the lowest orders (on the multiplicity of scatterings). In this case, the Pauli principle, which forbids the valence nucleons to fill the orbits already filled by the core nucleons, is disregarded. However, the formalism presented in Chapter 2 may be used to include the main features of the Pauli principle through the orthogonal projecting of the scattering operators in eq. (3.38) and thereafter. At the same time, such a procedure gives rise to some ambiguities due to the transition from the true interactions V_α to the optical potentials and from the initial t-matrices t_α to the t-matrices averaged over energy $\bar{t}_\alpha = t_\alpha(E + i\Delta)$. The ambiguities arise because the concept of orthogonality to the core states forbidden by the Pauli principle should also be modified when including the core excitations. If, for example, the standard phenomenological optical potential with the parameters fitted to the experimental cross sections of elastic scattering is used as optical potential (3.4) $W^{PP}(E + i\Delta)$ the imposition of additional orthogonality conditions will significantly alter the phase shifts of scattering [Am 70] thereby necessitating the corresponding *redefinition of all (or main) parameters of optical potential* to bring the new observables again in agreement with experiment. The same also relates to the projecting of the averaged t-matrix $\bar{t}_\alpha(E)$. One of the possible way of overcoming this difficulty is the separation of the contributions from the virtual excitations of the core and from the effects of the Pauli principle. In other words, since both types of the contributions appear to be corrections of the basic process, it will be expedient to neglect the effects of the Pauli principle in the terms describing the core exitation but include them in the operators and wave functions which are not contributed by the core excitations, for example when constructing the distorted waves.

Another way, which is more fundamental but, in turn, much more difficult, is to derive the interactions of clusters (or particles with core) and the *form of* the orthogonality conditions from the general N-particle integral equations, whereupon the approximate forms, which include correctly *the two types of the effects*, will be obtained from such equations. Some attempts to apply this procedure have already been made. In particular, the approximate integral equations of scattering (using the $^{16}O + 2N$ system as an example) have been obtained [Be 79a] from the exact N-body formalism. As to the form of the orthogonality conditions following from the rigorous N-body equations, this problem will be discussed in detail in Chapter 5 below.

Now, we shall briefly discuss the last effect associated with the composite structure of the particles involved in the reaction (of the core, in the given case), i.e. with the fact that the reduced widths of nucleons or clusters are different from unity. This difference is well known [Bo 69] to reflect the existence of other channels with other quantum numbers of fragments and other fragmentation in the system, i.e. to implicitly reflect the many-body nature of the process.

The t-matrix element corresponding to the three-body process of rearrangement $\alpha + (\beta + \gamma) \rightarrow \beta + (\alpha + \gamma)$ may be expressed, as usual through the transition operator $U_{\beta\alpha}$

$$T_{\beta n, \alpha m} = S_{\beta n} S_{\alpha m} \langle \Phi_{\beta n}, q_\beta | U_{\beta\alpha}(E + i0) | \Phi_{\alpha m}, q_\alpha \rangle , \tag{3.54}$$

where $\Phi_{\alpha m}$ and $\Phi_{\beta n}$ are the "effective" wave functions of two-body bound states; $S_{\beta n}$ und $S_{\alpha m}$ are the respective spectroscopic factors $(0 \leqslant S_{\alpha m}, S_{\beta n} \leqslant 1)$. After that, the two-body t-matrices t_γ forming part of the kernels of three-body equations can conveniently be presented in the form of the separable expansion

$$t_\gamma(E) \cong \sum_i S_{i\gamma} | g_{i\gamma} \rangle \tau_{i\gamma}(E) \langle g_{i\gamma} | S_{i\gamma} , \tag{3.55}$$

where $| g_{i\gamma} \rangle$ are the formfactors of the bound states and the propagators

$$\tau_{i\gamma} = \frac{-1}{E + \kappa_{i\gamma}^2} \int_0^\infty \frac{g_{i\gamma}^2(p)\, p^2\, dp}{(E - p^2 + i\epsilon)(p^2 + \kappa_{i\gamma}^2)}$$

($\kappa_{i\gamma}^2 = E_{i\gamma}$ is the binding energy of the bound state $i\gamma$). The spectroscopic factors $S_{i\gamma}$ give the weight of the effective two-body bound state $| i\gamma \rangle$ in the two-cluster wave function [Li 66].

Further, after introducing in the standard manner [Sc 74a] the operators $X_{\beta n, \alpha m}$ and $Z_{\beta n, \alpha m}$

$$X_{\beta n, \alpha m} = S_{\beta n} S_{\alpha m} \langle \Phi_{\beta n}, q_\beta | G_0 U_{\beta\alpha} G_0 | \Phi_{\alpha m}, q_\alpha \rangle$$

$$Z_{\beta n, \alpha m} = \bar{\delta}_{\alpha\beta} S_{\alpha m} S_{\beta n} \langle \Phi_{\beta n}, q_\beta | G_0 | \Phi_{\alpha m}, q_\alpha \rangle$$

we get the FE set [Be 73]

$$X_{\beta n, \alpha m} = Z_{\beta n, \alpha m} + \sum_{\gamma, r} Z_{\beta n, \gamma r} \tau_{\gamma r} X_{\gamma r, \alpha m} . \tag{3.56}$$

Owing to the structure of eq. (3.55) and eq. (3.56), the dependence of the three-particle reaction cross sections found by solving eq. (3.56) on the factors $S_{i\gamma}$ proves to be essentially *nonlinear*, in contrast to the standard DWBA-pattern. The calculations of the cross section of the reactions $^{16}O(d, p)$ or $^{16}O(d, n)$ [Be 73] carried out using the presented formalism have shown that the differential cross sections of the stripping reaction are strongly dependent on the spectroscopic factors of the $2S_{1/2}$ and $2D_{5/2}$ states of $^{17}O-^{17}F$ not only in value *but also in form*. The character of the dependence may be seen from Figs. 10 and 11.

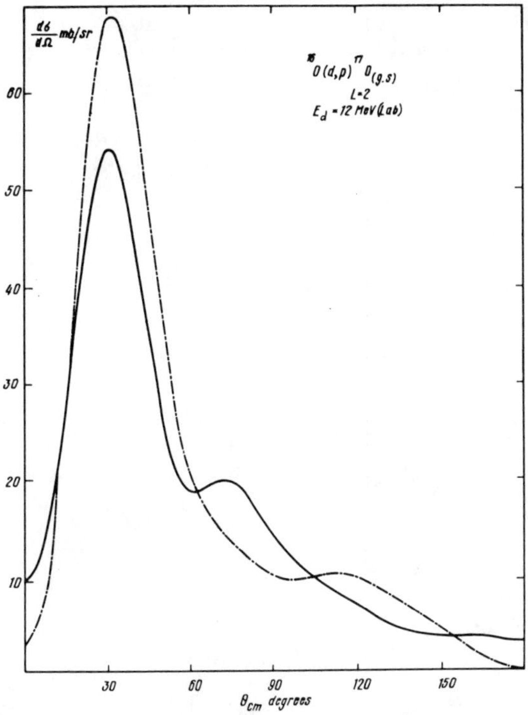

Fig. 10 The differential cross sections of the $^{16}O(d,p)$ reaction at $E_d = 12$ MeV (lab.) calculated on the basis of FE with separable potentials for different spectroscopic factors $S_{n,p}(l_j)$ of nucleon orbitals. The solid line is $S_p(d_{5/2}) = S_n(d_{5/2}) = S_p(s_{1/2}) = S_n(s_{1/2}) = 1$; the dash-dotted line is $S_p(d_{5/2}) = S_n(d_{5/2}) = 1$, $S_n(s_{1/2}) = S_p(s_{1/2}) = 0$, i.e. the interaction in the $s_{1/2}$ state is absent.

The comparison with the DWBA-pattern of stripping would have yielded more information if the spectroscopic factors of neutrons only were varied while the proton S-factors remained invariant. Such a procedure would correspond better to the simple Butler pattern of stripping (in which protons fails to penetrate into the inner region of nucleus) since the variations of the proton S-factors will change the t-matrix of proton scattering by the nucleus (in the model studied in [Be73]).

It is quite clear in case of heavier nuclei that the effect of the spectroscopic factors will be even stronger than in the case examined above and, therefore, the problem as a whole should be studied in more detail.

3.5 Conclusions

The present Chapter devoted to the presentation of the theoretical schemes developed in the core + two valence particles (for example, nucleons) problem is expedient to con-

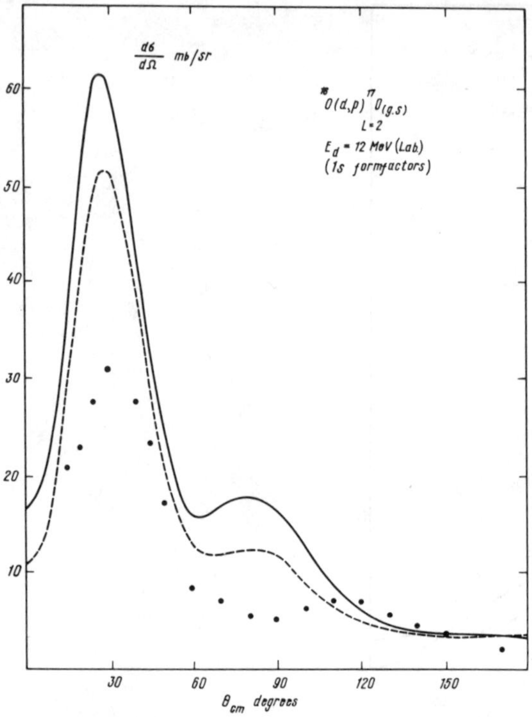

Fig. 11 The same as in Fig. 10 but with other spectroscopic factors of neutrons and protons. The solid line is $S_n(d_{5/2}) = S_p(d_{5/2}) = S_n(s_{1/2}) = S_p(s_{1/2}) = 1$; the dashed line is $S_n(d_{5/2}) = S_p(d_{5/2}) = 0.9$, $S_n(s_{1/2}) = S_p(s_{1/2}) = 0.8$. As a wave function of bound neutron, it is used one-node 1 s orbital. The dots represent the experimental data.

clude by formulating again the points where the theory has been elaborated and the points where further efforts should be made.

(1) In case of a core with only filled shells and subshells, OCM in combination with solution of the three-body problem will permit the Pauli principle to be adequately included in the integral Faddeev-approach and in the differential Schrödinger approach. It will be noted that the two approaches are *rather complementary than alternative to each other*, a fact that is due to the complementary nature of the approximation used in the approaches. In particular, the semiquantitative study carried out in terms of the differential approach has shown that the effects due to the Pauli principle give rise to an additional contribution to deuteron break-up which proves to be significant at low energies and localized mainly in the inner region of the nucleus.

(2) The appropriate three-body generalization of the Feshbach orthogonal projection formalism has been used to construct the theoretical scheme which makes it possible to include the virtual excitations of the core as the three-body rearrangement reaction proceeds. In this case, the equations derived [Lo75] allow a simple and clear physical interpretation.

(3) The further inclusion of the complex structure of the particles involved makes it necessary to take account of the spectroscopic factors of nucleons (and clusters) in two-body subsystems. This could be done without difficulties in case of a magic core because of the direct relationship of the bound state formfactors to the t-matrix of two-body scattering through the residue in the bound-state pole. Specific calculations have shown that the role of the two-body spectroscopic factors is very important and quite different from the role played by them in the DWBA-approach since, in case of the "accurate" treatment of three-body effects, the variations in the S-factors result in variation of not only value but also form of the differential cross sections of the rearrangement reactions.

Nevertheless, many problem have not been solved as yet. We shall discuss the most important ones of them.

(4) Even if the developed problems, which are described above in (1)–(3), are taken into consideration, there does not yet exist a unified formalism which would simultaneously include the Pauli principle, the virtual excitations of the core, and the difference of the reduced widths from unity, or any pair of the three effects.

(5) If we impose the requirement that the wave function of (A + 2) particles should be antisymmetric, then a consistent approach would give rise to not only two-body exchange forces but also three-body exchange forces due to triple exchanges of nucleons between all three constituents. The first estimates of such effects have been obtained in [Sc77, 79]. It is clear that the three-body forces will be significant in only the region of strong overlapping of all three cluster, i.e., in the inner zone of nuclei.

(6) The case of a core with partly filled shells gives rise to additional difficulties associated with the inclusion of semiforbidden states [Sa77] and, what is most important, with the rigorous derivation of the practically effective two- and three-body interactions from the general N-body equations. The formal aspect of such a derivation has been presented in [Kr66; Be71], however, none of the methods for calculating the effective interactions in practice is available at present.

(7) Finally, the basic difficulty with the entire domain of three-body rearrangement processes with composite particles is the absence (or, more strictly, weak mathematical substantiation) of the effective methods of solving the three-body integral or differential equations including all the complications as compared with the 3N-system accounted for by the complex structure of particles. The actual advances in the interesting and promising field of the three-body processes should be expected from the development of the methods which would fall outside the constraints of the simple separable models in the integral approach or include the complex boundary conditions in the differential approach. The works by Merkurjev et al. [Me76, 79] seem to justify optimistic expectations in this respect.

4 Interaction Between the Lightest Clusters

4.1 The Orthogonality Condition Model (OCM); Optical Potentials With Forbidden States

Although the techniques for calculating the RGM integral kernels and the resultant integro-differential equations have been much perfected since recently, so it proved possible even to program the derivation of the formulas for the RGM kernels (the review of the proposed approaches and the detailed description of the methods and the results obtained are presented in a special volume [P77]), the RGM calculations in the realistic cases (i.e. where the cluster functions are not of pure oscillator nature and the NN-interactions is treated in a realistic form) are nevertheless very tedious and physically obscure. It seems reasonable in this case to examine the simpler approaches which would, however, retain the main features of the complete microscopic approach. Saito's OCM (see the review [Sa77] and the references therein) is one of these approaches. In his model, which is formulated in terms of RGM, the Pauli principle is included by orthogonalizing the scattering functions to the forbidden states (determined in this Section below) in the variable of the relative motion of clusters. The brief history of the problem is as follows. In their pioneer study of the α–α scattering within the frame of RGM, Okai and Park [Ok66] have found that the relative motion function $\psi_{\alpha\alpha}(R)$ exhibits the characteristic shell-model oscillations in the inner region and that the localization of the nodes of $\psi_{\alpha\alpha}(R)$ almost coincides with that of the nodes of the shell-model wave functions of ^{8}Be in the relative variable $R_{\alpha\alpha}$. It was also found subsequently that the RGM-equations (4.2) for the relative motion function $\psi(R)$ have so called redundant (or spurious) solutions $\chi_i(R)$ which may be admixed at all energies to the scattering function $\psi(R)$ with arbitrary weights without changing the phase shifts. It is the exclusion of such spurious solutions from the scattering function (i.e. orthogonalization to $\chi_i(R)$) that results *automatically* in that the shell oscillations arise in $\psi(R)$. On the other hand, the spurious solutions were identified with the forbidden states in RGM corresponding to the unity EV's of the overlapping kernel K. It was noted after that [Sa68, 69] that the effect of the antisymmetrization operator was mainly to exclude such forbidden states from the function of relative motion or from the complete A-body wave function. Then the following step seems to be natural: to discard at all the antisymmetrization operator from the RGM-ansatz, and to replace its effect by the projecting operator onto the allowed subspace.

More formally we get the following. The RGM-equations are derived from the variational principle:

$$\langle \delta\Psi | H - E | \Psi \rangle = 0 \tag{4.1a}$$

where H is the total Hamiltonian, whereas the ansatz

$$\Psi = A(\phi_a \phi_b \psi) \tag{4.1b}$$

(where ψ is the function of relative motion of clusters) is taken for the complete wave function. Substituting the ansatz in eq. (4.1) and using the fact that ϕ_a and ϕ_b are the eigenfunctions of the internal Hamiltonians H_a and H_b, we find from eq. (4.1):

$$(T_R + V_D - \epsilon)(1 - K)\psi + V_{ex}\psi = 0\,, \tag{4.2}$$

where V_D is the direct (folding) potential of interaction of clusters "a" and "b":

$$V_D(\mathbf{R}) = \int d\tau_a\, d\tau_b\, \phi_b^*(1,\dots n_b)\phi_a^*(1,\dots n_a)\Big(\sum_{\substack{i\in a\\ j\in b}} v_{ij}\Big)\,\phi_a(1,\dots n_a)\phi_b(1,\dots n_b)\,,$$

T_R is the kinetic energy operator in the relative motion variable, K is the overlapping kernel; V_{ex} is the exchange (nonlocal) potential determined as

$$V_{ex}\psi \equiv \int dR'\, V_{ex}(R,R')\psi(R') = (1+\delta_{ab})^{-1}\,\langle\phi_a\phi_b\,|\sum_{\substack{i\in a\\ j\in b}} v_{ij} - V_D\,|\,A(\phi_a\phi_b\psi)\rangle\,. \tag{4.3}$$

Since the antisymmetrized function in the ket position in eq. (4.3) vanishes for the strictly forbidden states, we get

$$V_{ex}\Gamma = 0\,, \tag{4.4}$$

where $\Gamma = \sum_i |\chi_i\rangle\langle\chi_i|$ is the projector onto the forbidden subspace. Therefore, if the forbidden states are added with arbitrary weights to the solution of RGM-equation (4.2) at any energy, then eq. (4.2) remains valid. In other words, FS are the spurious solutions for the RGM-equation (4.2), and any operator acting in the relative motion variable should be independent of the weight of the spurious components, i.e. should vanish in the forbidden subspace. This means actually that, in the case of reduction of many-body problem to the relative motion of two clusters, this motion (and the corresponding operators) should be treated in a subspace orthogonal to FS. The forbidden states proper are defined as the eigenfunctions of the overlapping kernel

$$\int K(\mathbf{R},\mathbf{R}')\,\chi_i(\mathbf{R}')\,d\mathbf{R}' = \lambda_i\chi_i(\mathbf{R}) \tag{4.5}$$

corresponding to the unity eigenvalues $\lambda_i = 1$.

It is of importance to emphasize that the entire derivation of OCM given in [Sa68,69] is *purely heuristic* since the scheme for a consistent and rigorous derivation of OCM *with estimation of the discarded exchange terms* is hardly feasible even in terms of RGM. True, the semiforbidden and almost forbidden states for two- and three-cluster systems were consistently taken into account in the framework of the RGM approach [So77, 79]. However, the consistent derivation and rigorous substantiation of OCM should better be made in the framework of the complete N-body approach, not RGM. This problem is treated in Chapter 5.

Thus, if the EV spectrum of the overlapping exchange kernel K is step-like, where the first EV's equal unity and the remaining EV's are very small, the basic RGM-equation may be replaced, to within a high accuracy, by the following expression [Sa68, 69]:

$$\Lambda(T_R + V_D(R) - \epsilon_{rel})\Lambda\psi(R) = 0 \tag{4.6}$$

or in the form [Sa68, 69]:

$$(T_R + V_D - E)\psi(R) = \sum_i |\chi_i\rangle\langle\chi_i|\,T + V_D\,|\,\psi\rangle\,, \tag{4.6a}$$

where $\Lambda = 1 - \sum_{i=1}^{n} |\chi_i\rangle\langle\chi_i|$ is the projecting operator which projects onto the allowed subspace. The subsequent studies carried out mainly in Japan (see [Pr 77] and references therein) have shown that OCM is fairly fruitful for it permitted a comparatively simple method to be used in describing a large variety of experimental data. In particular, Fig. 12 shows how OCM describes the phase shifts of the $\alpha-\alpha$ scattering [Ho 77].

Later on, OCM was generalized in several directions, in particular for the many-channel scattering and for the few-cluster systems (see e.g. the detail reviews [Pro 77; Ho 77a, 78]), in particular, the systems

$$d + {}^3He \rightleftharpoons p + {}^4He$$
$$\alpha + {}^{16}O \rightleftharpoons {}^8Be + {}^{12}C$$
$$d + {}^4He \rightleftharpoons {}^3He + {}^3H$$

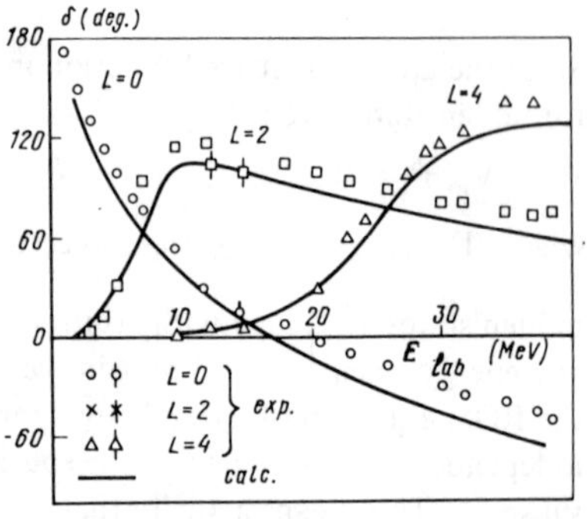

Fig. 12
The $\alpha-\alpha$ phase shifts in OCM for three lowest partial waves [Ho 77] with L = 0, 2, 4.

and others were studied. In all cases, a fairly good agreement was obtained between the OCM results and experiments or between the OCM and complete RGM results. This fact indicates that in many cases the simple and physically quite transparent OCM may effectively replace the tedious RGM formalism. However,further advances in the field are hampered by some factors, the most significant of which is that almost all real systems have a large spectrum of so called semiforbidden and almost forbidden states (AFS) which correspond to EV's of the operator K smaller than unity, $\lambda_n < 1$ ($\lambda_n \approx 1$ for AFS). Presented as an example in Table 1 are EV's of the operator $1 - K$ in the $\alpha + {}^{16}O$ system in the states with different angular momenta L (this table is borrowed from the review [Pr 77])

It follows from Table 1 that EV's with $\lambda_n \approx 1$ (and a number of EV with $\lambda_n \sim 0.5$) appear at different frequences of the oscillator functions of clusters instead of strictly unity EV's (or strictly zero EV) of the operator $1 - K$. In such cases, the meaning of OCM, as formulated in the works referred to above, gets obscure. If, however, AFS is used as rigorously forbidden states when constructing the projector Λ, there will appear at high energies the resonances of obscure nature [Cl 74; Sa 77] which are most probably spurious.

On the other hand, the role of the semiforbidden states is very important when studying the behaviour of the spectroscopic width amplitude

$$U_l(R) = \langle \phi_a \phi_b Y_{lm}(\hat{R}) | A(\phi_a \phi_b \psi(R)) \rangle$$

in the cluster overlapping region, especially in case of heavy ion scattering. It is the contribution from the semiforbidden states (in such systems as ${}^{16}O + {}^{16}O$, etc.) that results in

Tabelle 1 Eigen-values of the norm kernel of the system $\alpha + {}^{16}O$ in the two cases of the equal and unequal oscillator widths of clusters [Ho 77].

N	$\nu_\alpha = \nu_0$	$\nu_\alpha/\nu_0 = 1.47$				
		L = 0	L = 2	L = 4	L = 6	L = 8
0		0.3265 (– 6)	0.			
2		0.1117 (– 4)	0.6202 (– 5)			
4		0.6617 (– 3)	0.2997 (– 3)	0.1773 (– 3)		
6		0.1345 (– 1)	0.1234 (– 1)	0.9913 (– 2)	0.6197 (– 2)	
8	0.2292	0.2391	0.2384	0.2366	0.2339	0.2302
10	0.5103	0.5131	0.5127	0.5119	0.5105	0.5088
12	0.7185	0.7161	0.7161	0.7159	0.7154	0.7147
14	0.8459	0.8424	0.8423	0.8422	0.8420	0.8418
16	0.9178	0.9146	0.9146	0.9146	0.9145	0.9144
18	0.9568	0.9545	0.9545	0.9545	0.9545	0.9544
20	0.9775	0.9760	0.9760	0.9760	0.9760	0.9760
22	0.9884	0.9875	0.9875	0.9875	0.9875	0.9875
24	0.9941	0.9935	0.9935	0.9935	0.9935	0.9935
26	0.9970	0.9966	0.9966	0.9966	0.9966	0.9966
N		L = 1	L = 3	L = 5	L = 7	L = 9
1		0.1341 (– 5)				
3		0.4780 (– 4)	0.2784 (– 4)			
5		0.1431 (– 2)	0.1111 (– 2)	0.8070 (– 3)		
7		0.2726 (– 1)	0.2597 (– 1)	0.2370 (– 1)	0.2041 (– 1)	
9	0.3438	0.3518	0.3513	0.3505	0.3494	0.3479
11	0.6196	0.6179	0.6177	0.6174	0.6170	0.6164
13	0.7900	0.7861	0.7860	0.7858	0.7856	0.7854
15	0.8871	0.8834	0.8834	0.8833	0.8832	0.8831
17	0.9403	0.9375	0.9375	0.9375	0.9374	0.9374
19	0.9688	0.9669	0.9669	0.9669	0.9669	0.9669
21	0.9839	0.9827	0.9826	0.9826	0.9826	0.9826
23	0.9917	0.9910	0.9910	0.9910	0.9910	0.9910
25	0.9958	0.9953	0.9953	0.9953	0.9953	0.9953
27	0.9979	0.9976	0.9976	0.9976	0.9976	0.9976

practically total suppression of the inner oscillations in the scattering function. At the same time, it is physically obvious that the minor variations of the eigen frequencies of the clusters should not change the system spectrum significantly. The more comprehensive and detailed study of OCM (see Chapter 5) has shown that many of the difficulties with OCM can be eliminated by making the appropriate generalizations.

A more phenomenological approach to the composite-particle interactions was proposed in our works [Ne71, 72; Ku75]. The approach is much like OCM in its concepts but differs in actual realization. In the proposed approach the effective optical interaction potential between the clusters is directly constructed instead of calculations of the direct interaction potential V_D of clusters a and b in OCM. The effective potential is constructed in such a way that all the strictly forbidden states in the complete system a + b are the bound eigenstates of the potential. Therefore, the above mentioned optical potential (here, we mean the construction of only the real part of the potential, as in OCM, whereas the imaginary part may be merely fitted to the corresponding experimental data) should be sufficiently deep to 'contain" all the forbidden states. In this case, the basic requirement of the Pauli principle of orthogonalization of the scattering functions to the functions of the occupied states is automatically satisfied due to the hermiticity of the Hamiltonian. Since, however, not a single semiforbidden state exists in this scheme, the structure of the pure forbidden states can most simply be found directly from the shell model avoiding the calculations of RGM-kernels. After the structure of the forbidden states has been found, the rough values of parameters of the potential well can be found directly, whereupon their eventual values are specified by fitting to experimental data. In such a way, we obtain the microscopically substantiated optical potential (MSOP) whose eigenstate structure is in general coordinated with the shell model. Let the MSOP approach be illustrated using the canonical case of the $\alpha-\alpha$ scattering as an example. Consideration will be given to eight particles in the common oscillator well. Obviously, the configuration $(0s)^8$ with the total number of quanta $N = 0$ will be the lowest forbidden configuration. On dividing this configuration into two α-particles with configurations $(0s)^4$, the relative motion function will be of the form 0s. The configurations $\{(0s)^6\ (1p)^2\}$ and $\{(0s)^7\ (2d-2s)\}$ which lead to the relative motion functions of the form 2s and 2d will be the forbidden configurations with $N = 2$. At $N = 4$, the lowest allowed configuration $\{(0s)^4\ (1p)^4\}$ appears, which leads to the relative motion functions of the form 4s, 4d, and 4g. At the same time, admixtures of the forbidden configurations (for example, of the type $\{(0s)^k (N_1\, l)^{8-k}\}$ with $k > 4$) exist at both $N = 4$ and all higher N. As in RGM, such admixtures will reduce the wave function amplitude in the inner region without changing the on-shell properties of the function, i.e. without violating the phase shift fittings. This supression of the scattering functions in the inner region can readily be included by introducing the damping factor $A(E, R) < 1$, so that $A(E, R) \underset{\substack{R\to\infty \\ (E\to\infty)}}{\longrightarrow} 1$, i.e. the same as in the general case of transition from the nonlocal potential to the phase-equivalent local potential. Similarly to works [Bu75, 77a, b] the damping factor $A(E, R)$ may be constructed using the operator $\sqrt{1-K}$. However, in case of light cluster scattering, like $\alpha + \alpha$ such complications are insignificant and $A(E, R) \approx 0.85 - 0.90$.

Thus, we have to construct a local potential comprising the forbidden states as the bound states 0s, 1s, 1d (the spectroscopic notation is used here) and the allowed 2s-level near the threshod, the ground state ^{8}Be with two-node radial function.

In accordance with the above, the $\alpha-\alpha$ MSOP can readily be constructed in the form of the standard local static Saxon-Woods potential [Ne71; Ku75] with the parameters

$$V_0 = 125 \text{ MeV}, \quad R = 1.78 \text{ fm}, \quad \alpha = 0.66 \text{ fm} \tag{4.7}$$

i.e. *independent on both angular momentum and energy* (if the range $E_{\alpha,\mathrm{lab}} < 60$ MeV is only considered). The quality of the phase shift fittings is illustrated in Fig. 13 [1]); the corresponding $\alpha-\alpha$ differential cross sections are presented in Fig. 14. It can be seen from the figures (and especially from Fig. 13) that the simple *static* three-parameter potential eq. (4.7) permits the successful description of *four phase shift curves* $\delta_l(E)$ within a

---

1) When calculating the phase shifts, the Coulomb $\alpha-\alpha$ potential was added to the potential eq. (4.7).

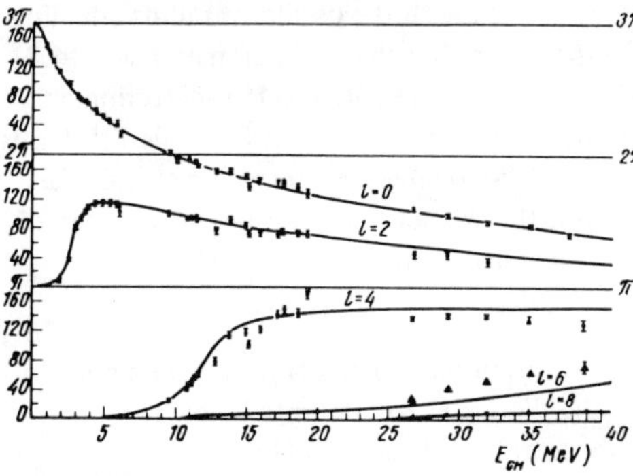

Fig. 13 The α–α phase shifts in MSOP for the partial waves with l = 0, 2, 4, 6, 8. The ordinate scale is taken in accordance with the generalized Levinson theorem (4.10). The dots show the experimental data.

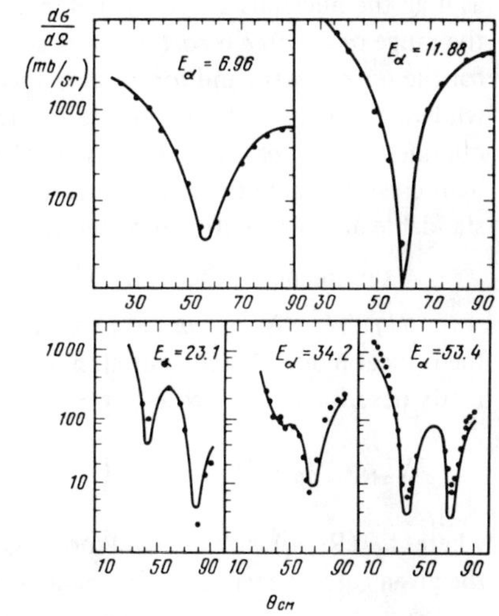

Fig. 14 The α–α differential cross sections at various energies, as found with MOSOP (4.7). At the highest energy used (E_α = 53.5 MeV) the real part given by (4.7) is supplemented by the absorbing potential of volume char acter with intensity $W_0 \cong 5$ MeV and with the same geometric parameters as in the real part.

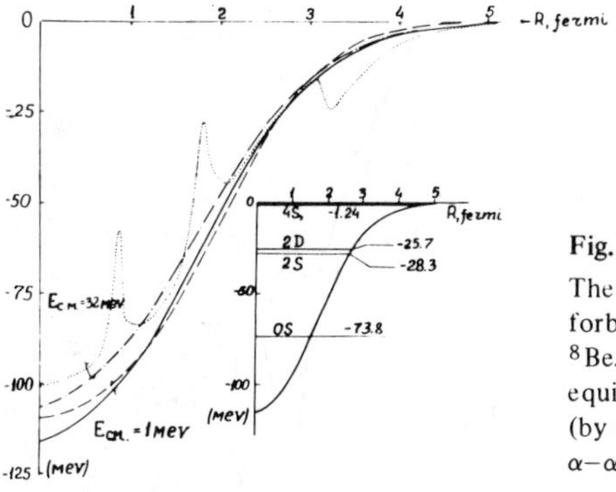

Fig. 15
The α – α MSOP scheme with the positions of the forbidden levels. The 4s-state is the ground state of ^{8}Be. It is shown for comparison the phase-shift equivalent potentials deduced from RGM-calculations (by dotted, dashed and dash-dot lines) at the various α–α relative motion energies (see [Ku75b]).

relatively wide energy range¹) $E_{lab} \lesssim 60$ MeV. The $\alpha-\alpha$ potentials, eq.(4.7) and its levels are shown in Fig. 15 from which the quasioscillator nature of the deep forbidden states (in particular, the 1s and 1d states are almost degenerate) is easily seen. Besides that, the distance between 0s and 1s – 1d states proves to be close to 36 MeV, which is just $2\hbar\omega_\alpha$, i.e. the doubled value of the oscillator frequency $\hbar\omega_\alpha$ characteristic of *free* α-particle. Quite the same situation takes place also for other pairs of light clusters, namely ^{4}He + n, ^{4}He + d, ^{4}He + ^{3}He(^{3}H) (see below). In other words, the pattern as a whole looks

¹) The range of proper description can be even more enlarged by introducing the energy-dependent depth $V_0(E) = V_0 - \frac{E_{c.m.}}{3}$ (MeV).

as if all the nucleons are contained in an oscillator well with the oscillator radius equaling the range of the *free α-particle*. It is also of importance that the well parameters eq. (4.7) for the α − α system and for other pairs of light clusters were fitted to the scattering phases, while the structure of the discrete spectrum, including the structure of forbidden states, is obtained as a *by-product*. This shows that the MSOP-model as a while gives (at least for light clusters where the reduced widths are about the Wigner limit) a physically understandable and inherently consistent pattern of the cluster interactions.

Examine now other pairs of clusters.

The $^4He + N$ system has been studied [Sa69b] in most detail from the viewpoint of the optical model. Satchler et al. have found the Woods-Saxon potential which can perfectly describe the n − α and p − α phase shifts up to $E_{c.m.} \simeq 20$ MeV:

$$V_{\alpha N} = V_0\, f_{ws}(R_0, a_0) + (\vec{l}\,\vec{s})V_1\, \frac{d f_{ws}(R_1, a_1)}{dr} \quad ^{1)}$$

where $f_{ws}(R_i, a_i)$ is the radial Woods-Saxon formfactor of radius R_i with diffusion a_i. In the given case, the values of the parameters are:

$$R_0 = 2.0 \text{ fm}, \quad R_1 = 1.5 \text{ fm}, \quad a_0 = 0.70 \text{ fm}, \quad a_1 = 0.35 \text{ fm}.$$

This potential contains the forbidden OS state with binding energy $\epsilon_b \simeq 18$ MeV whose wave function is almost 100% overlapping with the Os-orbital of the free α-particle [Ku80a].

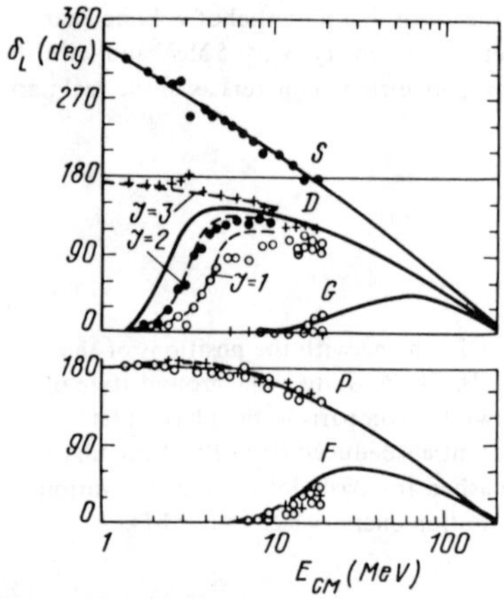

Fig. 16 The α–d phase shifts for MSOP (4.8).

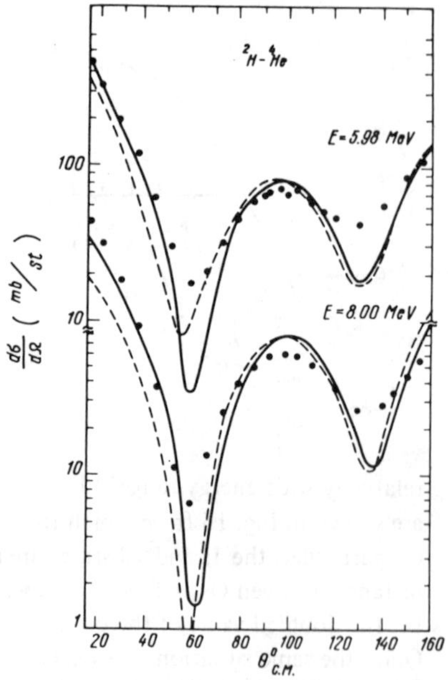

Fig. 17
The α–d differential cross sections at various energies found with MSOP (4.8) [Ko72, 73]. The experimental data is also shown.

[1]) Presented here is the static modification of the optical-model potential, which is found in [Ba78].

The $\alpha + d$ system in the MSOP approach was studied in [Ko72, 73]. The optical α+d potential obtained in [Ko72, 73] exhibits, in addition to the $\alpha - \alpha$ case, a dependence on spin and on state parity (the odd-even splitting [Wi 77; Ko72, 73]. The physical interpretation of the phenomena is given in [Ne72; Ko72; Wi 78]. The found optical α – d potential may be written as

$$V_l(r) = (V_0 + (-1)^l \Delta V)\, f_{ws}(R, a), \tag{4.8}$$

with parameter's values:

$$V_0 = 75.5 \text{ MeV}, \quad \Delta V = 2.5 \text{ MeV}, \quad R = 1.85 \text{ fm}, \quad a = 0.71 \text{ fm}.$$

In the $I^\pi = 1^+$ state, the potential contain the deep-lying forbidden 0s-state, whereas the ^{6}Li ground state is of 2s structure. Figs. 16 and 17 show the quality of the fittings of the phase shifts and differential cross sections of the ^{4}He-^{2}H scattering respectively. It is of great interest to note that the ^{6}Li wave function in the α – d relative coordinate found from the potential (4.8) coincides very well (taking account of the spectroscopic factor) with the analogous wave function found from the much more microscopic three-body model $^6\text{Li} \Rightarrow \alpha + 2\text{N}$ [Ku 80a; Ban 78] with realistic interactions. This means that the MSOP approach permits highly reliable calculations of *the form* of the exact wave function in the relative motion variable.

The $\alpha + t\,(\tau)$ system was studied in [Ne 72][1]) . Here again we have to include the significant odd-even splitting of the optical potential (for details, see [Ne 72; Wi 77]). The found MSOP can be written as

$$V^{(l)} = V_0(1 + (-1)^l \Delta V)\, f_{ws}(R, a) + (\vec{l}\,\vec{s})\, V_1\, \frac{d f_{ws}(R, a)}{dr} \tag{4.9}$$

with the parameter's values:

$$V_0 = 98.5 \text{ MeV}, \quad \Delta V = 11.5 \text{ MeV}, \quad V_1 = 4.2 \text{ MeV}, \quad R = 1.80 \text{ fm}, \quad a = 0.70 \text{ fm}.$$

Fig. 18 illustrate the quality of the fitting of the phase shifts $\delta(J^\pi)$ of the $\alpha + t(\tau)$ scattering. The positions of allowed and forbidden states of the α+ t potential are shown on the Fig. 19. It is clear that the importance of MSOP will decrease as still heavier clusters are examined, the spectroscopic factors decrease correspondingly, and the role of the imaginary part of the optical potential gets more important. In particular, the wave function in the ^{16}O + ^{16}O system found from RGM calculations [An 79] is so strongly supressed in the inner region that the shell-model oscillations are actually of but minor significance and, therefore, the MSOP-model with inclusion of a very small damping factor A(E, R) can be conveniently replaced by a smooth and shallow pseudopotential the introduction of which, as in the theory of solids, permits an easy parametrization of the scattering phase shifts and cross sections. Moreover, because of the comparatively easy deformability of heavy ions, even the RGM approximations may prove to be doubtful and, therefore, it would probably be more helpful to use a self consistent description of TDHF and similar types [We 77, Gi 77]. Nevertheless, the MSOP approach still retains its heuristic value even in case of the systems of the α + ^{40}Ca type.

Comparison between the two approaches discussed above (i.e. OCM and MSOP) has shown the following.

---

[1]) Here again we mean the optical-model approach. This system was studied in terms of RGM in many works (see the monograph [Wi 77] for references).

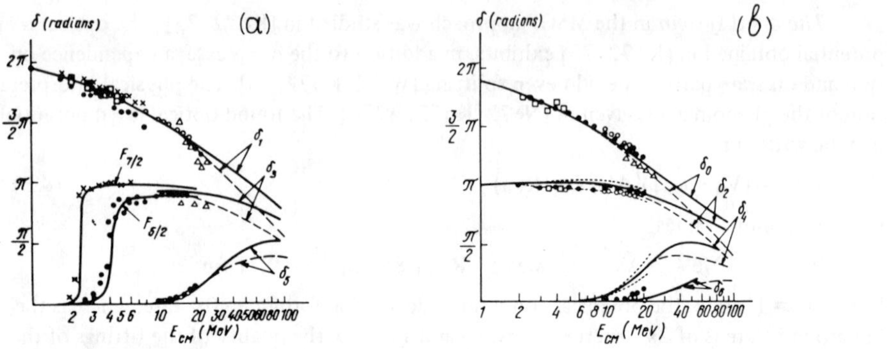

Fig. 18 The ^{4}He-^{3}H (and ^{4}He-^{3}He) phase shifts found with MSOP (4.9) (see ref. [Ne72]) as compared with experimental data of different works. The a-part corresponds to odd partial waves, b-part – to even waves.

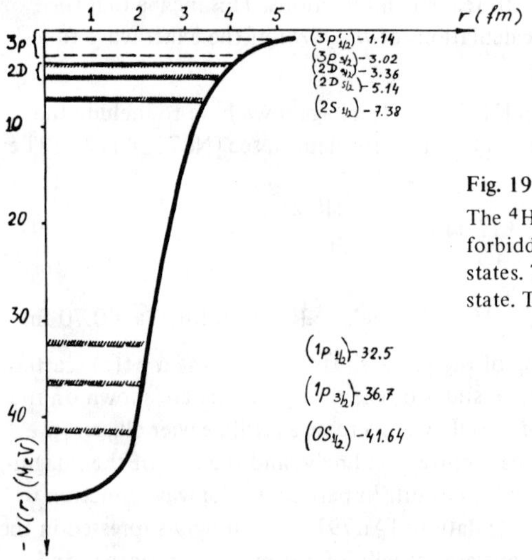

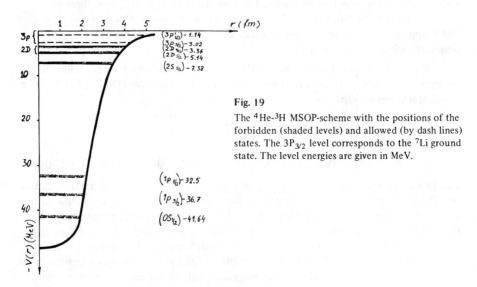

Fig. 19
The ^{4}He-^{3}H MSOP-scheme with the positions of the forbidden (shaded levels) and allowed (by dash lines) states. The $3P_{3/2}$ level corresponds to the ^{7}Li ground state. The level energies are given in MeV.

Though the MSOP approach is undoubtedly more phenomological than OCM, it has some important advantages, namely.

1. MSOP is more visual and simpler than OCM, retaining at the same time the main features of the latter.

2. As it was first shown by Swan [Sw55], the scattering of two clusters (or, generally, more composite particles) can be described by the generalized Levinson theorem

$$\delta_l(0) - \delta_l(\infty) = \pi(n_l + m_l)\,, \qquad (4.10)$$

where n_l and m_l are respectively the numbers of the allowed and forbidden states in a given partial wave l. In OCM, the theorem (4.10) was proved by Glöckle and Le-Tourneux [Gl78] and by Pomerantsev [Po78] (see the next Section for the detailed proof of the generalized Levinson theorem). It is a remarkable fact that in

MSOP the generalized Levinson theorem is *automatically* satisfied since in this case the forbidden states *are of equal validity* with the allowed states. It will be to the point to note that not a single semiforbidden state is contained in (4.10).

3. It will be seen from the discussion in Chapter 5 that the definition of the forbidden states as the *eigenstates* of the cluster interaction Hamiltonian, i.e. the dynamic definition of the forbidden states, in the MSOP approach agrees better with the complete microscopic pattern than the pure structural definition in OCM through the eigenfunctions of the overlap kernel K. In OCM, the forbidden states are treated as some *alien* states to which the solution for a given potential problem should be orthogonalized.

4. Finally, MSOP has another important advantage which is the reverse side of its partly phenomenological nature, namely MSOP permits a better and more accurate description of experimental data as compared with OCM. It will be noted that this advantage is of decisive importance when calculating the many-cluster systems where accurate fitting to two-body data is obligatory.

4.2 Generalization of the Levinson Theorem for the Case of Composite Particle Scattering

In any microscopic approach to the composite particle interactions, the potential of these interactions is nonlocal and, generally speaking, energy-dependent. Since the most general case of a nonlocal and energy-dependent interaction is too complex to be studied analytically, the results presented below relate, strictly speaking, to some models in which analytical examination is possible. Swan [Sw55] was the first to propose the generalization of the Levinson theorem for the case of composite particles in the atomic and nuclear physics. He has shown (by using a certain particular form of the exchange potentials as an example) in terms of the RGM approach that the forbidden states should be treated in the Levinson theorem on equal grounds with the physically observable allowed bound states, so that the generalized Levinson theorem must be of the form (4.10). Several versions of the proof of the theorem have been proposed [Gl76; En74; Po78]. The proof [Gl76] is the direct generalization of the standard proof of the Levinson theorem based on examination of the analytical behaviour of the Jost function of the problem in the upper half-plane $\mathrm{Im}\,k \geqslant 0$ [New66]. The scheme of the proof is in brief as follows. Let G(E) be GF corresponding to the left part of eq. (4.6a), $f_0(k)$ be the appropriate Jost function. Then the function

$$D(k) = f_0(k) \det [G_{\alpha\beta}(E)] \equiv f_0(k)\, g(k)$$

(where $[G_{\alpha\beta}(E)]$ is the matrix composed of the elements $\langle \chi_\alpha | G(E) | \chi_\beta \rangle$) is the Fredholm determinant of the integral equation corresponding to the Schrödinger equation (4.6a). It may be shown further that the zeroes in D(k) are directly related to the poles of the total S-matrix and that the argument of D(k) equals the scattering phase $\delta(k)$ in OCM, i.e.

$$D(k) = |D(k)|\, e^{-i\delta(k)}.$$

In this case, D(k) fails to coincide with the complete Jost function of the problem and differs from this function by a certain factor $\Delta(k^2)$. It has been show then in [Gl76] by examining the analytical properties of the matrix elements $G_{\alpha\beta}(k)$ in the upper half-plane that the factor

$$g(k) \underset{k\to\infty}{\Rightarrow} \frac{1}{k^{2n}}\left(1+0\left(\frac{1}{k^2}\right)\right) \quad , \quad \mathrm{Im}\, k>0\,, \tag{4.11}$$

where n is the number of the states forbidden by the Pauli principle. Since, however, the initial Jost function $f_0(k)\to 1$ at $k\to\infty$, then $D(k)\to g(k)$ in this limit. If now on the upper semiaxis there exist m zeroes of D(k) corresponding to the physical bound states, the standard contour integral [Ne66] is

$$\frac{1}{2\pi i}\int \frac{1}{D(k)}\frac{dD}{dk}\,dk = m\,. \tag{4.12}$$

Due to the property (4.11), however, the integral over the upper semicircle is

$$\int_C -\frac{2n}{k^{2n+1}}\,k^{-2n}\,dk = -2\pi i n\,.$$

Further, by using the properties $|D(-k)| = |D(k)|$ and $\delta(-k) = -\delta(k)$, we can readily find:

$$\int_{-\infty}^{\infty} dk\,\frac{d\ln D(k)}{dk} = 2i\,(\delta(0)-\delta(\infty))$$

and, with account of eqs. (4.11) and (4.12), we get eventually the theorem (4.10). Presented below will be another and simpler proof of the theorem given in [Po78] and based fully on the properties of completeness.

The expression (2.27) may be used to show that
$\langle\widetilde{\Psi}_E|\widetilde{\Psi}_{E'}\rangle = \langle\Psi_E|\Psi_{E'}\rangle \sim \delta(E-E')$,

i.e. that the normalization to the δ-function is conserved after orthogonalization. The completeness relation for the functions $\widetilde{\Psi}_E$ can be obtained if the basis functions is noted to coincide with the eigenfunctions of the Hamiltonian QHQ, where $Q = 1-\Gamma$. Therefore,

$$\int|\widetilde{\Psi}_E\rangle\langle\widetilde{\Psi}_E|\,dE + \widetilde{P}_B = 1-\Gamma\,, \tag{4.13}$$

where $\widetilde{P}_B = \sum_\alpha|\widetilde{\chi}_\alpha\rangle\langle\widetilde{\chi}_\alpha|$ is the projector onto the bound states of the pseudohamiltonian $\widetilde{H} = H+\lambda\Gamma$ (being $|\widetilde{\chi}_\alpha\rangle$ are also the eigenfunctions of QHQ). Since $\Gamma\widetilde{P}_B = \widetilde{P}_B\Gamma = 0$ then $\widetilde{P}_B+\Gamma$ is the total projector on the entire discrete spectrum of the operator QHQ. After that, the Möller operator should be introduced:

$$\widetilde{\Omega} = \int_0^\infty dE\,|\widetilde{\Psi}_E\rangle\langle\Psi_{0E}|$$

which is the isometric reflection of the continuum of the Hamiltonian H_0 on the continuum of $\widetilde{H}$ (or QHQ, which is the same). The completeness relation (4.13) gives the equality

$$\widetilde{\Omega}^*\widetilde{\Omega} - \widetilde{\Omega}\,\widetilde{\Omega}^* = \widetilde{P}_B + \Gamma \,. \tag{4.14}$$

Further, following the method due to Jauch [Ja58] we take the trace from the equality (4.14) and express the trace of the left part through the on-shell t-matrix:

$$i\pi \int_0^\infty dE\, \mathrm{Sp}\left\{\tilde{t}^* \frac{\partial \tilde{t}}{\partial E} - \tilde{t}\, \frac{\partial t^*}{\partial E}\right\} = \widetilde{N}_B + N_\Gamma \,, \tag{4.15}$$

where $\widetilde{N}_B$ is the number of bound states χ_α, N_Γ is the dimension of the projector Γ. It will be noted that N_Γ in the right hand part is obtained because the trace and dimension of the projector are equal. If use is made now of the unitarity of the t-operator and to write the latter in the form

$$\tilde{t}\,(E) = -\frac{1}{\pi} \sum_l e^{i\widetilde{\delta}_l(E)} \sin \widetilde{\delta}_l(E) |\chi_l\rangle\langle\chi_l| \tag{4.16}$$

we shall find, after substitution in eq. (4.15) and integration, that

$$\frac{1}{\pi} \sum_l (\widetilde{\delta}_l(0) - \widetilde{\delta}_l(\infty)) = \widetilde{N}_B + N_\Gamma \,.$$

If the potential is spherically symmetrical, the expansion (4.16) reduces to the ordinary expansion in partial waves and $|\chi_l\rangle\langle\chi_l| = 1$, for the energy and the moment exhaust all the variables. If now the projector Γ is invariant in rotations and is the sum of the projectors with definite l, i.e. $\Gamma = \sum_l \Gamma_l$, it will be easy to obtain eventually:

$$\frac{1}{\pi} (\widetilde{\delta}_l(0) - \widetilde{\delta}_l(\infty)) = \widetilde{N}_{B_l} + N_{\Gamma_l} \,. \tag{4.17}$$

It can be seen that the proof presented above has been made disregarding the explicit analytical properties of the solutions at the complex k and is based completely on the t-matrix unitarity and on the completeness properties of the basis in the truncated space. The significance of the generalized Levinson theorem for practical purposes is that it predicts considerable elastic phase shifts at high energies in the systems with numerous forbidden states in which the number of the allowed bound states is usually small. In particular, considerable elastic phase shifts at $E_{rel} \sim 200-400$ MeV in c.m.s. were predicted in [Ne71] for the $\alpha-\alpha$ scattering. (It will be noted that the work [Ne71] was one of the first to make physical application of the generalized Levinson theorem). The predictions were been verified in [Fo76] by experimentally measuring the elastic scattering in the $\alpha-\alpha$ system at $E_\alpha \sim 600-800$ MeV (lab.). A noticeable elastic cross section of nondiffractive nature was found. The final conclusions will be drawn, however, only from the still unavailable experimental data in the energy range $150\text{ MeV} \lesssim E_{lab} \lesssim 600\text{ MeV}$, to make if only for qualitative theoretical interpretation.

5 Ghost States as Generalization of the Concept of Forbidden States in the Resonating Group Method

5.1 Ghosts in the Theory of Low-Energy n–d and d–d Scatterings

Though the theory of forbidden states in nuclear systems developed in [Fe62; Au72] has proved to be very helpful, the subsequent studies have shown that this approach has noticeable disadvantages associated with the appearance of almost forbidden ($\lambda_n \approx 1$) and semiforbidden states. A much more significant drawback, however, is that the theory developed in [Fe62; Au72] is based exclusively on RGM or on the cluster approximation for the complete wave function of the system of the form of (4.16), i.e. essentially on the adiabatic model with exchange, whereas the concept of the states forbidden by the Pauli principle is much more exhaustive and, generally speaking, should not be associated with special approximations for wave functions.

It is, therefore, of great interest to examine what *in the rigorous theory of many-body systems* is in correspondence with the forbidden states in RGM. Luckily, such a possibility does exist and is based on the rigorous Faddeev-Yakubovsky theory of many-body systems. Namely, one is to take the problem of three-or-four nucleons and to find out, in the frame of the rigorous theory, what type of the forbidden states appears in such systems. The conclusions found in such a (i.e. rigorous) way may be applied after that to other nuclear systems. The relevant study was carried out in [Ku77a, 78a] in the cases of n–d scattering in the channel with $S = \frac{3}{2}$ and d–d scattering in the channel with $S = 2$. The main results are presented in brief below.

Examine first the n–d scattering in the quartet channel. Since the spins of all three nucleous are parallel, the orbital wave function should be antisymmetric under transposition of two neutrons, i.e. only odd partial waves should be taken into account in two-neutron interaction. It is well-known, however, that such interactions may be neglected at low energies. The remaining two n–p interactions will be the triplet ^{3}S. Under such circumstances, the complete orbital wave function is divided into the sum of two Faddeev components (according to two n–p interactions)

$$\Psi_{s,a} = \psi^{(1)}_{s,a} \pm \psi^{(2)}_{s,a}\,, \tag{5.1}$$

where the plus and minus signs correspond respectively to the spatially symmetric and antisymmetric orbital functions (it will be noted that the symmetric function cannot be realized physically) and are denoted with the subscripts s and a. The important feature of the Faddeev approach is the fact that the resultant equations for the symmetric $\psi^{(i)}_s$ and antisymmetrical $\psi^{(i)}_a$ Faddeev components are *different* (the kernels for the s- and a-cases differ in sign).

The physically non-realizable channel with the symmetric orbital function will be called the ghost channel and the bound state in this channel – the ghost bound states, or simply ghosts. It seems expedient to note the specific features of the problem of the permutational symmetry in terms of the Faddeev-Yakubovsky theory (and any other mathematically correct theory of many-body scattering). The fact is that, although the complete Schrödinger function Ψ exhibits one or another permutational symmetry its Faddeev components $\psi^{(i)}$ (a partition into channel components similar to (5.1) is characteristic of any "good" theory) fail to exhibit definite permutational symmetry. In a "good" theory, the type of the permutational symmetry of solution is determined by the structure of integral kernel and the procedure of finding a kernel corresponding to the solution of a given symmetry includes the partition and reduction of the matrix integral kernel into submatrices corresponding to the irreducible representations of the permutation group. This analysis has not been made yet in the general case (see, however, the detailed study [Be 79]); but in the special cases examined here it may be readily carried out [Ku 77a].

Thus, FE set for the components $\psi^{(i)}_{s,a}$ is of the form

$$\psi^{(i)}_{s,a} = \varphi^{(i)} \pm G_0 T_i P_{12} \psi^{(i)}_{s,a}, \qquad i = 1, 2 \tag{5.2}$$

where $\psi^{(2)}_{s,a} = \pm P_{12} \psi^{(1)}_{s,a}$ and the plus and minus signs correspond to the ghost and physical channels respectively; P_{12} is the transposition operator of two identical neutrons; $G_0 = (E - H_0)^{-1}$ is the free GF; T_i is the t-matrix of the triplet interaction in channel i (in this case $T_2 = P_{12}\ T_1\ P_{12}$). The function of the initial state is $\varphi^{(i)} = \varphi_{23}(\mathbf{p})\, \delta(\mathbf{q} - \mathbf{q}_0)$ where $\varphi_{23}(\mathbf{p})$ is the wave function of deuteron; $\mathbf{q}_0$ is the momentum of the incident neutron. Since the equations (5.2) are the same for any i, the subscript i in eq. (5.2) is "dumb" and is used only to discriminate the Faddeev component $\psi^{(i)}$ from the complete Schrödinger wave function $\Psi_{s,a} = (1 \pm P_{12})\, \psi^{(i)}_{s,a}$.

Besides that, the sign and norm of the kernel $K = G_0 T_i P_{12}$ are of great importance to the scattering problem, of which it is the norm that determines the intensity of interaction. In its turn, the kernel norm depends upon the kernel EV's which are determined by the matrix equation

$$\mathbb{K}_2(E)\, \chi_n^{s,a} = \pm\, \alpha_n(E)\, \chi_n^{s,a}, \tag{5.3}$$

where the matrix kernel $\mathbb{K}_2(E)$ is known [Sc 74a] to be the operator

$$\mathbb{K}_2(E) = \begin{pmatrix} 0 & G_0(E) T_1(E) \\ G_0(E)\, T_2(E) & 0 \end{pmatrix}, \tag{5.4}$$

$\chi_n^{s,a}$ are the two-component columns: the plus and minus signs in eq. (5.3) correspond to the superscripts s and a respectively. The bound states in the physical (i.e. a-) and ghost (i.e. s-) channels are in correspondence with EV's $\alpha_n(E) = -1$ and $\alpha_n(E) = +1$ respectively. And now the specific calculations [Ku 77a, 78a] have permitted a bound state with $E_G \approx -5.5$ MeV in the ghost channel to be found, which is nothing but the *quartet triton* with spin $\frac{3}{2}$ forbidden by the Pauli principle and with the *symmetric* spatial part of the wave function. In the physical channel (where the kernel sign corresponds to repulsion), bound states are absent, i.e. the kernel $K = G_0 T_1 P_{12}$ does not have high negative EV's. Since, however, the FE set (5.2) is equivalent to the input Schrödinger equation, the existence of a unity-valued EV's of the Faddeev kernel in the ghost channel means that the complete Hamiltonian of the system will contain a bound eigenstate of forbidden symmetry, i.e. a ghost, the fact that may be predicted without calculations. Indeed, if consideration is given to the $p + 2n$ system in a symmetric spatial state with two S-triplet np-interactions, it will become absolutely clear that this system will contain a bound state with energy $E_G \sim 2\,\epsilon_d$, where ϵ_d is the binding energy of deuteron (the strict equality is $E_G = 2\,\epsilon_d$ would have held if the proton had been infinitely heavy). From the viewpoint of mathematics, the relationship between the physical and ghost channels can be seen even more explicitly. Namely, since the FE kernels (5.2) for the two channels differ only in signs, the resolvents of the corresponding equations are the special values of the resolvent $(1 - \gamma K)^{-1}$ at $\gamma = \pm 1$. At the same time, the resolvent is a *meromorphic* function of γ over the *entire* plane. Therefore, the solution in the ghost channel is determined by the straightforward analytical continuation of the physical channel solution.

Now the question arises what is the importance of the ghost states to the scattering theory? First of all, it is clear that from the viewpoint of the effect on the scattering problem the ghosts play the same role as the physical bound states. In particular, the ghosts give rise to the same divergence of the Born series (for the Faddeev-Yakubovsky equations) as the physical bound states, while the wave function of scattering in the physical channel must be orthogonal to the wave functions of *both physical bound states and ghosts.*

Besides that, it was shown as long ago as [Sw55] using qualitative arguments that the quartet s-phase of the n–d scattering begins (similarly to the doublet s-phase) from π (for the normalization $\delta(\infty) = 0$), and that, while the bound state in the doublet channel exists (triton) it is absent in the quartet channel, but instead (as it was noted above) a ghost bound state is present. Moreover, it has been confirmed [Sl71] by means of direct, numerical claculations of the particular three-body model made on the basis of FE that the total interval of variations of the quartet and doublet s-phase of n–d scattering is *the same and equals* π. In other words, we have drawn an important conclusion that the contribution from the ghost to the generalized Levinson theorem for *three-body scattering* is the same as that from the physically observable bound state. Though this conclusion generalizes to the true three-body scattering the results which has been proved for OCM (see Chapter 4, Section 4.5), the difficulties with interpretation of the three-body Levinson theorem [Wr71] have still precluded any rigorous proof. Nevertheless, we make sure that the formalism presented below in this Section is a step in the right direction.

Return now to the convergence of the FE iterations for the n–d scattering. We shall follow the works [Ku77a, 78a; Po78]. To provide for the convergence of FE (5.2) iterations, we shall make orthogonal projecting in eq. (5.2) and obtain

$$\widetilde{\psi}_a^{(i)} = \widetilde{\varphi}^{(i)} - \widetilde{K}_i\,\widetilde{\psi}_a^{(i)} \tag{5.5}$$

$$\widetilde{\psi}_s^{(i)} = \widetilde{\varphi}^{(i)} + \widetilde{K}_i\widetilde{\psi}_s^{(i)}, \tag{5.6}$$

where $\widetilde{K}_i = (1 - P(E))K_i$, $\widetilde{\varphi}^{(i)} = (1 - P(E))\,\varphi^{(i)}$ and the energy-dependent nonhermitian projector $P(E) = 1 - G_i\Gamma(\Gamma G_i\Gamma)^{-1}\Gamma$. As applied to FE (5.2) such projecting may be called semieigenprojecting since the complete (i.e. Schrödinger-type) wave function does not change after such projecting.

$$\widetilde{\Psi} = \Psi - P(E)\Psi \equiv \Psi$$

as $\Gamma\Psi = 0$ for both eigen- and ghost-projecting, but the Faddeev components

$$\psi^{(i)} = G_0 V_i \Psi \quad \text{and} \quad \widetilde{\psi}^{(i)} = \widetilde{G}_0 V_i \Psi$$

do not coincide with each other:

$$\widetilde{\psi}^{(i)} = G_0 V_i \Psi - G_0 \Gamma(\Gamma G_0 \Gamma)^{-1} G_0 V_i \Psi = \psi^{(i)} - G_0 \Gamma(\Gamma G_0 \Gamma)^{-1} \Gamma \psi^{(i)} . \tag{5.7}$$

In case of eigen-projecting onto the functions of *the same symmetry*, we get

$$P\Gamma = \pm \Gamma, \quad \Psi = \psi^{(1)} \pm \psi^{(2)}$$

whence

$$\Gamma\psi^{(i)} = 0, \quad \widetilde{\psi}^{(i)} = \psi^{(i)} . \tag{5.8}$$

At the same time, in case of semieigen-projecting the equalities (5.8) are invalid and, therefore, to restore $\psi^{(i)}$ with the help of $\widetilde{\psi}^{(i)}$, one should invert the relation (5.7). Simple transformations made using the input FE (5.2) give for the case of $P\Gamma = +\Gamma$ examined here:

$$\psi_a^{(i)} = \widetilde{\psi}_a^{(i)} + G_0 \Gamma(\Gamma G_i \Gamma)^{-1} \Gamma(\varphi^{(i)} - K_i \widetilde{\psi}_a^{(i)}) . \tag{5.9}$$

We shall write now the off-shell amplitude of elastic scattering in the physical channel:

$$F_a = -\frac{\mu}{2\pi} \langle \varphi^{(i)} | G_i^{-1} | \psi_a^{(i)} - \varphi^{(i)} \rangle , \tag{5.10}$$

where the contribution of the second Faddeev component on the energy shell is zero due to the factor G_i^{-1}. In this case, however, the addend $G_0 \Gamma(\Gamma G_0 \Gamma)^{-1} \Gamma \psi^{(i)}$ in (5.10) also fails to contribute on-shell and, therefore, relations (5.7) and (5.9) may be used to write the amplitude F_a in the form

$$F_a = \frac{\mu}{2\pi} \langle \varphi^{(i)} | G_i^{-1} K_i | \widetilde{\psi}_a^{(i)} \rangle + \frac{\mu}{2\pi} \langle \varphi^{(i)} | \Gamma(\Gamma G_i \Gamma)^{-1} \Gamma | \varphi^{(i)} - K_i \psi_a^{(i)} \rangle . \tag{5.11}$$

For the eigen-projecting, the second term in eq. (5.11) vanishes due to eqs. (5.2) and (5.8), but nevertheless this term can readily be included since in this case the relation (5.11) at iterations (5.5) will give a *symmetric* series whose first term will never contain the three-body kernel K_i

$$\widetilde{F}_0 = \frac{\mu}{2\pi} \langle \varphi^{(i)} | \Gamma(\Gamma G_i \Gamma)^{-1} \Gamma | \varphi^{(i)} \rangle \tag{5.12}$$

whereas the first term of eq. (5.11) alone gives, after iterations, an asymmetric series whose first term corresponds to the Born approximation with orthogonalized plane wave:

$$\widetilde{F}_1 = -\frac{\mu}{2\pi} \langle \varphi^{(i)} | V | \widetilde{\varphi}^{(i)} \rangle .$$

For the ghost-projecting, the series must always begin from the term (5.12), i.e. from $\widetilde{F}_0$. And for the function Ψ_s of the symmetric ghost channel the same projecting (i.e. the equation (5.6)) will be the eigen-projecting and the relation (5.11) is valid if K_i is replaced by $-K_i$, and the index a by s. We shall simultaneously iterate eqs. (5.5) and (5.6) and substitute the resultant approximations in eq. (5.11). This procedure gives the series for the amplitudes F_a and F_s so the convergence of the series is to be verified at various energies and coupling constants. For our purpose, it is sufficient to examine the simple three-body models in which the two-body interactions is taken to be of the form of the Yamaguchi potential

$$V(k, k') = \lambda\, g(k)\, g(k') , \qquad (5.13)$$

where $g(k) = (k^2 + \beta^2)^{-1}$. The sole dimensionless parameter in such model (note that we include only *two* triplet n−p interactions, whereas the n−n interaction in the triplet state is only allowed for the odd orbital momenta and, therefore, is insignificant to our purposes) is the deuteron wave vector α related to the deuteron binding energy ϵ_d as $\epsilon_d = (\alpha\beta)^2/m$. After that, by fitting the parameters of (5.13) to low-energy N−N scattering in the triplet state, we get $\alpha = 0.16$. Table 2 shows the convergence of FE iterations for the amplitudes F_a and F_s and for the relevant phase shifts δ_a and δ_s as a function of the number of iterations.

Table 2: Convergence of the orthogonal iterations of FE for n−d scattering in symmetric and antisymmetric channels at $E_{n,\,lab} = 1.4$ MeV ($\alpha = 0.16$).

Number of iterations	Re F_a	Im F_a	δ_a (rad)	Re F_s	Im F_s	δ_s (rad)
0	− 3.42	6.55	2.052	− 3.42	6.55	2.052
1	− 2.79	7.45	1.929	− 4.05	5.64	2.193
2	− 3.71	5.87	2.134	− 4.97	4.07	2.456
3	− 3.98	6.33	2.131	− 4.70	3.61	2.487
4	− 4.01	5.49	2.201	− 4.73	2.76	2.673
5	− 4.20	5.74	2.202	− 4.54	2.51	2.637
...	...	...	...	...	...	...
...	...	...	...	...	...	...
19	− 4.003	5.316	2.216	− 3.455	1.84	2.653
20	− 4.004	5.325	2.216	− 3.454	1.85	2.651

It can be seen from Table 2 that the convergence of the orthogonalized iterations of FE for the scattering phases in the physical a-channel and the ghost s-channel is very good. A somewhat better convergence for the a-case may be explained by the sign-changing iteration series in the a-case (the kernel will be $-K_i$) and the sign-constant series in the s-case (the kernel is $+K_i$). The situation as the whole is very similar to the potential scattering on the pure attracting potential V and on the repulsive potential $-V$ which is a mirror reflection of the former [New66].

Now, we shall briefly discuss the role of the ghost states in the four-body scattering theory [Po77]. The most simple case here is the d + d scattering in the state with the total spin $S = 2$. For the same approximations as those used in the n−d scattering problem, we find that the n−n and p−p nuclear interactions may be neglected, so that four triplet n−p interactions will remain in the total Hamiltonian. Analysis of the four-body Yakubovsky equations [Po77; Ya67], similar to that presented above for the three-body problem but much more tedious has shown that the quintet channel involves as many as *six* ghost channels, namely two symmetric channels and four channels of mixed symmetry. For the sake of brevity, consideration will be given to a symmetric ghost channel. The subsequent analysis has shown that, although the kernels in the spatially antisymmetric and symmetric ghost channels will not be the same to within a sign as in the n–d case, they still are directly interrelated as

$$\hat{K}_{ph} = \frac{\hat{B} - 1}{\hat{B} + 1} \hat{K}_G$$

where $\hat{K}_G$ and $\hat{K}_{ph}$ are the kernels in the ghost and physical channels respectively; $\hat{B}$ is an operator, such as the large EV's in the ghost channels give rise to also large negative EV's in the physical channel. It can be found using the variational method that the symmetric ghost channel contain a bound state (*a ghost α-particle*) with binding energy ≈ 20 MeV. Using again the orthogonal projecting and the projector onto this ghost bound state, we can rearrange the Yakubovsky four-body equations in such a way that they become convergent (in the S-wave) at low energies. A similar situation will also take place in the n-^{3}H or n-^{3}He scatterings with quantum numbers ST = 01, 10, 11; in this case, however, the effective Hamiltonians will be somewhat different from the case of the d–d scattering.

5.2 Substantiation of the Generalized Orthogonality Condition Model (GOCM)

We shall return again to the ghosts in the n–d scattering and analyze the contribution of the zero-order approximation amplitude $\widetilde{F}_0$ at various α. It can be seen from Table 2 that the zero-order approximation phase shift $\widetilde{\delta}_0$ at low energies is close to the exact quartet n–d phase. Fig. 20 presents a comparison between the exact and zero-order approximations for the phase shifts depending on the dimensionless momentum of bombarding particle p_0 in c.m.s. at $\alpha = 0.16$ ($\epsilon_d = 2.2$ MeV) and $\alpha = 0.32$ ($\epsilon_d = 8.8$ MeV). It can easily be seen that, as the deuteron binding energy increases, the quality of the zero-order approximation $\widetilde{F}_0$ (and, correspondingly, $\widetilde{\delta}_0$) improves rapidly. The same

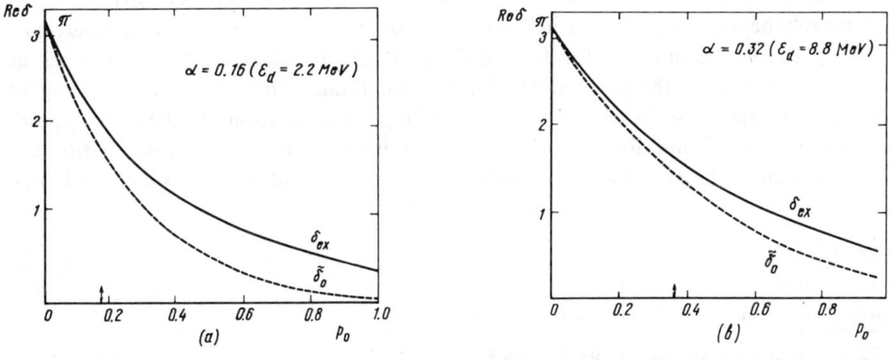

Fig. 20 Comparison between the accurate (δ_{ex}) and zero-order approximation (δ_0(E)) of phase shifts for the quartet channel of n–d scattering at various binding energies of model deuteron.
(a) The case $\epsilon_d = 2.2$ MeV ($\alpha = 0.16$), (b) the case $\epsilon_d = 8.8$ MeV ($\alpha = 0.32$).

relates naturally to the scattering length. In particular, at $\alpha = 0.16$, $\tilde{a}_0 = 9.85$, and $^4a_{exact} = 9.09$, the error $\sim 6.0\%$. Hence, in case of d + d scattering in quintet channel, the quality of the zero-order approximation for exact amplitude should be even better than in the quartet n−d scattering.

Examine now in more detail the expression for $\tilde{F}_0(E)$ in case of a single ghost state Φ_G:

$$\tilde{F}_0(E) = \frac{\mu}{2\pi} \frac{\langle \varphi^{(i)} | \Phi_G \rangle \langle \Phi_G | \varphi^{(i)} \rangle}{\langle \Phi_G | G_i(E) | \Phi_G \rangle}, \tag{5.14}$$

where Φ_G is the spatially symmetric function. If the three-body GF G_i is replaced by corresponding pole term $2\mu |\varphi_d\rangle\langle\varphi_d| / (p_0^2 - p^2)$ (where p_0 is the momentum of the incident particle), the expression (5.14) turns out to be the expression for the one-body orthogonality scattering amplitude (cf. the amplitude $\tilde{A}_0$ in the theory of potential scattering, see Chapter 2):

$$\tilde{f}_0(p_0) = \frac{\mu}{2\pi} \frac{u(p_0') u(p_0)}{\langle u | g_0 | u \rangle}, \tag{5.15}$$

where $g_0 = 2\mu/(p_0^2 - p^2)$ is the one-body free GF;

$$u(p) = \langle \varphi_d(k) | \Phi_G(k, p) \rangle \tag{5.16}$$

is the formfactor of the ghost state. It will be noted that the one-particle amplitude $\tilde{f}_0$ is unitary *at all energies*, while the initial three-body amplitude $\tilde{F}_0(E)$ is unitary *below the three-body threshold only*. Therefore, it may hardly be expected that the one-body orthogonality scattering amplitude $\tilde{f}_0(p_0)$ will give a proper description of the phase shift *above* the three-body threshold. At low energies, however, $\tilde{f}_0$ gives a very good approximation to $\tilde{F}_0$ and, thereby, to the *complete three-body amplitude*, and the agreement between $\tilde{F}_0$ and $\tilde{f}_0$ improves as the binding energy increases. Some idea about the correspondence between the exact three-body, orthogonality *three-body* scattering (F_0), and orthogonality *one-body* scattering ($\tilde{f}_0$) amplitudes can be inferred from Table 3 which presents the respective scattering lengths, namely the exact quartet length ^4a and the lengths $\tilde{a}_0$ and $\tilde{a}_0^{op}$ corresponding to $\tilde{F}_0$ and $\tilde{f}_0$ respectively, as functions of binding energy.
It can easily be seen that $\tilde{a}_0$ and $\tilde{a}_0{}^{op}$ are in a good mutual agreement. The comparison between the expression (5.15) for $\tilde{f}_0$ and (2.28) of Chapter 2 has permitted the conclusion that $\tilde{f}_0$ is nothing but the pure OCM (one-body) amplitude of orthogonality scattering, i.e. the scattering amplitude in OCM when $V_D = 0$. In this case, contrary to the Saito approach, we have strictly dynamically determined the functions to which the orthogonalization should be made, namely they one merely *formfactors of the ghost states* $u(p)$ (see eq. (5.16)).

Table 3

α	0.04	0.08	0.16	0.32	0.64
$\tilde{a}_0$	27 83	16.43	9.85	5.81	3.57
$\tilde{a}_0{}^{op}$	28.35	16.75	9.91	5.83	3.58
$^4a_{exact}$	31.20	16.94	9.10	5.48	3.81

In itself eq. (5.14) may be treated as the generalization of OCM (with $V_D = 0$) which can approximately include the deuteron break-up (it will be emphasized that the denominator of eq. (5.14) at energies above the break-up threshold contains a contribution from deuteron break-up). We can get OCM only by using the pure pole term from the channel GF G_i. Obviously, the corresponding generalization of the formula (5.14) for the general case of scattering of two clusters, a and b is

$$\widetilde{F}_{0,a+b}(E) = \frac{\mu_{ab}}{2\pi} \cdot \frac{\langle \phi_a \phi_b | \Phi_G \rangle \langle \Phi_G | \phi_a \phi_b \rangle}{\langle \Phi_G | G_a * G_b | \Phi_G \rangle} , \tag{5.17}$$

where $G_a * G_b$ is the convolution of two channel GF; ϕ_a and ϕ_b are the inner functions of the scattered clusters. Using again the pole terms of G_a and G_b, we get, as before, the one-particle OCM in which we have only the variable of relative cluster motion.

Thus, the ghosts, i.e. the bound states with forbidden symmetry make it possible not only to construct a convergent iteration series for many-particle scattering but also to obtain a generalization of the OCM approach. It will be noted that we obtained above the special generalization of OCM at $V_D = 0$. However, on rewriting the input FE on the basis of distorted waves [Ku78, 80] and making orthogonal projecting in them, we get the OCM generalization of general form [Ku80b].

Presented below will be another comparison between the ghost states in the many-body scattering theory and the forbidden states in RGM. The comparison between the definition of the ghost states (5.3) and forbidden states in RGM (4.5) shows that they are essentially analogous. In particular, they both are eigenstates of integral kernels with unit EV's. There exist, however, significant differences. While forbidden states in RGM are determined algebraically through the pure structural overlapping kernel, i.e. in a way which is actually independent of the interaction forces *between* the clusters, the ghosts (in the exact scattering theory) are determined in a *dynamical way*, i.e. their energies and wave functions *will be essentially dependent* on the forces between the particles.

In the complete N-particle scattering theory, the ghosts are the precise bound states of the N-particle Hamiltonian, i.e. when the symmetry of spin part of the wave function is not related unambigously to the symmetry of the orbital part of the wave function. Such bound states, however, either do exist or do not. This means that, if the *forbidden* states are *accurately treated, none of the semiforbidden or almost forbidden states* appears. In other words, the appearance of such states is due to the special approximations of RGM and OCM. In particular, if the analogy with the Faddeev pattern of three-body scattering is prolonged, then AFS with $\lambda \sim 1$ correspond to EV's of the Faddeev kernel K $\alpha(E) \lesssim 1$ (at all energies from the region of pure discrete spectrum) in the ghost channel. However, such values of $\alpha(E)$ are known to be due, as a rule [Ne66], to the presence of the virtual or resonance states of the complete Hamiltonian with forbidden symmetry above the threshold. On the other hand, such-states belong, strictly speaking, not to the discrete spectrum but to the continuum. To get a deeper insight into the situation at a whole, let us examine the conventional variational approach to solving the scattering problem and, with this purpose, write the ansatz for the wave function of the form

$$\Psi_t = \sum_i a_i \varphi_i + \Psi_{asympt} , \tag{5.18}$$

where φ_i are the square integrable basis functions; a_i are the linear variational parameters; Ψ_{asympt} is the asymptotic part of the trial wave function. Further, it is well-known that, when the Kohn, Hulten, or Rubinov variational principles [Ji74; Sc74a] are used, the scattering matrix exhibits singularities at the energies satisfying the condition

$$\det \| \langle \varphi_i | H - E | \varphi_j \rangle \| = 0 . \tag{5.19}$$

In particular, if the Kohn technique is used, we get the following set of algebraic equations:

$$\sum_{j=1}^{N} M_{ij} a_j = R_i \tag{5.20}$$

where $M_{ij} \equiv -2 \langle \varphi_i | H - E | \varphi_j \rangle$, $R_i = 2 \langle \varphi_i | H - E | \Psi_{asympt} \rangle$ and the scattering matrix will be found from the equality

$$\frac{\tan \delta}{k} = \lambda + \sum_{ij} M_{ij} a_i a_j - 2 \sum_{i} a_i R_i - \langle \Psi_{asympt} | H - E | \Psi_{asympt} \rangle$$

where λ is the trial value of $\tan \delta / k$. It is clear that the matrix M_{ij}^{-1} at energies E_k satisfying (5.19) is singular, so the solution for the basic equation (5.20) will also be singular at such energies.

Returning now to AFS, we can conclude that they themselves and the spurious resonances closely associated with them, similarly to the above mentioned singularities appearing in case of the variational treatment of the scattering problem, are probably due to the constraint to the functional space of the problem [1]). At the same time, far from all the semiforbidden states are the reflection of the true singularities of the accurate Hamiltonian onto a narrower subspace; most of them are probably the mere finitely-dimensional image of smooth continuum in the ghost (or disregarded other) channels of scattering. It is quite possible, therefore, that, although each of these states fails to carry useful information, some mean-integral derivatives, such as, say, the functions $F = \sqrt{1 - \hat{K}}\, \phi$ (where $\hat{K}$ is the overlapping kernel; ϕ is the function of relative cluster motion) can be in a general agreement with their accurate analogues in the rigorous scattering theory.

Therefore, the applicability of OCM is probably associated with the small number of particles in a system. In more complex systems (of the $^{16}O + \alpha$ type), the appearance of semiforbidden states is the unavoidable pay for the neglection of the great number of the degrees of freedom in the system. In such cases, the only alternative is to use again the complete RGM-pattern. Thus, the appearance of numerous semiforbidden states is probably an indication of the ever increasing hazard for the applicability of OCM.

The content of the present chapter may be summarized in brief as follows. We have shown that the ghost bound states, which are nothing but than the bound states of the complete Hamiltonian with a symmetry forbidden by the Pauli principle, appear instead of the model forbidden states in RGM and OCM when the rigorous many-body approach to cluster scattering is applied. The many-body amplitudes of orthogonality scattering,

---

[1]) The fact that OCM involves the singularities in the form of the spurious resonances, and not phase discontinuities, is due to the treatment of scattering on the continuous and not discrete, basis in at least a single channel, i.e. for relative motion variable.

which can approximately include the cluster breaup-up and generalize OCM for the case of true *many-body scattering*, may be obtained as zero-order approximation by applying the technique of orthogonal projecting onto ghost states. After that, using the pole parts of the many-body GF contained in such amplitudes, we get the standard OCM with the overlapping formfactors of the exact ghost functions with the inner functions of clusters taken as the wave functions of the forbidden states. The obtained pattern, apart from its heuristic value in the many-body scattering theory, permits the cluster scattering pattern given by RGM and OCM to be clarified and specified.

It is also of interest to use the ghost states and to apply the appropriate technique of orthogonal projecting (onto the ghost states) in the atomic and molecular physics (electron scattering by atoms and molecules). In particular, the e – H scattering in the triplet state is almost completely analogous, from this viewpoint, to the quartet n – d scattering. The said channel also contains a ghost bound state (and possible not a single) of H^- ion with symmetric spatial function of two electrons. And in just the same manner as in the n–d scattering, the amplitude of pure orthogonality scattering $\widetilde{F}_0$ (5.12) (including the action of the long-range polarization potential) should be a good approximation for the e – H triplet scattering at low energies. It is useful to note that the OCM concept was first proposed as long ago as 1961 [Li61] just for the problem of e – H scattering. Besides that, the success of the separable model for the triplet e – H scattering constructed on the ghost state formfactor [De75] is quite understandable from this viewpoint, although the standard physical substantiation of the applicability of the separable model (i.e. domination of the near pole of the bound state) is absolutely inapplicable to the triplet S-wave e – H scattering.

Besides that, the orthogonal projecting onto the ghost states in the field of electron scattering by molecules (e–N_2 and others) is extensively used in practice to approximately describe the exchange effects [Bu70, 72, 77c; Ch76]. In this case, the application of ghost projecting instead of the complete inclusion of the exchange effects (which results inavoidably in a very complex calculational technique for molecules) makes it possible to significantly simplify the calculations and to physically clarify the picture as a whole.

6 The Systems of Three Composite Particles

6.1 The ^{12}C Nucleus as a 3α System With Forbidden States

6.1.1 The oscillator basis for a three-boson system with eliminated forbidden states

As is known, there are two different approaches to the three body problem – one using the Faddeev equation (FE) and the other using the variational method. In the preceding chapter, we have discussed a modification of FE for three composite particles. In this chapter, taking the ^{12}C nucleus as an example, we shall discuss a second approach, in which the ^{12}C nucleus will be treated as a system 3α. As the variational basis, we use the oscillator basis for the translation invariant shell model (TISM), while the $\alpha\alpha$ interaction has been chosen in the form of an optical potential with forbidden states (see Chapter 4).

Attempts to treat the ^{12}C nucleus in the ground state with $T = 0$ as a system 3α have repeatedly been made for a long time (see, the review [HO77]). However in the overwhelming number of these works, the Pauli exclusion principle for the interaction was taken into account in the rough a manner by introducing into the potential a l-dependent local repulsive core. Naturally, none of these potentials leads to a correct, from the microscopic standpoint, wave function for the relative motion in the $\alpha\alpha$ subsystem.

Although the 3α model for the ^{12}C nucleus considered here fails to take into account a number of factors (such as the triple exchange forces, spin-orbital interaction in the system of twelve nucleons), it still seems very interesting to try OPFS in the 3α system and see what will be the off-shell behaviour of these interactions, whether they yield reasonable binding energies, etc.

The specificity of work with a complete TISM basis is, in our case, that it must be purified from forbidden states in each particle pair. This, not quite elementary, requirement can readily be taken into account in the frame of the oscillator basis, when we remember the easiness of particle recoupling allowed by it. Here we shall consider a very simple case of constructing a projected TISM basis corresponding to the assumption that the forbidden states of the $\alpha + \alpha$ Hamiltonian are approximated by the oscillator wave functions. The best approximation of the wave functions for the forbidden states is reached for the oscillator-parameter value $\hbar\omega = 22.5$ MeV, which we have used.

The states of the TISM basis for the three composite boson will, in accordance with [Kr69; Mo69, Va71, Sm63, Ku70] be numbered by the quantum numbers N, [f], $(\lambda\mu)$, K, Ω, L, M. Since the ^{12}C nucleus has the lowest-lying shell-model configurations $s^4 p^8$, we shall restrict ourselves to a set of states with the total number of oscillator quanta $N \geqslant 8$. The permutation symmetry of 3α particles $[f] = [3]$ places the following limitations [Kr69] upon the number $K = \lambda, \lambda - 2, \lambda - 4, ..., 1$ or 0: $K = 0 \{\mathrm{mod}\ 3\}$; if $K = 0$, μ should be even. The complete list of the TISM basis vectors for $N \leqslant 12$ is given in Table 4. However, not all states with $N \geqslant 8$ are free from the forbidden components,

Table 4: Forbidden and allowed states for the system of 3α in the oscillator basis

N	Basis vectors of the TISM $(\lambda\mu)$	forbidden states $(\lambda\mu)$	allowed states $(\lambda\mu)$
0	(00)	(00)	
1			
2	(20)	(20)	
3	(30)	(30)	
4	(40) (02)	(40) (02)	
5	(50) (31)	(50) (31)	
6	$(60)^2$ (22)	$(60)^2$ (22)	
7	(70) (51) (32)	(70) (51) (32)	
8	$(80)^2$ (61) (42) (04)	$(80)^2$ (61) (42)	(04)
9	$(90)^2$ (71) (52) (33)	$(90)^2$ (71) (52)	(33)
10	$(10{,}0)^2$ (81) $(62)^2$ (24)	$(10{,}0)^2$ (81) (62)	(24) (62)
11	$(11{,}0)^2$ $(91)^2$ (72) (53) (34)	$(11{,}0)^2$ (91) (72)	(91) (53) (04)
12	$(12{,}0)^3$ (10,1) $(82)^2$ (63) (44) (06)	$(12{,}0)^2$ (10,1) (82)	(12,0) (82) (63) (44) (06)

which can be constructed as follows. The oscillator function $|\,\mathrm{nlm}\,(\mathbf{r})\rangle$ for the forbidden state (nl = 00, 20, 22) of the subsystem $\alpha+\alpha$ where $\mathbf{r}=\mathbf{r}_{12}=\mathbf{r}_1-\mathbf{r}_2$ is the coordinate of the relative motion of two α clusters must be multiplied by an oscillator function $|\,\mathrm{N-n},\Lambda\mathrm{M}_\Lambda(\boldsymbol{\rho})\rangle$ dependent on the second Jacoby coordinate $\boldsymbol{\rho}=\boldsymbol{\rho}_3=\frac{1}{2}(\mathbf{r}_1+\mathbf{r}_2)-\mathbf{r}_3$. Using the Clebsch-Gordan coefficients for the group SU(3), we shall construct from these products the states with a definite SU(3) symmetry $(\lambda\mu)$:

$$|\,\mathrm{n,N-n}\,(\lambda\mu)\Omega\mathrm{LM}\rangle = \sum_{l\Lambda}\langle(\mathrm{n}0)l\times(\mathrm{N-n},0)\Lambda\,|\,(\lambda\mu)\Omega\mathrm{L}\rangle\times|\,\mathrm{n}l(\mathbf{r}_{12}),\ \mathrm{N-n}\Lambda(\boldsymbol{\rho}_3):\mathrm{LM}\rangle\,. \tag{6.1}$$

When n = 0, only $(\lambda\mu)=(\mathrm{N}0)$ is allowable. At n = 2, $(\lambda\mu)=(\mathrm{N}0)$, (N − 2, 1) and (N − 4, 2) are possible. Then we symmetrize the vector (6.1) with respect to the permutation of α particles. Since the vector (6.1) is symmetrical relative to the permutation P_{12}, the symmetrization operator S takes on the form:

$$\mathrm{S}=(1+\mathrm{P}_{13}+\mathrm{P}_{23}) \tag{6.2}$$

and we arrive at the following expression for the forbidden-state function

$$|\,\mathrm{A}=3\,\mathrm{N}\,[3]\,(\lambda\mu)\Omega\mathrm{LM}\rangle_{\mathrm{forbid}}=\frac{1}{\mathrm{W}}\mathrm{S}\,|\,\mathrm{n,N-n}\,(\lambda\mu)\Omega\mathrm{LM}\rangle\,, \tag{6.3}$$

where, using the results obtained in ref. [Kr69], we can write the normalizing factor in the following form

$$\mathrm{W}^2=3\,[1+2\,\langle\mathrm{n,N-n}\,(\lambda\mu)\Omega\mathrm{LM}\,|\,\mathrm{P}_{23}\,|\,\mathrm{n,N-n}\,(\lambda\mu)\Omega\mathrm{LM}\rangle]=$$

$$=3\,[1+2\,(-1)^{\mathrm{N-n}}\,\mathrm{D}^{\lambda/2}_{\mathrm{n-N/2,\,n-N/2}}(2\pi/3)]\,. \tag{6.4}$$

The condition for a certain symmetrical function $\overline{\psi}_{LM}$ to be orthogonal to the function (6.3) has the form

$$\langle \overline{\psi}_{LM} \mid A = 3, N, [3] (\lambda\mu)\Omega LM \rangle =$$

$$= \frac{3}{W} \langle \overline{\psi}_{LM} \mid n, N - n (\lambda\mu)\Omega LM \rangle = 0 , \tag{6.5}$$

where $n \leqslant 2$, and the quantum numbers N, $(\lambda\mu)$, Ω assume all possible values. Since the transformation (6.1) is unitary, one can rewrite the condition (6.5) in the form

$$\langle \overline{\psi}_{LM} \mid nl\,(\mathbf{r}_{12}), N - n\ \Lambda(\boldsymbol{\rho}_3): LM \rangle = 0 \ \text{ at } nl = 00,\ 20,\ 22\,. \tag{6.6}$$

Here the values $N - n$ and Λ are arbitrary; therefore, we obtain the condition

$$\langle \overline{\psi}_{LM} \mid nl\,(\mathbf{r}) \rangle = 0, \qquad nl = 00,\ 20,\ 22\,. \tag{6.7}$$

Due to the symmetry of the function $\overline{\psi}_{LM}$ relative to the permutations of α particles, the condition of orthogonality to the forbidden states (6.3) of the three-body system reduces to the condition (6.7), which deals with only one degree of freedom $\mathbf{r} = \mathbf{r}_{12}$ with fixed enumeration of α particles.

The direct calculation shows that at $N \geqslant 8$ all states (6.3) with $n = 0, (\lambda\mu) = (N0)$ or $n = 2, (\lambda\mu) = (N0), (N - 2, 1)$ and $(N - 4, 2)$ are linearly-independent nonvanishing vectors. Hence we come to the following procedure for constructing the projected TISM basis which is "purified" from forbidden states: for each value of N one should construct at first a complete TISM basis $\overline{\psi}_{NLM}$, then the total set of forbidden states (6.3) and finally, orthogonalize the basis states $\overline{\psi}_{NLM}$ to the forbidden states by using, for example, the Gram-Schmidt procedure. The overlap integrals between the original basis vectors $\overline{\psi}_{NLM}$ and forbidden states are given by formulae (6.4), (6.5). Specifically, for the ^{12}C nucleus, this leads to the following results. In the ground state the given nucleus has the spin 0^+; therefore, only the vectors TISM with the even N and μ may be contained in its wave function. Consequently, at a fixed N, there are three types of linearly independent vectors of the forbidden states, namely

$$\begin{array}{ll} \text{two sets of vectors with} & \mu = 0, \\ \text{one set of vectors with} & \mu = 2. \end{array} \tag{6.8}$$

All vectors with $\mu > 2$ are allowed. Therefore, in practice, the above procedure for orthogonalization to the forbidden states should be carried out only for the TISM basis vectors with $\mu = 0$ and 2 (see Table 4).

From the condition (6.7) there follows still another possibility of projecting forbidden states in a three-body system. It reduces to the diagonalization in the frame of a complete TISM basis of the operator

$$P = P_1 + P_2 + P_3\,. \tag{6.9}$$

where $P_1, P_2\ P_3$ are the projectors upon the forbidden states for the degrees of freedom $\mathbf{r}_{12}, \mathbf{r}_{13}, \mathbf{r}_{23}$ respectively. Owing to the symmetry of the three-boson states relative to the permutations of the particles, it is sufficient to diagonalize a part of the operator P, for example

$$P_1 = |00\rangle\langle 00| + |20\rangle\langle 20| + |22\rangle\langle 22|\,. \tag{6.10}$$

The allowed states correspond to zero eigenvalues of this operator. It is clear that the latter approach is also applicable to systems composed of four and larger number of α clusters. It enables one to generalize the procedure of projection also to the case when one uses functions of the general type rather than oscillator wave functions.

In the literature, there have been suggested also other methods for separation allowed and forbidden states in systems of three [Ho74, 75] and four [Ho76; Su77] clusters, and also for other multicluster systems, which are based on studying the properties of eigenvalues and eigenfunctions of three-cluster and more complicated integral overlap-kernels. Naturally, for the system 3α they lead to the same results as those described above.

Let us now turn to the results of specific calculations.

6.1.2 Levels of the ^{12}C nucleus in the 3α model with forbidden states

In the variational calculation [Sm74] the trial function was represented as a linear combination of allowed TISM basis vectors:

$$\psi_{\text{trial}} = \sum_{i=1}^{i_{max}} C_i | A = 3\,N\,[3]\,(\lambda\mu)\Omega LM \rangle_{\text{allowed}} . \qquad (6.11)$$

Here index i denotes the total set of the TISM quantum numbers: N, $(\lambda\mu)$, Ω; i_{max} is the number of the vector upon which the basis is cut off. The unknown coefficients C_i and the level energies E were obtained by solving a set of linear equations:

$$\frac{\partial}{\partial C_i} (\psi_{\text{trial}} \,|\, H - E \,|\, \psi_{\text{trial}}) = 0, \quad i = 1, 2, \dots, i_{max} \qquad (6.12)$$

and the corresponding secular equation.

The potential of $\alpha\alpha$ interaction, which enters into the Hamiltonian H, has the following form (see Chapter 4)

$$V_{\alpha\alpha}(r) = V_0 \,\{1 + \exp[(r - r_0)/a]\}^{-1} + V_{\text{Coulomb}}(r) . \qquad (6.13)$$

where $V_0 = -125$ MeV, $r_0 = 1.78$ fm, $a = 0.66$ fm. It will be noted that in the given calculation it is impossible to vary, as usual, the parameter $\hbar\omega$ because its value has been predetermined ($\hbar\omega = 22.5$ MeV) from the condition of best approximation for the forbidden states. Therefore, the class of the trial functions (6.11) used in such a calculation is less flexible than in the conventional variational calculations on an oscillator basis. As a result, to obtain good convergence, one has to employ a basis of sufficiently high dimension. The calculation was performed up to $N_{max} = 34$, the matrix of dimension 68×68 was diagonalized. The energy of the ground state 0_1^+ and the positions of first two excited states 0_2^+ and 0_3^+ were determined:

$$E_{0_1^+} = -15.06 \text{ MeV}, \quad E_{0_2^+} = 9.12 \text{ MeV}, \quad E_{0_3^+} = 11.75 \text{ MeV} . \qquad (6.14)$$

For comparison, a calculation was made of the 3α system with interaction (6.13) on a complete TISM basis, i.e., without orthogonalization to forbidden states and it has been found that

$$E'_{0_1^+} = -217.3 \text{ MeV}, \quad E'_{0_2^+} = -164.9 \text{ MeV}, \quad E'_{0_3^+} = -116.9 \text{ MeV} . \qquad (6.15)$$

Thus the orthogonalization to forbidden states in the absence of repulsion in the $V_{\alpha\alpha}$ potential does keep the $3\,\alpha$ system from collapsing. This fact again confirms our initial assumption that it is quite unneccessary to introduce the repulsive core into the $\alpha\alpha$ interaction.

It can also be seen that the scales of the kinetic and potential energies in the $3\,\alpha$ system are large. In comparison with these scales the obtained overestimation by 7.8 MeV[1]) of the binding energy of the ^{12}C nucleus may be regarded as a relatively small deviation if we remember that the model under discussion is merely a "first approximation", because we leave out of account the following aspects:

a) The ground-state ^{12}C nucleus appears to have the structure intermediate between the shell-model and $3\,\alpha$ cluster ones. The orthogonalization to the two particle forbidden states takes into account only the contribution from the exchanges of identical nucleons between two different α clusters, while in the $3\,\alpha$ system a definite contribution should come also from the nucleon exchanges between three α clusters. This will lead to the renormalization of the $V_{\alpha\alpha}$ interaction in the $3\,\alpha$ system.

b) As is indicated by RGM calculation [Wi77], the interaction between two α clusters associated with nucleon exchanges between them is nonlocal and energy-dependent. The potential (6.13) is equivalent to such an interaction in the sense of describing the phases of $\alpha\alpha$ scattering and elimination of certain forbidden states.

Therefore the t matrix for the potential (6.13) may still be different from the t matrix for the "true" nonlocal interaction.

c) If we proceed from the shell model of the nucleus (this in reality corresponds to the central region, where the nucelon density is high), then the probability will be high that in the ^{12}C nucleus there are (virtual) α particles in the excited states [Go76], which may already be present, for example, in the 2 S state of the relative motion, etc., i.e., they should be treated in a special way. The calculations of Jackson et al. [Ja70] show however that their contribution to the binding enery of the ^{12}C nucleus is not large.

Thus a "second approximation" should include the effects listed above.

As regards the excited states 0_2^+ and 0_3^+, these probably have the $3\,\alpha$-cluster nature. This inference is consistent with the results of microscopic calculations [De72], which definitely indicate that the ^{12}C nucleus in the 0_2^+ state represents a vibrating chain of α clusters.

We have dwelt at length on the 0^+ levels of the ^{12}C nucleus. A similar calculational scheme based on the orthogonality-condition model described in Chapter 4 was applied also to states with other angular momenta [Ho74, 75; Ka77] Fig. 21 compares with experiment the results of these calculations published by Horiuchi [Ho74, 75] who used a deep attractive interaction potential of the folding-potential type and carried out the elimination of the forbidden states in the $3\,\alpha$ system by the method described above in this section. The energy matrix for the $3\,\alpha$ system was diagonalized on a basis that in-

[1]) It will be noted that calculations with the $\alpha\alpha$ potential containing the repulsive core usually underestimate the binding energy of ^{12}C nucleus [Ag77].

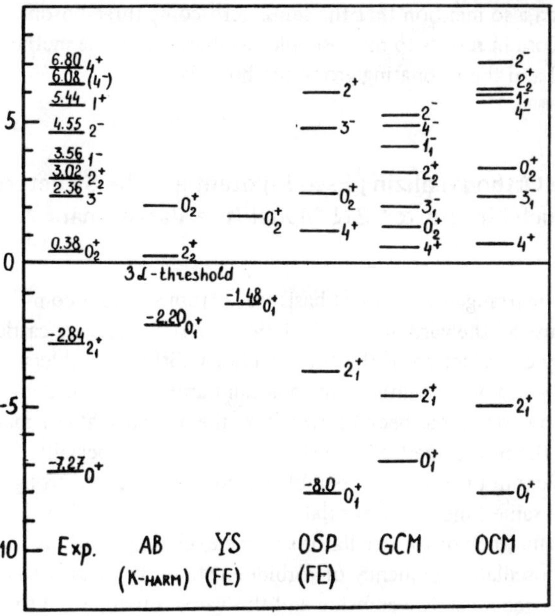

Fig. 21
The level spectra of ^{12}C-nucleus in the various theoretical 3α-models: first at the left – experimental data; AB – the result of variational calculation with the $\alpha-\alpha$ Ali-Bodmer potential; YS – the result of solution of FE with $\alpha-\alpha$ Yamaguchi potential; OSP – FE solution with $\alpha-\alpha$ oscillating separable potential; GCM – the result of application of generator coordinate method; OCM – orthogonality-condition-model solution.

cluded all the allowed states with the quantum number $N \leqslant 30$. The figure also presents the results of a full microscopic calculation for the ^{12}C nucleus by the generator coordinate method [Ka 77]. It can be seen that all the observed levels whose excitation energy is below 15.11 MeV where the first level $J = 1^+$ with $T = 1$ lies, are well reproduced by the theory, with the exception of the 1^+ level with the energy 12.71 MeV, which represents spin excitation and does not have the $3\,\alpha$ structure. The level with an energy of 13.35 MeV has the negative parity, but the value of its spin J has not been identified as yet. The theory predicts the appearance of the 4^- level from the rotational band with $K = 3$ based on the 3^- level with the energy 9.64 MeV. The results have also shown (see [Ho 74, 75, 77]) that the second excited states 0^+ and 2^+ have the structure of $^8Be\,(0^+) + \alpha$, where α particle moves with the orbital angular momentum 0 or 2, respectively, at a distance of about 5 fm from ^8Be. These levels are characterized by large α-particle reduced widths, which is in good agreement with experiment. Thus the calculated total widths Γ_α for the levels 0^+, 2^+, 1^-, and 3^- are equal to 8.7, 1700, 370, and 29 KeV, respectively, which is close to the experimental values 8.7, 3000, 315 and 34 KeV. The peripheral character of the 0^+, 2^+ levels is manifested also by the fact that their wave functions are "spread", in the sense of expansion in the oscillator basis, over many states with a large number of quanta N. At the same time, the "spread" of the wave function for the compact ground state ^{12}C nucleus is much less, here the low lying allowed oscillator state with $N = 8$ is presented with a large weight and is dominant. OCM calculations reproduces well a correct value $6.1\ fm^2$ for the matrix element for a E0 transition between the states 0_2^+ and 0_1^+, which is close to the experimental value $5.8\ fm^2$ [Ho 75,77]. Thus one can state that the assumption about the $3\,\alpha$ structure with forbidden states for ^{12}C describes correctly the properties of the levels of this nucleus lying below the first excited state with T = 1. One can also mention that the semimicroscopic three-boson model with forbidden states is close in results to microscopic calculations by the method of generator coordinates [Ka 77] and the resonating group method [Fu 77].

6.2 Projecting Through the Orthogonalizing Pseudopotentials; the Structure of the ^{6}He–^{6}Li–^{6}Be Nuclei in the $\alpha + 2N$ Model (the Variational Analysis)

Though the application of rearranged variational basis freed from spurious components is very convenient because all the variational calculations on such basis are carried out in the standard way, the very construction of the basis is a fairly difficult problem (see subsection 6.1 above) and, which is most important, one can hardly avoid the constraints of the oscillator basis. However, it has been known from the 3N-calculations made on this basis [Ja 71; St 74b] that the results are but very slowly convergent, especially where the interaction potentials exhibit certain undersirable features (hard core, strong tensor components, etc.). At the same time, it is essential to the cluster physics that the variational calculations in the framework of the oscillator scheme excluding spurious states is only practical when the oscillator frequency of forbidden states, i.e. $\hbar\omega \approx 18$ MeV and the oscillator frequency of a free α-particle coinside and this value is used as the total oscillator frequency for all basis functions. On the other hand, most states of light nuclei (say, in the systems of the type ^{6}Li, ^{9}Be, etc.) lie near or above the threshold and are loose in structure. Obviousely, the value of $\hbar\omega$ inherent to such states should be much smaller, otherwise the variational expansion convergence gets notably slow. In such cases, therefore, it is reasonable that another approach based on orthogonalizing pseudopotentials [Kr 74] be used. Though the OPP concept is a whole the same as described in the previous Chapters, its realization will be *quite different*. Namely, let us find the eigenenergies and eigenfunctions of the N-body Hamiltonian

$$H(1, 2, \dots, N) = \sum_{i=1}^{N} t_i + \sum_{i>j} V_{ij} + \sum_{ijk} V_{ijk} + \dots,$$

where t_i are the operators of kinetic energy; V_{ij} are the two-body forces; V_{ijk} are the three body forces and etc., i.e. we have to solve the Schrödinger equation

$$H(1, 2, \dots, N)\,\Psi(1, \dots, N) = E\Psi(1, \dots, N) \tag{6.16}$$

with the additional conditions of orthogonality of the general form

$$\Gamma_1\Psi = 0, \quad \Gamma_2\Psi = 0, \dots, \quad \Gamma_r\Psi = 0\,. \tag{6.17}$$

The solution $\Psi(1, \dots, N)$ will be expanded, as usual, in a certain complete basis $\{\Phi_j\}$

$$\Psi(1, 2, \dots, N) = \sum_{i=1} C_i \Phi_i(1, \dots, N)\,.$$

After that, the following procedure may be conveniently applied, instead of the rearrangement of the basis $\{\Phi\}$ described in subsection 1 above, to exclude the spurious components.

The pseudo-Hamiltonian

$$\widetilde{H}_\lambda(1, \ldots, N) = H(1, \ldots, N) + \lambda \sum_{k=1}^{r} \Gamma_k$$

will be determined and then the initial complete basis will be used to minimize the functional

$$[E] = \min \frac{\langle \Psi_t | \widetilde{H}_\lambda | \Psi_t \rangle}{\langle \Psi_t | \Psi_t \rangle} \tag{6.18}$$

by taking a sufficiently high (but not infinite) value of λ. It can easily be proved using the results presented in the previous Chapters that, for the limit $\lambda \to \infty$ and *when the dimension of the basis gets infinitely high*, the solutions of eq. (6.18) will exhibit the required properties of orthogonality (6.17). In this case, the basis $\{\Phi\}$ and the functions contained in the projectors Γ_i may be of arbitrary forms (not only of the oscillator form). The major difference of the scheme described here from the approach using the projected basis (see Subsect. 1) is that the trial function in the present scheme is strictly speaking, non-orthogonal to the forbidden states (due to the finiteness of the basis and the constant λ) but, as the dimension of the basis increases, the orthogonality conditions are better and better satisfied (see below, Table 5.1).

This approach was used in [Ku 80a] together with the variational approach [Kr 77, Ku 75a] based on the expansion in non-minimal randomized Gaussian basis to calculate the spectrum of the low-lying states in the ^{6}He$-^6$Li$-^6$Be nuclei in terms of the $\alpha - 2N$ model. The calculational scheme may be outlines in brief as follows. The complete wave function of three-body system Ψ^I is presented as the expansion in LS-basis:

$$\Psi^I(1, 2, 3) = \sum_{LSM_LM_S} \langle LM_L SM_S | IM \rangle \Phi_{LM_L}(\mathbf{r}_1, \mathbf{r}_2, \mathbf{r}_3)\, \chi_{SM_S}(s_1, s_2, s_3)$$

where $\Phi_{LM_L}(\mathbf{r}_1, \mathbf{r}_2, \mathbf{r}_3)$ is the orbital wave function with total orbital moment L and with its projection M_L, $\chi_{SM_S}(s_1, s_2, s_3)$ is the spin wave function corresponding to the total spin S and to its projection M_S. The spin wave function is constructed using the standard procedure while the orbital function will be represented through the appropriate normalized Jacobi coordinates (see Fig. 22) which, in their turn, are expressed through the particle coordinates as

$$\mathbf{x}_i = \left(\frac{2\, m_j m_k}{\hbar^2 (m_j + m_k)}\right)^{1/2} (\mathbf{r}_j - \mathbf{r}_k) \equiv \tau_{jk}\, \mathbf{r}_{jk}$$

$$\mathbf{y}_i = \left(\frac{2\, m_i (m_j + m_k)}{\hbar^2 (m_i + m_j + m_k)}\right)^{1/2} \left(\frac{m_j \mathbf{r}_j + m_k \mathbf{r}_k}{m_j + m_k} - \mathbf{r}_i\right) \equiv k_{ijk}\, \rho_i$$

$ijk = 1, 2, 3$ and their cyclic transpositions. In these coordinates, the three-body Hamiltonian takes the form

$$H = H_0 + V \;\text{ with }\; H_0 = \frac{\partial^2}{\partial x_i^2} + \frac{\partial^2}{\partial y_i^2} \;\text{ and }\; V = \sum_{i=1}^{3} V_i(r_{jk}^{-1} x_i) + V^{(3)}\,.$$

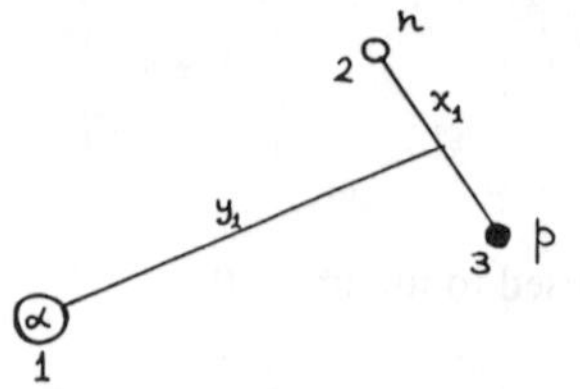

Fig. 22
The choice of the coordinates in the three-body model of the ^{6}He-^{6}Li-^{6}Be nuclei.

Henceforth, the three-body forces $V^{(3)}$ will be neglected (in the case of as loose systems as $^6\text{He}-^6\text{Li}-^6\text{Be}$, the contribution from three-body forces must be minor). Now, the orbital function $\Phi_{LM_L}(\mathbf{x}_1, \mathbf{y}_1)$ (the choose of coordinates is shown in Fig. 22) will be expanded in the nonorthogonal Gaussian basis:

$$\Phi_{LM_L}(\mathbf{x}_1, \mathbf{y}_1) = \sum_{\lambda_1 l_1} \Phi^L_{\lambda_1 l_1}(x_1, y_1)\, I^{LM_L}_{\lambda_1 l_1}(\hat{x}_1, \hat{y}_1) =$$

$$= \sum_{\lambda_1 l_1} \left(\sum_{i=1}^{n} C_{\gamma i}\, \phi_{\gamma i}(x_1, y_1) \right) I^{LM_L}_{\lambda_1 l_1}(\hat{x}_1, \hat{y}_1)\,, \tag{6.19}$$

where the spherical tensor

$$I^{LM_L}_{\lambda_1 l_1}(\hat{x}_1, \hat{y}_1) = \sum_{m_{\lambda_1} m_{l_1}} \langle \lambda_1 m_{\lambda_1} l_1 m_{l_1} | LM_L \rangle\, Y_{\lambda_1 m_{\lambda_1}}(\hat{x}_1)\, Y_{l_1 m_{l_1}}(\hat{y}_1)$$

is expressed through the superposition of products of two spherical harmonics $Y_{\lambda_1 m_{\lambda_1}}(\hat{x}_1)$ and $Y_{l_1 m_{l_1}}(\hat{y}_1)$. The basis functions are taken to be of the form of Gaussians with the appropriately selected scale parameters

$$\phi_{\gamma i}(x, y) \equiv N_i\, x^\lambda y^l \exp(-\alpha_{\lambda i} x^2 - \beta_{l i} y^2) \tag{6.20}$$

and the normalization constants are

$$N_i \equiv 2^{\lambda + l + 3} \sqrt{\frac{2\alpha_{\lambda i}^{\lambda + 3/2}\, \beta_{l i}^{l + 3/2}}{\pi (2\lambda + 1)!!\, (2l + 1)!!}}\,.$$

The index $\gamma \equiv \lambda, l$; $C_{\gamma i}$ are the linear variational parameters. It may be proved [Kac 35] that the basis (6.20) is *complete* provided the scale coefficients $\alpha_{\lambda i}$ and $\beta_{l i}$ satisfy certain conditions, for example of the type $\sum_{i=1}^{\infty} (\alpha_{\lambda i})^{-1} = \infty$. The essence of the stochastic variational method (SVM) is the concept of random search for the nonlinear parameters $\alpha_{\lambda i}$ and $\beta_{l i}$ (or the analogous parameters for the system of any number of particles) by the trial and error method with subsequent selection of optimal components. The various combinations of the random search with the known procedures of deterministic search were proposed [Kr 77]. It has been shown in [Kr 77, Ku 77b; Bar 77, 78] using numerous systems (in particular 3N, 4N, 3 α, four-quark system $qq\bar{q}\bar{q}$ etc.) as examples that SVM gives a highly effective calculational scheme which is especially suitable to study the cluster structure of nuclei [Kr 77]. In particular the rate of convergence on the many-dimensional randomized Gaussian basis proves to be much in excess (by an order of magnitude) of that in the oscillator scheme. Presented below will be the results of the calculations of the low-lying state structure of nuclei with $A = 6$ obtained in the above presented formalism using the OPP procedure. Since the aim of the calculation was essentially to give an illustration, i.e. to demonstrate the effectiveness of the calculational scheme developed, we disregarded the spin-orbital $\alpha - N$ interaction and the noncentral components of the N–N forces[1]). In this case, the pure L–S coupling takes place. The following two-particle interactions were used.

In the $\alpha - N$ subsystem, the potential given in [Sa54] was used without the spin-orbital component: $V_{\alpha N}(r) = -V_0 \exp(-\eta r^2)$ with the parameters:

$$V_0 = 47.32 \text{ MeV}, \quad \eta = 0.18923 \text{ fm}^{-2} \tag{6.21}$$

The potential was properly fitted to the s- and p-wave phase shifts.

In the N–N subsystem, the rectangle well potential

$$V_{NN}^{s}(r) = \begin{cases} -V_1^s, & r < b_s \\ 0, & r > b_s \end{cases}$$

was used with the parameters [Be70a; Ac72]:

$$V_1^1 = 34.4012 \text{ MeV}, \quad b_1 = 2.0719 \text{ fm} \tag{6.22}$$

for the triplet state and

$$V_1^0 = 16.85 \text{ MeV}, \qquad b_0 = 2.37 \text{ fm} \tag{6.23}$$

for the singlet state. The $\alpha - N$ potential (6.21) comprises a low-lying OS-state (with binding energy $\epsilon_b \simeq 12$ MeV) forbidden by the Pauli principle in a five-nucleon system. Therefore, to exclude the contribution of this state, one has to require orthogonalization of the total function (6.19) to both OS-states (in both $\alpha - N$ coordinates), i.e. we get two additional conditions (6.17). Since, however, both projectors (i.e. Γ_2 and Γ_3) do not commute, the total projector will be an infinite sum of the products of Γ_2 by Γ_3 (see Chapter 2). Unlike all the previous works based on this scheme, the projecting approach used here *does not require* that a combersome total projector be constructed but, instead, includes the matrix elements of the projectors Γ_2 and Γ_3 on the chosen variational basis. It is of importance to emphasize that, because of the Gaussian basis, *all* the matrix elements of the Hamiltonian and projectors in the $\alpha - N$ subsystem can be calculated on this basis in the completely analytical form. The above discussed procedure provides us with a rapidly realizable calculational scheme which leaves but simple algebraic operations to computer (namely, the calculations of determinants and the solution of the set of linear equations).

Let now the results be discussed. The convergence of the low-lying states of ^{6}He-^{6}Li-^{6}Be is presented in Table 6.1 as a function of the number of terms.

Table 6.2 shows the convergence of the energies of the ^{6}Li ground state at a fixed number of terms n (= 4) with increasing λ.

---

[1]) The results obtained including the spin-orbital forces have been available and will be presented in a separate publication.

Table 6.1 [1])

n	4	6	8	10
$E(^6Li, 1^+)$ MeV	– 0.468	– 2.138	– 2.554	– 2.561
$E(^6Li^*, 0^+)$ MeV		3.126	3.027	3.000
$E(^6He, 0^+)$ MeV	2.565	2.248	2.151	2.148
$E(^6Be, 0^+)$ MeV	4.910	4.646	4.507	4.498

[1]) The data shown in Table 6.1 have been obtained at the value of the orthogonalizing constant $\lambda = 900$ MeV. A very rapid convergence can be seen (the same relates to the energies of the ground states of ^{6}He and ^{6}Be.

Table 6.2

λ (MeV)	0	100	500	900
$E_{gr.\,st}(\lambda)$ MeV	– 5.387	– 1.019	– 0.547	– 0.468
$P_\xi(\lambda)$	0.336	0.0024	$2 \cdot 10^{-4}$	$7 \cdot 10^{-5}$

The third line of Table 6.2 lists the values of the admixture $P_\xi(\lambda)$ of the forbidden state in the ground state function. The admixture was determined as

$$P_\xi(\lambda) = \int [\varphi^*_{0s}(x_2) \cdot \Psi^I(\lambda; x_1, y_1)]^2 \, dx_1 \, dy_1$$

It can easily be seen that at already $\lambda = 500$ MeV the admixture of the forbidden configuration becomes very small. Thus, the orthogonality projecting method presented here gives a very simple and universal calculational scheme.

Fig. 23 shows a comparison between the calculated and experimental energies of the ^{6}He-^{6}Li-^{6}Be levels. It can be seen that the distance between the levels of isobaric triplet $I^\pi T = 0^+ 1$ in the ^{6}He-^{6}Li-^{6}Be nuclei is reproduced extremely well, but the absolute position of the theoretical levels is shifted upwards by ~ 3 MeV as compared with experiment (see also in [Wa62]). It is known [Ba68], however, that the inclusion of the spin-orbit forces depresses the energies of all the low-lying states. Therefore, such an inclusion will improve the agreement with experiment for absolute energies too.*)

The found wave function of the ^{6}Li ground state may be used to calculate other interesting observables, such as the r.m.s. charge radius and the electromagnetic (charge) formfactor of ^{6}Li (see ref. [Ku80a]).

It is also of great interest to consider the weight P of $\alpha - d$ cluster component determined as

$$\chi(y_1) = \int \varphi_d(x_1) \Psi^I(x_1, y_1) \, dx_1 \, dy_1 ; \quad P = \int_0^\infty |\chi(y_1)|^2 \, y_1^2 \, dy_1 .$$

*) *Added in proof.* Indeed, the subsequent calculations with inclusion of P- and D-components in the trial wave functions and of more realistic forces have shown the excellent agreement of theoretical results with a great body of experimental data.

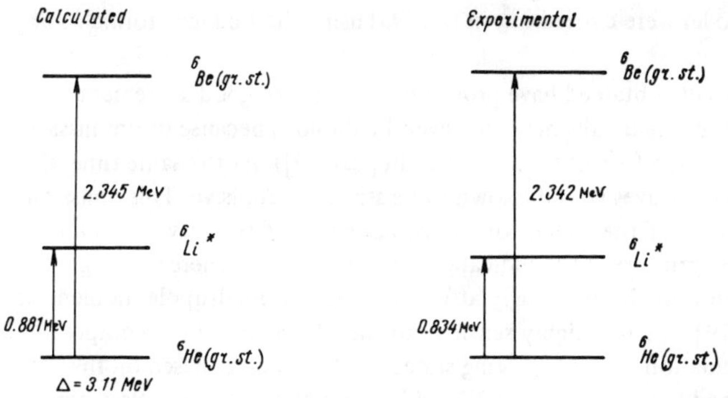

Fig. 23 Comparison between the theoretical and experimental energies of levels in the ^{6}He-^{6}Li-^{6}Be nuclei.

The calculations in our model give P = 0.68, which is in a reasonable agreement with the results published elsewhere [Ba 79a].

It seems expedient now to briefly describe the results of other calculations of the structure of nuclei with A = 6 made in terms of three-body model [1]) with special emphasis to the variational results. Only a number of a great variety of works (for detailed references, see the recent work [Ba 79a]) will be mentioned here. Wackman and Austern [Wa 62] were the first to overcome the constraints of the shell model when studying in detail the spectrum of the low-lying states of nuclei with A = 6 including the spin-orbital and tensor forces. The Pauli principle was approximately taken into account by introducing the projecting operator $\Gamma = 1 - \Gamma_n - \Gamma_p$, which is not the exact total projector for the $\alpha + 2N$ system, but merely its main part. Unfortunately, the results of [Wa 62] may be treated as but semiquantitative because of the error in definition of the operators of the angular momentum in the $\alpha - n$ and $\alpha - p$ subsystems. The error was eliminated in the work of the Italian group [Ba 78b] in which, unfortuantely, only the $I^\pi = 0^+1$ levels of isobaric triplet were calculated. Besides that, the Pauli principle was taken into account using also but a part of the total projector

$$\Gamma = 1 - \Gamma_n - \Gamma_p + \Gamma_n\Gamma_p + \Gamma_p\Gamma_n \tag{6.24}$$

which, though, exceeds that used in [Wa 62]. Nevertheless, quite resonable results were obtained for the absolute and relative energies of the isobatic triplet levels, so that the terms of the total projector discarded in (6.24) seem to make a small contribution. All the same, it is desirable that the total projector should be included to obtain more reliable results. The most accurate and realistic calculations of the structure of the ^{6}Li ground state

[1]) It will be emphasized that, in accordance with the conceptualism adopted in the present review and described in the Introduction, we are interested in the results obtained just in the cluster model with *effective* interactions between clusters, and not in the completely microscopic calculations which are still impossible ab initio for nuclei with A = 6 with the modern computers.

in terms of three-body model were carried out in [Ba 79a] using the Faddeev formalism in coordinate representation.

As the whole, the results obtained have proved to be in a very good agreement with experimental data. The fine details may, however, be dubious because of ommission of singlet P-wave NN interaction (it was set to be zero in [Ban 79]). At the same time, the NN interaction in odd singlet waves is well known to be strongly repulsive. This omission has resulted in some increase of the weight of the components of the wave function $\Psi(J^\pi T)$ with L = 1 (~ 5 % against ~ 2.5 % indicated in most cases elsewhere (see e.g. [Wa 62])). The same reason is likely to strongly affect the value of quadrupole moment of $^6Li_{gr.st.}$ obtained in [Ban 79] which is highly sensitive to the admixture of the components with L = 1. The calculation of the low-lying states for A = 6 nuclei based on the variational scheme described here with correct NN odd potential is published elsewhere [Ku 81 a, Wor 81].

In summarizing the results presented in this Chapter, it will be emparized again that the variational method on non-orthogonal many-dimensional Gaussian basis combined with the pseudo-potential method for including the Pauli principle gives a simple and highly effective calculational scheme for practical purposes of studying the cluster states in nuclei.

7 The NN-System in the Quark Model

7.1 Introductory Comments

In the present chapter, nucleons are treated as compound particles, "3 q-clusters", and the consequences for the NN system ensuring from the identity of quarks as Fermi particles are studied. Here appear interesting possibilities of connecting the properties of the 6 q-system symmetry with the character of the wave function of mutual motion of nucleons as is shown by experience of studying clusters in atomic nuclei discussed in the previous chapters. Hence, many useful physical conclusions may be drawn.

The identity of quarks in different nucleons implies complete antisymmetrization of the NN wave function $\Psi_{NN}(q^6)$ with respect to quark permutations. In a rough approximation, disregarding the dynamics of the six-quark system, it can be written that

$$\Psi_{NN}(q^6) = \frac{1}{N_A} A \{\Phi^l_{NN}(\mathbf{R})\Psi_{N_1}(q^3)\Psi_{N_2}(q^3)\}_{LCST} \tag{7.1}$$

Here $\Phi^l_{NN}(\mathbf{R})$ is the orbital part of an ordinary nuclear wave function $\Psi_{Nucl}(\mathbf{R}) = \Phi_{NN}(\mathbf{R}) \{\chi^{S_1T_1}_{N_1} \chi^{S_2T_2}_{N_2}\}_{ST}$. Now let R be expressed in terms of the quark coordinates $\mathbf{R} = \frac{1}{3}(\mathbf{r}_1 + \mathbf{r}_2 + \mathbf{r}_3) - \frac{1}{3}(\mathbf{r}_4 + \mathbf{r}_5 + \mathbf{r}_6)$ and the spin-isospin parts $\chi^{S_iT_i}_{N_i}$ be replaced by the quark wave functions of nucleons

$$\Psi_{N_1}(q^3) = A_1 \{\varphi^{L_1}_{N_1}(\mathbf{r}_1\mathbf{r}_2\mathbf{r}_3)\chi^{C_1S_1T_1}_{N_1}\} \tag{7.2}$$

where φ_{N_1} is the orbital and $\chi^{C_1S_1T_1}_{N_1}$ the spin-charge parts of the function (the superscript C means three colour quark charges which can be treated as three projections of "the colour moment" $C_z = 0, \pm 1$). The antisymmetrizers A_i and A in (7.2) and (7.1) are, respectively, equal to

$$A_1 = \frac{1}{6}(I - P_{12} - P_{13} - P_{23} + P_{12}P_{13} + P_{13}P_{12}),$$

$$A = \frac{1}{10}\left(I - \sum_{i=1}^{3}\sum_{j=4}^{6} P_{ij}\right); \quad A^2 = A, \quad A_i^2 = A_i.$$

The factor N_A takes into account the renormalization of the wave function when affected by the antisymmetrizer A. For the discrete spectrum function, the value N_A is determined by the relation

$$N_A^2 = \langle \{\Phi_{NN}\Psi_{N_1}\Psi_{N_2}\} | A \{\Phi_{NN}\Psi_{N_1}\Psi_{N_2}\}\rangle$$

The braces $\{...\}_{LCST}$ in (7.1) means that the addition is made of all moments

$$L = l + L_1 + L_2, \quad C = C_1 + C_2, \quad S = S_1 + S_2, \quad T = T_1 + T_2\,'$$

in which case

$$L_i = 0, \quad C_i = 0, \quad S_i = \frac{1}{2}, \quad T_i = \frac{1}{2}.$$

What is the advantage of the many-particle approach to the NN system over the two-particle approach? The case is that in the region of small NN distances $R \lesssim 0.5$ fm the form of the wave function $\Phi_{NN}(R)$ is known with a considerable uncertainty. A theoretical description in terms of NN + mesons is here inapplicable and we can proceed only from the phenomenological NN interaction models. At the same time a description of this region in terms of quarks possesses a number of advantages. Experience of studying the nuclear many-particle systems (such as $\alpha\alpha$) shows that the function $\Psi_{NN}(q^6)$ in the 3q-cluster overlap must resemble the functions $\Psi_n(q^6)$ of quark shell model states (n is the numeration of levels of the quark shell model; by the latter we shall mean either the MIT bag model [CH74] or more simple oscillatory models [Mi78; Li77; Ce77]). As will be seen, the Pauli principle fixes rather strictly the form of the lowest shell model states in the 6q-system which will enable us to describe the properties of the NN system at small distances in the many-particle approach more definitely than in the case of the NN phenomenology of core type in the NN forces.

We may proceed from the cluster expansion of the lowest shell model states in the 6q-system

$$\psi_n(q^6) = \sum_{B_1B_2} \Phi^n_{B_1B_2}(R)\,\Psi_{B_1}(q^3)\,\Psi_{B_2}(q^3), \tag{7.3}$$

where B_1B_2 = NN, NΔ, $\Delta\Delta$, N*N, N*N*,... etc., which will define the form of the wave function $\Phi_{NN}(R)$ at small distances (just as the shell model function of the ^{8}Be nucleus in the s^4p^4 configuration defines the form of the wave function of the $\alpha\alpha$-system at small distances).

A rigorous consideration of the relationship between representations of the type (7.1) and (7.3) for the case of many-particle nuclear systems was made in the book [Wi77a]. According to [Wi77a] a good interpolation of many-particle (e.g. q^6) wave-function can be obtained by the formula

$$\Psi(q^6) = \sum_n C_n\Psi_n(q^6) + A\left\{\sum_i \Phi_i^{(+)}(R)\,\Psi_{B1i}(q^3)\,\Psi_{B2i}(q^3)\right\}, \tag{7.4}$$

where $\Phi_i^{(+)}(R)$ has the asymptotics of scattering and Ψ_n, the asymptotics of bound states. Variation in the functions $\Phi_i^{(+)}$ and the amplitudes C_n leads to a set of the coupled integrodifferential and algebraic equations

$$\delta\,\langle\Psi_{B1i}\Psi_{B2i}\,|H-E|\,\Psi(q^6)\rangle = 0 \tag{7.5}$$

Here $B_{1i}B_{2i}$ correspond to the open channels only, i.e. B_1B_2 are the colour singlet barions. The terms $C_n\Psi_n$ are responsible for the inclusion of polarization of the 3q-clusters and if a limited number of open channels is included in the form (7.4) the items $C_n\Psi_n$ account effectively for the effect of all remaining channels which are not included explicitly in the calculation. Solutions to a set of equations (7.5) have the property: with increasing number of the items $C_n\Psi_n$ the amplitude of the functions $\Phi_i^{(+)}(R)$ in the cluster overlap region tends to zero and the behaviour of the many-particle system in this region is represented entirely by the terms $C_n\Psi_n$.

In what follows, however, we shall exploit the simplest ideas of the resonating group method connected with formula (7.1). In particular, the shell-model behaviour of a many-particle wave function in the cluster overlap area is reproduced here by a big values of the overlap integrals between the shell model and resonating group wave functions (see formulae (7.17) and (7.18) below). In fact, the cluster representation (7.1) is possible for the shell-model wave function itself.

The first steps in realizing this approach have already been undertaken. In the papers [Ri78; Ok80] the RGM calculations for the six-quark system were performed. True, these calculations did not include the polarization terms. Namely, the paper [Ri78] treated the single-channel (NN) problem with the oscillatory forces of a pairwise qq-interaction and the paper [Ok80], the two channel problem (NN + ΔΔ) with the qq-forces as the string potential [Ru75; Gu75; Li75]. The results of these calculations show

i) independence of the $\Phi_{NN}(R)$ function at small distances $R \lesssim 0.5$ fm upon the energy of the system E_{NN} within a wide range $0 \lesssim E_{NN} \lesssim 1$ GeV;

ii) nodal character of the $\Phi_{NN}(R)$ wave function with a stable position of the node at the point $R \simeq 0.4$ fm;

iii) nonresonance shape of the S-wave phase shifts of NN scattering in the energy range under consideration.

A similar situation was observed in studying the $\alpha\alpha$-scattering in the RGM in which case it was a consequence of limitations imposed by the Pauli principle (orthogonality to forbidden states 0S and 2S, see chapter 4). However, the node of the NN wave function at small distances is of a different origin. In order to investigate the present case it is necessary to analyse the structure of the principal terms $\Psi_n(q^6)$ in the wave function (7.4).

The properties of symmetry of the quark-quark forces play a decisive role in this analysis. From atomic physics (the Hund rule [Hu29]) and nuclear physics (the Wigner supermultiplets [Wi37; Hu37]) it is known that the spectrum of a many-particle system depends on what states of a pair of fermions (symmetrical $[2]^X$ or antisymmetrical $[1]^X$) the energy proves to be the lowest in. What is an analog of the Hund rule for the quark system? To begin with, it should be noted that the exchange forces become more important in the quark system in which case they are associated with a new charge – the colour C and a new symmetry corresponding to unitary transformations in the colour space (the SU_3^C group).

Considerations based on the quantum chromodynamics (QCD) show (see the next section 2) that the splitting of hadron levels is determined by the colour magnetic quark-quark interaction of the type

$$\sum_{i<j}^{N} \sum_{a=1}^{8} \lambda_i^a \vec{\sigma}_i \lambda_j^a \vec{\sigma}_j v_{qq}(r_{ij}). \tag{7.6}$$

In this formula λ_i^a and λ_j^a are Gell-Man's eight SU_3 matrices which operate on the colour variables of particles i and j. By analogy with the nuclear case of the SU_4^{ST} spin-isospin symmetry [Wi37] it is here convenient to introduce the colour-spin symmetry SU_6^{CS} [Ja77]. We may think of the SU_6^{CS} as being an approximate symmetry and of the Young scheme $[f]^{CS}$, as being a good quantum number for the multiquark states. It is also convenient to introduce a group U_{12}^{CST} which corresponds to the direct product of the colour, spin and isospin spaces, for the Young scheme $[f]^{CST}$ is unambigously related with the symmetry of the orbital (X) part of the wave function $[f]^{CST} = [\tilde{f}]^X$. As a result, many more quantum numbers than simply the spin S, isospin T and colour C, must be defined in classifying the states in the multiquark systems. Here, we use the quantum numbers (the Young schemes or Casimir invariants) of the following reduction chain

$$U_{12}^{CST} \supset U_6^{CS} \times U_2^T \supset SU_6^{CS} \times SU_2^T \supset SU_3^C \times SU_2^S \times SU_2^T \tag{7.7}$$

In the colour space use is made of the reduction

$$SU_3^C \supset O_3^C \supset O_2^C \tag{7.8}$$

which introduces the moment C (Casimir invariant of the group O_3^C) and its projection C_z.

Let us now pass to the language of the shell model for quarks which is actually used in the MIT bag model [Ch74] and the oscillatory models [Ce77; Mi78] as well. The basis set of the $\Psi_n(q^N)$ states of the multiquark system in the $s^{N_s} p^{N_p}$ configurations ($N=N_s+N_p$ is the total number of quarks) will be written as

$$\Psi_n(q^N) \equiv |s^{N_s} p^{N_p} [f]^X \omega LM, [g_C] CC_z SS_z [h_{CS}] TT_z [\tilde{f}]^{CST}\rangle =$$

$$= \sum_{i=f}^{n_f} \frac{\Lambda_i}{\sqrt{n_f}} |s^{N_s} p^{N_p} [f]^X \omega LM(r_i)\rangle |[g_C] CC_z SS_z [h_{CS}] TT_z [\tilde{f}]^{CST} (\tilde{r}_i)\rangle \tag{7.9}$$

Here (r_i) is the Yamanouchi symbol defining the row of the irreducible representation [f] of the permutation group; n_f is the dimension of the representation; $\frac{\Lambda_i}{\sqrt{n_f}}$ is the Clebsch-Gordan coefficient of the symmetrical group [Ha64] for constructing completely anti-symmetric representation $[1^N]$ in the product of representations $[f]^X \circ [\tilde{f}]^{CST}$; $\Lambda_i = \pm 1$ in accordance with the choice of the phase factors in ref. [Ha64]. The orbital parts of the present wave function

$$\phi_{ni}^X \equiv |s^{N_s} p^{N_p} [f]^X \omega LM(r_i)\rangle$$

depend on the sort of model of quark confinement. The wave functions in the CST-space

$$\chi_{ni}^{CST} \equiv |[g_C] CC_z SS_z [h_{CS}] TT_z [\tilde{f}]^{CST} (\tilde{r}_i)\rangle$$

are of universal character independent of the model. These can be constructed of the colour, spin and isospin wave functions, possessing a definite permutation symmetry ($[g_C](r_C)$, $[\varphi_S](r_S)$, $[\rho_T](r_T)$) with the help of the Clebsch-Gordan coefficients of the S_N group [Ha64]. For these functions χ_{ni}^{CST} to be corresponding to a certain Young scheme $[h_{CS}]$ in the spin-colour space, it is first necessary to combine $[g_C](r_C)$ and $[\varphi_S](r_S)$ into the resulting irreducible representation $\{h_{CS}\}$

$$|[h_{CS}](r_{CS})\rangle = \sum_{(r_C),(r_S)} \left(\begin{array}{cc|c} [g_C] & [\varphi_S] & [h_{CS}] \\ (r_C) & (r_S) & (r_{CS}) \end{array}\right) |[g_C](r_C)\rangle |[\varphi_S](r_S)\rangle \tag{7.10}$$

and then, after combining the representations $[h_{CS}]$ and $[\rho_T]$, we obtain

$$|[f_{CST}](r_{CST})\rangle = \sum_{(r_{CS}),(r_T)} \left(\begin{array}{cc|c} [h_{CS}] & [\rho_T] & [f_{CST}] \\ (r_{CS}) & (r_T) & (r_{CST}) \end{array}\right) |[h_{CS}](r_{CS})\rangle |[\rho_T](r_T)\rangle . \tag{7.11}$$

The quantum numbers $[g_C]$ C, T, L + S of the basis (7.9) are the integrals of motion. The remaining quantum numbers are approximate and the degree of their accuracy depends on the qq-forces symmetry. The set of quantum numbers used in (7.9) corresponds to the QCD forces to the atmost extent (7.6). So, we shall calculate here only on the basis (7.9) in which the index n replaces the following set of quantum numbers:

$$n = \{N_s N_p, [f]^X \omega, [h_{CS}]\} \tag{7.12}$$

In contrast to the atomic and nuclear shell models where the orbital quantum numbers and the integral of motion (L + S, T) completely define the state Ψ_n, in the quark system one should define, in addition to those, the Young scheme $[h_{CS}]$ (or other quantum numbers). It will be noted that the degeneracy of shell levels in quantum numbers $[h_{CS}]$ is removed by the QCD forces of single-gluon exchange (7.6). The Pauli principle imposes restrictions upon a set of permissible values of $[h_{CS}]$ defined by the Clebsch-Gordan series for the inner product of the Young schemes

$$[g_C] \circ [\varphi_S] = \sum_h \nu_h [h_{CS}] . \qquad (7.13)$$

Namely, only those $[h_{CS}]$ are permissible which at the next stage of multiplication of the Young schemes

$$[h_{CS}] \circ [\rho_T] = \sum_f \nu_f [f]^{CST} \qquad (7.14)$$

lead to the Young scheme $[f]^{CST} = [\tilde{f}]^X$, satisfying the Pauli principle. In respect to the 6q-system under consideration the restrictions (7.13) and (7.14) lead to the following. The S-shell corresponds to the configuration $s^6\ [6]^X\ L = 0$. Let us consider the state with quantum numbers of a deuteron. Then, (S, T) = (1, 0), $[\varphi_S] = [42]$, $[\rho_T] = [3^2]$. Colour singlet 6q state has the Young scheme $[g_C] = [2^3]$. We have

$$[2^3]^C \circ [42]^S = [42]^{CS} + [321]^{CS} + [2^3]^{CS} + [31^3]^{CS} + [21^4]^{CS}$$

in which case the Pauli principle fixes the Young scheme $[f_{CST}] = [1^6]$ which should be contained in the product $[h_{CS}] \circ [3^2]^T$. This requirement is fulfilled by $[h_{CS}] = [2^3]$ only. Thus, in the S-shell the Pauli principle unambigously fixes the single shell model state with quantum numbers of a deuteron:

$$\Psi_0(q^6) \equiv | s^6 [6]^X\ L = 0,\ [2^3]^C\ C = 0\ \ S = 1\ [2^3]^{CS}\ T = 0\ [1^6]^{CST} \rangle . \qquad (7.15)$$

The following shell model level with the same quantum numbers (S, T) = (1, 0), $[g_C[= [2^3]$ should be looked for among the p-excitations of the same parity, i.e. in the S^4P^2 configuration [Ne 77; Ob 79a].

It is clear that the same set of quantum numbers will be assigned to the state with the symmetry $S^4P^2\ [6]^X$ but it is also clear that the present level will not be the lowest-lying state in the configuration S^4P^2. Really, the Young scheme $[2^3]^{CS}$ is not the most symmetrical among all possible schemes in the product $[2^3]^C \circ [42]^S$. The most symmetrical Young scheme is here $[42]^{CS}$ (the weight of symmetrical pairwise states $[2]^{CS}$ in this state is the largest). The QCD forces (7.6), removing the degeneracy in $[h_{CS}]$, act as attraction in the symmetric pairwise states $[2]^{CS}$ and as repulsion in the antisymmetric pairs $[1^2]^{CS}$ (see section 7.2). Thus, the state with the Young scheme $[42]^{CS}$ is the lowest-lying state in the S^4P^2 configuration.

In the S^4P^2 configuration the Young scheme $[42]^{CS}$ is consistent only with the orbital symmetry $[42]^X$. Then $[f]^{CST} = [2^2 1^2]$ and the product $[42]^{CS} \circ [3^2]^T$ contains this Young scheme. So, we obtain that in the case of the unbroken symmetry SU_6^{CS} the first nearest energy level to the (7.15) is the shell model level

$$\Psi_1(q^6) \equiv | s^4 p^2\ [42]^X\ L = 0,\ [2^3]^C\ C = 0\ \ S = 1\ [42]^{CS}\ T = 0\ [2^2 1^2]^{CST} \rangle \qquad (7.16)$$

An accurate calculation in the MIT bag model will be presented in Section 7.2. It will be noted that in the nonrelativistic quark model with the oscillatory forces of the pair-wise qq-interaction [Li77] we obtained [Ob79a] that the level (7.16) is lower-lying than the level (7.15) due to the colour magnetic forces. If the shell state (7.16) in the non-relativistic model is the ground state of the 6q-system it is not difficult to understand why the RGM calculations [Ri78; Ok80] show the nodal character of the wave function $\Phi_{NN}(R)$ at small distances. Really, the function $\Phi_{NN}(R)$ in the two 3q-cluster overlap region should resemble the shell model function (7.16). But the s^4p^2 shell model state, when rewritten as the expansion (7.3) in the TISM functions [Sm63; Ne69] (see sec. 7.3, this Chapter), leads to the function $\Phi_{NN}(R) = C\,\Phi_{20}(R)$ (Φ_{nl} is the oscillatory function with n excitation quanta). This function has a node at the point $R = \Omega^{-1/2}$ where Ω is the TISM oscillatory parameter (by different papers [Li70, Ce77; Dr72; Ho73; Ke74; Mi78] the value of $\Omega^{-1/2}$ for nucleons is equal to $0.36 \div 0.48$ fm).

It is interesting to mention one peculiarity of the quark shell model that limits the range of its validity. It is known that the MIT bag model predicts a rich spectrum of dibarion resonances [Ae78] which were not observed in experiment. (Yet, a number of experiments [Mi76 Ka77a, Al70, Au77] do not rule out the existence of resonances; these cases were studied in the papers [Sw71; Ka77b; Ho77, Hi77; Sw78].) Moreover, this model predicts not resonances but levels of zero width.

This is due to the fact that all models of quark confinement are rather rough idealizations, which disregard the coupling between resonances and the channels of decay. The condition of confinement is formulated either as the boundary condition [Ch74] or as the infinite potential barrier [Ce77] (see Sect. 7.2), for each quark *individually*. Here, the fact that quarks may escape the region of confinement in *triplets* producing colour singlet barions (this is significant for the six-quark system), and that the escape of single quarks may be followed by production of a $q\bar{q}$ pair (this is significant for the 3q and $q\bar{q}$ hadrons) is by no means taken into account. Fig. 24 shows the corresponding diagrams. The channel of the 6q-system decay given in Fig. 24(c) is the same as the $^8\text{Be} \to \alpha + \alpha$ decay and must be studied in an analogous manner.

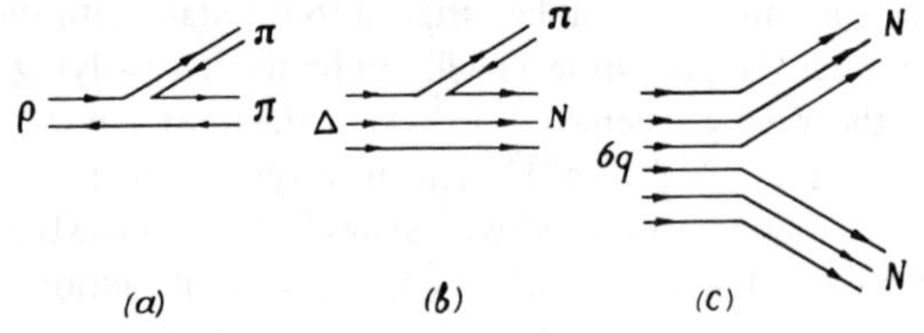

Fig. 24
The decay channels of quark shell model states.

It will be noted that in the recent papers [Ch75; Br79] in the bag model a connection with the external π-meson field was included by the boundary condition. This connection leads to the finite width of the Δ-resonance [Ch75]. But the introduction of this connection should affect substantially the bag itself. In particular, in the paper [Br79] the parameters of the bag, surrounded by meson cloud, differ substantially from those of the MIT bag. Consideration is here necessarily restricted to spectroscopy of the six-quark system and its symmetry and the problem of mixing of various shell states of the type

(7.15) and (7.16) in the quark-cluster overlap. This is the way to account for the imperfectness of the models of quark confinement. It will be noted that in the papers, concerning the quark bags, consideration was made of the s^N-type states only. Here we study at length the states of the type $s^{N_s} p^{N_p}$ $(N_s + N_p = N)$ which is a rather interesting problem both in the sence of its physical interpretation and of development of the technique of fractional parentage expansions in the multiquark systems.

7.2 Spectroscopy of the Six-Quark System in the MIT Bag Model

The 6q-system state spectrum in the S^6 configuration was studied in the MIT bag model in a number of papers [Ae78; Ja77; Mu78; Ho79]. Since in the system of non-strange quarks (p- and n-quarks) the complete orbital symmetry $[6]^X$ leads to the colour magnetic repulsion, in the papers [Ae78; Ja77; Mu78; Ho79] special attention was given to the systems with additional inner degrees of freedom: the strangeness (S) and the charm (C). Addition of s- and c-quarks, i.e. extention of the isospin group SU_2^T to the groups SU_3^F and SU_4^{FC}, permits one to consider a wider set of the Young schemes in the charge space. Really, in the case of p- and n-quarks, the total isospin of 6q-system T = 0 fixes unambigously the Young scheme $[\rho_T] = [3^2]$, but if there is one more charge degree of freedom the total isospin T = 0, for example, in the system of p-, n- and s-quarks, may also be consistent with the Young scheme $[\rho_F] = [2^3]$ in flavour space (F). As a result, in these systems the Pauli exclusion principle can be satisfied by fixing more symmetric Young schemes in the CS-space and we obtain the colour magnetic attraction. This was first shown by Jaffe [Ja77], who obtained in the s^6 configuration a dibarion resonance with quantum numbers of the ΛΛ-system. The resonance of Jaffe has the Young scheme $[3^2]^{CS}$ in the spin-colour space.

In the system of nonstrange quarks the $s^{N_s} p^{N_p}$ configurations are as interesting cases as the s^6 configurations. Here the colour magnetic attractive forces act which could arise to resonances in the higher partial waves in the NN-system. Besides, the $s^4 p^2$ configuration may be related to the problem of a core in the NN forces in the $^3S_1$- and $^1S_0$-states.

The $s^4 p^2$ configuration was suggested in [Ne77] and treated in our paper [Ob79a]. The present chapter gives a detailed account of the calculation technique in the excited multi-quark configurations of the type $s_{1/2}^{N_s} p_{1/2}^{N_p}$. Insofar as the magnetic moments of the $s_{1/2}$ and $p_{1/2}$ Diràc states in the bag are different one cannot use the technique of Casimir operators which substantially simplify the calculations in the case the $s_{1/2}^N [N]^X$ configurations [Ja77, Ch74]. Therefore, the technique of fractional parentage expansions is developed which is described at length in Sect. 3. In the present section the Jaffe mass formula [Ja77] is generalized to the case of an arbitrary configuration $s_{1/2}^{N_s} p_{1/2}^{N_p}$ in the MIT bag.

7.2.1 Connection with the problem of a core in the NN forces

The problem of the level spacing in the 6q-system is directly connected with the problem of a core in the NN forces.

In the papers [Li77; Ta78] the NN potential v(R) was calculated from the qq forces in the adiabatic approximation. In the adiabatic approach the point $R = 0$ corresponds to the shell configurations s^6 of 6q system. Therefore an idea about the NN forces at small ranges is extracted here from the s^6 configuration. Indeed, the calculations performed in the papers [Li77; Ta78] have shown that in this approximation colour magnetic repulsive forces predominate. This result agrees well with the phenomenological potentials of the NN interaction (see e.g. [Ha62; Re68]), which contain the repulsive core at the distances $R \simeq 0.5$ fm.

But a real situation in the 3q-cluster overlap region may have little in common with the adiabatic picture what is definitely shown by the results of the RGM calculations [Ri78; Ok80], in particular, by the node character of the wave function. It is not unlikely that, just as in the $\alpha\alpha$ system, the repulsive core in the NN forces is actually a phenomenological representation of a node of the wave function $\Psi_{NN}(q^6)$ in the point $R \simeq 0.4$ fm. A simple interpretation of such a node implies that the quark dynamics (combined with the Pauli principle) fixes the shell model state $s^4p^2\,[42]^X$ as energetically favourable as compared with the remaining shell model states. Nodal behaviour of the function $\Phi_{NN}(R)$ (as contrasted to the dying-out of the function in the model with a core) does not prevent nucleons from approaching one another arbitarily close ($\Phi_{NN}(0) \neq 0$). A choice between these two conceptions can be made only on the basis of data on the processes which are highly sensitive to small internucleonic distance (see e.g. [Ba73] and the discussion in Sect. 7.4). As to the experimental data on the NN scattering phase shift and electromagnetic form factors of a deuteron at nonrelativistic momentum transfer, the model with the wave function with one node is here as good as the standard model with a core (this was shown in the paper [Ne75] in the model of the attractive NN potential with one forbidden state).

In calculations of the effects that are sensitive only to small internucleonic distances one may expand the wave function (7.1) in the shell model basis $\psi_n(q^6)$

$$\Psi_{NN}(q^6) = \sum_{n=0,1,\dots} |\Psi_n\rangle \langle \Psi_n | \frac{1}{N_A} A\{\Phi_{NN}\Psi_{N_1}\Psi_{N_2}\}\rangle . \qquad (7.17)$$

Let us denote the wave functions $\Phi_{NN}(R)$ of the two phenomenological models as Φ_{NN}^{core} and Φ_{NN}^{node} respectively. Then the representation

$$\Psi_{NN}(q^6) \simeq \begin{cases} C_0\,|\Psi_0\rangle, & \text{if } \Phi_{NN} = \Phi_{NN}^{core}; \\ C_1\,|\Psi_1\rangle, & \text{if } \Phi_{NN} = \Phi_{NN}^{node}; \end{cases}$$

where

$$C_n = \langle \Psi_n \,|\frac{1}{N_A}\, A\{\Phi_{NN}\Psi_{N_1}\Psi_{N_2}\}\rangle , \qquad (7.18)$$

is a reasonably good approximation to expansion (7.17) at small distances if the function $|\Psi_0\rangle$ and $|\Psi_1\rangle$ are the lowest-lying shell model states (7.15) and (7.16), respectively. (In Sect. 7.3 these functions are constructed on the basis TISM.)

7.2.2 The MIT bag. The equations of motion and the conservation laws

The hadron spectroscopy has been well studied in the quark model [Ch74; Wi77b; Ce77], since the old problem with the quark statistics [Ko69] was solved in the coloured theory. The spin and isospin splitting of hadron levels which was previously described semi-phenomenologically by mass formulas of the SU_3- and SU_6-models, received now an explanation (in the first-order in the constant of QCD α_s) as the splitting of levels with certain values of the Young schemes $[f]^{CS}$ by the colour magnetic forces. The problem of the nature of forces ensuring confinement of quarks is still open. That is why, the hadron spectroscopy uses, different models of confinement which contain phenomenological parameters. But the MIT bag model has a number of advantages:

i) it is a relativistic model in which the energy of a field ensuring confinement of quarks is a constituent of the hadron mass. This is important in extrapolating the model from the case of well-studied systems 3q and $q\bar{q}$ to the multiquark system ($N \geqslant 6$);

ii) it takes into account the dependence of Diràc magnetic moment of a quark on the quantum numbers of the state which is essential in calculations of the contribution of the colour magnetic forces of QCD in the $s^{N_s}p^{N_p}$ type excited configurations.

The MIT bag model has a number of drawbacks [Wo77] too (the problem of spurious excitations of center of mass is not solved, use is made of the quadruplicated values of the constant $\alpha_s = 2.2$ and other drawbacks noted in Sect. 7.4), which is a stimulating motive for building up more realistic models [Br79]. But these models are not yet developed sufficiently well to give the mass formula.

There exist good survey papers concerning both the MIT bag model [Ha78; Gr76] and the quantum chromodynamic [Ma78; Sl78]. Following the basic papers [Ch74; Ja77; Ma78; Ae78] we present here at short only the data we are concerned in. The model is based on the notions appeared in the development of non Abelian gauge-field theories. In particular, the idea that the confinement of coloured fields (quarks and gluons) will ultimately be explained as an effect of the QCD vacuum instability [Fr73; Ko74; Ca79], seems very attractive. The stable vacuum phase, which is outside the hadrons, is not consistent with the existence of the coloured electric field. On the contrary, inside the hadrons the phase-transition takes place. Here we have the ordinary vacuum of QCD with an excess of volume energy B(MeV/fm^3). B is the only constant that characterizes the interhadronic phase of vacuum in which case it is assumed that B can be calculated from the first principles using only the value of constant α_s.

In the MIT bag model the constant B is considered as a phenomenological parameter. The following scheme of confinement of coloured fields is proposed. A hadron in dynamic sence is like a "bubble" (or a "bag") whose "skin" is unpenetrable to colour electric fields. The bag is considered to be a spherical cavity filled with the quanta of colour quark and vector-gluon fields. Pressure of the fields, confined in the cavity, counterpoises an excess of volumetric energy $E = BV_{bag}$ (V_{bag} is the cavity volume). The equations of motion and the boundary conditions for the present system of fields may be obtained on the basis of the variational principle for the action

$$W \equiv \int L(x)\, d^4x =$$

$$= \int_{bag} \left(-\frac{1}{4} F_a^{\mu\nu} F_{a\mu\nu} + \frac{i}{2} \bar{q}_\alpha \gamma^\mu \partial_\mu q_\alpha + g\bar{q}\gamma_\mu A_a^\mu \frac{\lambda_a}{2} q - B \right) d^4x . \qquad (7.19)$$

Integration in (7.19) goes only over the four-dimensional volume of the bag in which case the boundary of the bag is a dynamic variable, which should be varied in the same manner as the field amplitudes (in the case of a spherical cavity one should vary the bag radius $R \to R + \delta R$).

The first and second items in the integrand (7.19) are the Lagrangians of free massless fields [gluons (A_a^μ) and quarks (q_α)]; $\alpha = 1, 2, 3$; $a = 1, 2, \ldots, 8$; $\mu = 0, 1, 2, 3$. The third item is the field interaction term proportional to the constant QCD $g(\alpha_s = g^2/4\pi)$. The constant B may be treated as the confining field which is introduced in the simplest way: it has no field degrees of freedom proper (amplitudes of the field), its single dynamical variable is the boundary of a cavity. (The radius R in the case of a spherical bag.)

The gluon field strength

$$F_a^{\mu\nu} = \partial^\mu A_a^\nu - \partial^\nu A_a^\mu - g f_{abc} A_b^\mu A_c^\nu \tag{7.20}$$

and the Lagrangian in (7.19) are invariant under "non Abelian" gauge transformation

$$q(x) \longrightarrow e^{i\frac{1}{2}\lambda_a\varphi_a(x)} q(x),$$

$$A_a^\mu(x) \longrightarrow A_a^\mu(x) + \partial^\mu \varphi_a(x) + g f_{abc} A_b^\mu(x) \varphi_c(x),$$

where the matrices $\frac{1}{2}\lambda_a$ are the generators of eight independent rotations in the coloured space,

$$\left[\frac{1}{2}\lambda_a, \frac{1}{2}\lambda_b\right] = i f_{abc} \frac{1}{2}\lambda_c . \tag{7.21}$$

f_{abc} are the structural constants of the Lie algebra of the SU_3 group. On the basis of the principle of stationarity of the action $\delta W = 0$ we obtain for a spherical bag of radius R:

i) equations for the fields inside the bag

$$i\gamma^\mu \partial_\mu = -g\gamma_\mu A_a^\mu \frac{\lambda_a}{2} q, \quad r < R, \tag{7.22}$$

$$\partial^\mu F_{a\,\mu\nu} = g f_{abc} F_{b\,\mu\nu} A_c^\mu + g\bar{q}\frac{\lambda_a}{2}\gamma_\nu q, \quad r < R; \tag{7.23}$$

ii) linear boundary conditions in the bag surface

$$i\gamma^\mu n_\mu q_\alpha = q_\alpha, \qquad r = R, \tag{7.24}$$

$$n_\mu F_a^{\mu\nu} = 0, \qquad r = R, \tag{7.25}$$

where n_μ is the covariant normal to the bag surface, $n = \{0, -\hat{r}\}$, $\hat{r} = \frac{R}{R}$;

iii) a nonlinear boundary condition which has the meaning of balance of pressures produced by fields in the bag surface

$$-\frac{1}{4} F_{a\,\mu\nu} F_a^{\mu\nu} + \frac{1}{2} n_\mu \partial^\mu(\bar{q}q) - B = 0, \quad r = R. \tag{7.26}$$

The boundary condition (7.25) for $F_a^{\mu\nu}$ implies that all colour charges of the bag turn into zero

$$Q_a \equiv \int_{V_{bag}} j_a^0(x)\, d^3r ,$$

where

$$j_a^\mu = g(\bar{q}\gamma^\mu \frac{\lambda_a}{2} q + f_{abc} F_b^{\nu\mu} A_{c\nu})$$

is the colour current occuring in eq. (7.23) in the same way as the electromagnetic current occurs in the Maxwell equations. By analogy with electrodynamics, the field strength tensor $F_a^{\mu\nu}$ may be divided into the electric ($\mathbf{E}_a$) and magnetic ($\mathbf{B}_a$) components

$$F_a^{\mu\nu} \equiv \begin{pmatrix} 0 & F_a^{01} & F_a^{02} & F_a^{03} \\ F_a^{10} & 0 & F_a^{12} & F_a^{13} \\ F_a^{20} & F_a^{21} & 0 & F_a^{23} \\ F_a^{30} & F_a^{31} & F_a^{32} & 0 \end{pmatrix} = \begin{pmatrix} 0 & E_a^1 & E_a^2 & E_a^3 \\ -E_a^1 & 0 & B_a^3 & -B_a^2 \\ -E_a^2 & -B_a^3 & 0 & B_a^1 \\ -E_a^3 & B_a^2 & -B_a^1 & 0 \end{pmatrix} \quad a = 1, 2, \dots, 8. \qquad (7.27)$$

Then the boundary condition (7.25) reflects the turning into zero the value of the normal component of electric field and the tangential component of magnetic field in the bag surface

$$\hat{\mathbf{r}}\,\hat{\mathbf{E}}_a = 0, \; r = R \qquad (7.28)$$

$$\hat{\mathbf{r}} \times \mathbf{B}_a = 0, \; r = R. \qquad (7.29)$$

Applying the Gaussian theorem to the condition (7.28) we obtain

$$Q_a = 0. \qquad (7.30)$$

The dynamics of quarks and gluons inside the bag may be studied in terms of the perturbation theory. Using the smallness of the constant QCD $\alpha_s < 1$, we first obtain the solution of the boundary problem (7.22) and (7.24) for the quarks only, neglecting the gluon fields. Then we include the gluon contribution in the second order in g.

7.2.3 The quark eigenmodes. The quantization

The boundary problem for free Diràc fields $\Psi(x)$ in the bag

$$(i\,\partial_0 \gamma^0 + i \quad \boldsymbol{\gamma})\,\Psi(x) = 0, \quad r < R \qquad (7.31)$$

$$i\hat{r}\gamma\Psi = \Psi, \quad r = R \qquad (7.32)$$

has a solution as a discrete spectrum of states

$$\Psi_{nkjm}(x) = \frac{N(\omega_{nk})}{\sqrt{4\pi}} \begin{pmatrix} i\,j_k\left(\frac{\omega_{nk}}{R} r\right) U_{jm}^k(\hat{r}) \\ j_{k-1}\left(\frac{\omega_{nk}}{R} r\right) \sigma\hat{r}\,U_{jm}^k(\hat{r}) \end{pmatrix} e^{-i\frac{\omega_{nk}}{R}t} \equiv \psi_{nkjm}(\mathbf{r})\,e^{-i\frac{\omega_{nk}}{R}t}, \qquad (7.33)$$

where $\frac{\omega_{nk}}{R}$ are the eigenmodes determined from the boundary condition (7.32) as solutions to the equation

$$j_k(\omega_{nk}) = j_{k-1}(\omega_{nk}). \qquad (7.34)$$

$j_k(x)$ are the spherical Bessel functions, $j_{-k}(x) = (-1)^{k+1} j_{k-1}(x)$, $j_0(x) = \frac{1}{x}\sin x$,

$j_1(x) = \frac{1}{x}\left(\frac{1}{x}\sin x - \cos x\right)$, $j_k(x) = \sqrt{\frac{\pi}{2x}}\, J_{k+\frac{1}{2}}(x)$.

$U^k_{jm}(\hat{\mathbf{r}})$ is the Pauli spinor which is a state with a certain value of the total momentum $\mathbf{j} = \boldsymbol{l} + \frac{1}{2}\boldsymbol{\sigma}$ and of the spin-orbital operator $\boldsymbol{l\sigma}$ [Bj64]

$$j^2 U^k_{jm} = j(j+1) U^k_{jm},$$

$$l\sigma\, U^k_{jm} = -(1+k)\, U^k_{jm},\quad k = \pm(j+\tfrac{1}{2}),\ \text{if } j = l \pm \tfrac{1}{2},$$

$$\frac{1}{\sqrt{4\pi}} U^k_{jm}(\hat{\mathbf{r}}) = \begin{pmatrix} \sqrt{\dfrac{l+\frac{1}{2}\pm m}{2l+1}}\; Y_{lm-\frac{1}{2}}(\hat{\mathbf{r}}) \\ \sqrt{\dfrac{l+\frac{1}{2}\mp m}{2l+1}}\; Y_{lm+\frac{1}{2}}(\hat{\mathbf{r}}) \end{pmatrix},\quad k = \mp(j+\tfrac{1}{2}). \tag{7.35}$$

The normalizing factor $N(\omega_{nk})$ for classical fields is arbitrary. We shall consider the solutions (7.33) as the basis functions normalized to 1,

$$\int\limits_{V_{bag}} \Psi^+_{nkjm}(x)\Psi_{nkjm}(x)d^3\mathbf{r} = \int\limits_{V_{bag}} \psi^+_{nkjm}(\mathbf{r})\psi_{nkjm}(\mathbf{r})d^3\mathbf{r} = 1,$$

$$N^{-2}(\omega_{nk}) = \int\limits_0^{\omega_{nk}} [j_k^2(x) + j_{k-1}^2(x)]\, x^2 dx,$$

$$\int\limits_{4\pi} U^{k^+}_{jm}(\hat{\mathbf{r}})\, U^k_{jm}(\hat{\mathbf{r}})\, d\Omega = 1. \tag{7.36}$$

Let us consider six independent quark fields $q_{\alpha T_z}(x)$, [$\alpha = 1, 2, 3$ (the colours), $T_z = \pm\frac{1}{2}$ (the isospin projections)] which are a superposition of eigenmodes (7.33) of the spherical cavity

$$\bar{q}_{\alpha T_z}(x)\, q_{\alpha T_z}(x) = \sum_{nkjm} b_{\alpha T_z(nkjm)} \Psi_{nkjm}(x). \tag{7.37}$$

The nonlinear boundary condition (7.26) which, in the absence of gluon fields should be written as

$$\hat{\mathbf{r}}\, \nabla(\bar{q}q) = 2B,\ r = R, \tag{7.38}$$

where

$$\bar{q}q \equiv \sum_{\alpha, T_z} \bar{q}_{\alpha T_z}(x)\, q_{\alpha T_z}(x) = \sum_{\alpha T_z} \sum_{nkjm} \sum_{n'k'j'm'} b^*_{\alpha T_z(nkjm)} b_{\alpha T_z(n'k'j'm')} \times$$
$$\times\, \bar{\psi}_{nkjm}(\mathbf{r})\psi_{n'k'j'm'}(\mathbf{r})\, e^{i\frac{1}{R}(\omega_{nk} - \omega_{n'k'})t} \tag{7.39}$$

imposes certain limitations upon the value of the amplitudes $b_{\alpha T_z(nkjm)}$. Firstly, the stationarity of the right-hand side of the equation (7.38) implies the requirement

$$\omega_{nk} = \omega_{n'k'} \tag{7.40}$$

i.e. only one mode ω_{nk} may participate in the expansion (7.37) for each field $q_{\alpha T_z}(x)$ (the quark states $q_{\alpha T_z}$ if being a superposition of several modes of oscillations are prohibited). Second, by rewriting the condition (7.38) with the inclusion of (7.40) as

$$\frac{d}{dr} \sum_{\alpha T_z(nkjm)} b^*_{\alpha T_z(nkjm)} b_{\alpha T_z(nkjm)} \bar{\psi}_{nkjm}(\mathbf{r}) \psi_{nkjm}(\mathbf{r}) = 2B, \quad r = R \tag{7.41}$$

we simultaneously obtain the following requirement of the spherical symmetry

$$\sum_{\alpha T_z(nkjm)} b^*_{\alpha T_z(nkjm)} b_{\alpha T_z(nkjm)} U^{k^+}_{jm}(\hat{\mathbf{r}}) U^k_{jm}(\hat{\mathbf{r}}) = \text{const} \tag{7.42}$$

which is satisfied only by the states: $s_{1/2}$ and $p_{1/2}$. In what follows consideration is given to the states $s_{1/2}$ and $p_{1/2}$ that is why we write down the solutions (7.33)–(7.35) for the case $k = \pm 1$

$$\psi_{n-1\frac{1}{2}m}(\mathbf{r}) = \frac{N(\omega_{n-1})}{\sqrt{4\pi}} \begin{pmatrix} i j_0\left(\frac{\omega_{n-1}}{R} r\right) U_m \\ -j_1\left(\frac{\omega_{n1}}{R} r\right) \sigma \hat{\mathbf{r}} U_m \end{pmatrix}, \tag{7.43}$$

$$\psi_{n1\frac{1}{2}m}(\mathbf{r}) = \frac{N(\omega_{n1})}{\sqrt{4\pi}} \begin{pmatrix} i j_1\left(\frac{\omega_{n1}}{R} r\right) \sigma \hat{\mathbf{r}} U_m \\ j_0\left(\frac{\omega_{n1}}{R} r\right) U_m \end{pmatrix}, \tag{7.44}$$

$$\tan(\omega_{nk}) = \frac{\omega_{nk}}{1 + k\,\omega_{nk}}, \quad k = \pm 1, \tag{7.45}$$

$$N(\omega_{nk}) = \sqrt{\frac{\omega^3_{nk}}{R^3} \frac{\omega_{nk}}{\omega^2_{nk} - \sin^2 \omega_{nk}}}, \quad k = \pm 1. \tag{7.46}$$

In (7.44) use was made of the relations

$$U^1_{\frac{1}{2}m}(\hat{\mathbf{r}}) = \sigma \hat{\mathbf{r}} U_m, \quad (\sigma \hat{\mathbf{r}})(\sigma \hat{\mathbf{r}}) = 1, \quad \hat{\mathbf{r}}^2 = 1,$$

where U_m $(m = \pm \frac{1}{2})$ is the Pauli spinor.

The eq. (7.45) has positive and negative solutions

$$\omega_{1-1} = 2.043, \quad \omega_{11} = 3.812, \quad \omega_{2-1} = 5.396, \ldots$$

$$\omega_{-11} = -2.043, \quad \omega_{-1-1} = -3.812, \quad \omega_{-21} = -5.396, \ldots$$

Calculate the energy of the Diràc fields in the cavity by integrating the energy-momentum tensor

$$T^{\mu\nu}_{Dir} = -\frac{i}{2}\left[\bar{q}\gamma^\mu \partial^\nu q - (\partial^\nu \bar{q})\gamma^\mu q\right],$$

$$E_{Dir} = \int_{V_{bag}} T^{00}(x) d^3\mathbf{r} = \sum_{\alpha T_z(nkjm)} \frac{\omega_{nk}}{R} b^*_{\alpha T_z(nkjm)} b_{\alpha T_z(nkjm)} \tag{7.47}$$

In (7.47) the summation goes over positive and negative eigenfrequencies. We remove the difficulties with negative energies by the standard method of quantum field theory. We replace the field amplitudes $b_{\alpha T_z(nkjm)}$ by the operators of particle annihilation $b_{\alpha T_z}$ at $n>0$ or by the operators of antiparticle creation $d^+_{\alpha T_z}$ at $n<0$. The operators b, b^+, d, d^+ satisfy the anticommutation relations

$$\{b_{\alpha T_z(nkjm)}, b^+_{\alpha T_z(n'k'j'm')}\}_+ = \delta_{nn'}\delta_{kk'}\delta_{jj'}\delta_{mm'},$$
$$\{d_{\alpha T_z(nkjm)}, d^+_{\alpha T_z(n'k'j'm')}\}_+ = \delta_{nn'}\delta_{kk'}\delta_{jj'}\delta_{mm'}. \quad (7.48)$$

Writing down the expression for energy of the field (7.47) through the operators b, b^+, d, d^+, we redetermine the zero point of the energy (the state of vacuum) by the substitution $dd^+ \to :dd^+: \equiv d^+d$.

The following operator of energy (Hamiltonian) is obtained

$$H_{Dir} = \sum_{\alpha T_z} \sum_{nkjm}^{n>0} \frac{\omega_{nk}}{R} [b^+_{\alpha T_z(nkjm)} b_{\alpha T_z(nkjm)} + d^+_{\alpha T_z(nkjm)} d_{\alpha T_z(nkjm)}] + E_0(R). \quad (7.49)$$

Summation in (7.2.33) is made only over the positive frequencies $\omega_{nk} > 0$. In conventional field theory, the energy of zero-point fluctuations is discarded since the volume filled by fields does not change in any process. However, in the bag model, there appears a finite term $E_0(R)$ depending upon R. The term $E_0(R)$ was estimated in the papers of the ref. [Ch74]

$$E_0(R) = -\frac{Z_0}{R}, \quad Z_0 \simeq 2.$$

In the MIT bag model Z_0 is the phenomehological constant.

When the operators b, b^+, d, d^+ are normalized by the relations (7.48), the non-linear boundary condition (7.41) may be treated as an equation for determining the equilibrium radius of the bag $\bar{R}$. Substituting the expressions (7.43)–(7.46) into (7.41) we obtain the equation for $\bar{R}$

$$\sum_{\alpha T_z(nkjm)}^{n>0} \frac{\omega_{nk}}{R} [b^+_{\alpha T_z(nkjm)} b_{\alpha T_z(nkjm)} + d^+_{\alpha T_z(nkjm)} d_{\alpha T_z(nkjm)}] - \frac{Z_0}{R} = 4\pi R^3 B. \quad (7.50)$$

The solution may be written down as:

$$\bar{R} = \left(\frac{\sum_{nk} N_{nk}\omega_{nk} - Z_0}{4\pi B} \right)^{1/4}, \quad (7.51)$$

where

$$N_{nk} = \sum_{\alpha T_z} \sum_{jm} [b^+_{\alpha T_z(nkjm)} b_{\alpha T_z(nkjm)} + d^+_{\alpha T_z(nkjm)} d_{\alpha T_z(nkjm)}]$$

is the summed number of quarks and antiquarks having the present excitation mode ω_{nk}. The total number of particles $N = \sum_{nk} N_{nk} = \int_{V_{bag}} :\bar{q}q: d^3r$.

The energy of the bag E_{bag} includes also the volume energy of the cavity:

$$T^{\mu\nu}_{bag} = T^{\mu\nu}_{Dir} + g^{\mu\nu}B,$$

$$E_{bag} = E_{Dir} + \int_{V_{bag}} B d^3\mathbf{r} = \sum_{nk} N_{nk} \frac{\omega_{nk}}{R} - \frac{Z_0}{R} + \frac{4\pi}{3} R^3 B . \tag{7.52}$$

Substitute the equilibrium value of the bag radius (7.51) into (7.52) and obtain finally

$$E_{bag} = \frac{16\pi}{3} \bar{R}^3 B = \frac{4}{3}(4\pi B)^{1/4} \left(\sum_{nk} N_{nk}\omega_{nk} - Z_0\right)^{3/4} . \tag{7.53}$$

Comparing (7.53) with (7.52) we see that field degrees of freedom of the bag in sum have the energy $4\pi \bar{R}^3 B$ which is three times the energy of the confining field. This is a result of the virial theorem.

7.2.4 Effective Hamiltonian in the second order in the constant g

In the MIT bag model the gluon fields are taken into account as the static fields produced by the quark currents. Eigenmodes of gluon correspond to the quarks bag excitation. The energy of the gluon field is quadratic in $\mathbf{E}_a$ and $\mathbf{B}_a$ so in order to obtain the expression in the second order in g one should substitute into the Hamiltonian the linear in g approximations for $\mathbf{E}_a$ and $\mathbf{B}_a$. Therefore we exclude from (7.20) and (7.23) the terms nonlinear in gluon fields and obtain the following boundary problem for $\mathbf{E}_a$ and $\mathbf{B}_a$

$$\nabla \mathbf{E}_a = g : \bar{q} \frac{\lambda_a}{2} \gamma^0 q : \ , \ r < R ,$$

$$\nabla \times \mathbf{B}_a = g : \bar{q} \frac{\lambda_a}{2} \boldsymbol{\gamma} q : \ , \ r < R ,$$

$$\hat{\mathbf{r}}\, \mathbf{E}_a = 0 , \quad r = R ,$$

$$\hat{\mathbf{r}} \times \mathbf{B}_a = 0 , \quad r = R , \tag{7.54}$$

which can be solved as the problem for the Maxwell field.

The Hamiltonian of a system of fields (quarks + gluons) H_{field} is equal to

$$H_{field} = \int_{V_{bag}} : \bar{q} \gamma^\mu \partial_\mu q : d^3r + \int_{V_{bag}} \left[\frac{1}{2}(\mathbf{E}_a^2 + \mathbf{B}_a^2) - \mathbf{j}_a \mathbf{A}_a\right] d^3r + E_0(R) \tag{7.55}$$

in which case $\mathbf{j}_a$ is the colour current which appears in eqs. (7.54), $\mathbf{j}_a = \nabla \times \mathbf{B}_a$, $\mathbf{B}_a = \nabla \times \mathbf{A}_a$ Using the latter relations and the boundary conditions (7.54) one may transform the term $-\mathbf{j}_a \mathbf{A}_a$ in the Hamiltonian (7.55) to the form

$$- \int_{V_{bag}} \mathbf{j}_a \mathbf{A}_a d^3\mathbf{r} = - \int_{V_{bag}} \mathbf{B}_a^2 \, d^3\mathbf{r} ,$$

$$H_{field} = \int_{V_{bag}} : \bar{q} \gamma^\mu \partial_\mu q : d^3\mathbf{r} + \frac{1}{2} \int_{V_{bag}} (\mathbf{E}_a^2 - \mathbf{B}_a^2) d^3\mathbf{r} + E_0(R) . \tag{7.56}$$

The effective Hamiltonian of quark interaction in the bag is obtained by substituting into the second term (7.56) the solution to the boundary problem (7.54) as explicit expressions of the fields $\mathbf{E}_a$ and $\mathbf{B}_a$ in terms of the quark field operators

$$q_{\alpha T_z}(x) = \sum_{nkjm} \left[b_{\alpha T_z(nkjm)} \psi_{nkjm}(r\hat{\mathbf{r}})\, e^{-i\frac{\omega_{nk}}{R}t} + d^{+}_{\alpha T_z(nkjm)} \psi_{nkjm}(r\hat{\mathbf{r}})\, e^{i\frac{\omega_{nk}}{R}t} \right],$$

$$q = \begin{pmatrix} q_{1\frac{1}{2}} \\ \vdots \\ q_{\alpha T_z} \\ \vdots \end{pmatrix} \tag{7.57}$$

To write down the solution to the boundary problem (7.54), let us determine the following functions

$$\mathbf{m}'_a(r') = g \int_{4\pi} \frac{1}{2} \mathbf{r}' \times : q(r'\hat{\mathbf{r}}', t)\, \boldsymbol{\gamma} \frac{\lambda_a}{2}\, q(r'\hat{\mathbf{r}}', t) : r'^2 d\Omega' ,$$

$$e'_a(r') = g \int_{4\pi} : \bar{q}(r'\hat{\mathbf{r}}', t)\gamma^0 \frac{\lambda_a}{2}\, q(r'\hat{\mathbf{r}}', t) : r'^2 d\Omega' . \tag{7.58}$$

They have a transparent physical meaning: $\mathbf{m}'_a(r')dr'$ is the colour magnetic moment of the spherical layer $(r', r' + dr')$, $e'_a(r')$ is the colour charge of the spherical layer. Then, using the standard formula of electrodynamics for the field of magnetic dipole the solution can be written down as

$$\mathbf{B}_a(r) = \frac{3\,\hat{\mathbf{r}}\,(\hat{\mathbf{r}}\,\mathbf{m}_a(r)) - \mathbf{m}_a(r)}{4\pi r^3} + \frac{\mathbf{m}_a(R)}{4\pi R^3} +$$

$$+ \int_r^R \left[\hat{\mathbf{r}} \times : \bar{q}(r'\hat{\mathbf{r}}, t)\, \boldsymbol{\gamma} \frac{\lambda_a}{2}\, q(r'\hat{\mathbf{r}}, t) : + \frac{3\,\hat{\mathbf{r}}\,(\hat{\mathbf{r}}\,\mathbf{m}'_a(r')) - \mathbf{m}'_a(r')}{4\pi r'^3} \right] dr', \quad r < R, \tag{7.59}$$

$$\mathbf{E}_a(r) = \hat{\mathbf{r}}\, \frac{e_a(r)}{4\pi r^2}, \quad r < R . \tag{7.60}$$

Here we have the following designations for the characteristics of the region $0 < r' < r$

$$\mathbf{m}_a(r) = \int_0^r \mathbf{m}'_a(r')\, dr' , \qquad \boldsymbol{\mu}_a = \mathbf{m}_a(R) ;$$

$$e_a(r) = \int_0^r e'_a(r')\, dr' , \qquad Q_a = e_a(R) .$$

The boundary condition for the magnetic field is satisfied by a special choice of the constant of integration $\mathbf{C} = \dfrac{\mu_a}{4\pi R^3}$. But we cannot act analogously in the case of the electric field since we should have to subtract from (7.60) the field $\hat{\mathbf{r}}\,\dfrac{Q_a}{4\pi R^2}$ dependent upon the directions. For the boundary condition for an electric field to be satisfied the summed colour charge of quarks in the bag is reguired to be equal to zero

$$\langle \mathrm{bag} | Q_a | \mathrm{bag} \rangle = 0\,. \tag{7.61}$$

As is shown by the calculations [Ch74] a contribution from the colour electric energy to the total energy of the bag for these states is very small, therefore the term $\frac{1}{2}\mathbf{E}_a^2$ is henceforth neglected in the Hamiltonian of the colour-neutral bag.

Substitute in (7.59) the expansion (7.57) of the field $q(x)$ in the spherically symmetrical ($s_{1/2}$ and $p_{1/2}$) eigenmodes (7.43) and (7.44). As a result the following operator is obtained

$$\begin{aligned}\mathbf{B}_a(\mathbf{r},t) = \frac{g}{4\pi}\sum_{T_z}\sum_{\alpha\bar{\alpha}}\sum_{m\bar{m}}\Big\{\frac{1}{2}\lambda_a^{\bar{\alpha}\alpha}\,\boldsymbol{\sigma}^{\bar{m}m}\sum_{nk}\Big[2M_{nk}(r) - \frac{\mu_{nk}(r)}{r^3} + \frac{\mu_{nk}}{R^3}\Big] \\ b^{+}_{\bar{\alpha}T_z(nkj\bar{m})}\,b_{\alpha T_z(nkjm)} + \frac{1}{2}\lambda_a^{\bar{\alpha}\alpha}\,3\,\hat{\mathbf{r}}\,(\boldsymbol{\sigma}^{\bar{m}m}\hat{\mathbf{r}})\sum_{nk}\frac{\mu_{nk}(r)}{r^3}\,b^{+}_{\bar{\alpha}T_z(nkj\bar{m})}\,b_{\alpha T_z(nkjm)} + \\ + \frac{1}{2}\lambda_a^{\bar{\alpha}\alpha}\cdot i\,[\hat{\mathbf{r}}\times\boldsymbol{\sigma}^{\bar{m}m}]\sum_{nk\neq\overline{nk}} J_{nk,\overline{nk}}\,b^{+}_{\bar{\alpha}T_z(\overline{nk}j\bar{m})}\,b_{\alpha T_z(nkjm)}\,e^{-i\frac{\omega_{nk}-\omega_{\overline{nk}}}{R}}\Big\},\end{aligned} \tag{7.62}$$

$$j = \frac{1}{2}\,,\ k = \pm 1,\ m = \pm\frac{1}{2},\ T_z = \pm\frac{1}{2},\ n > 0.$$

Here we followed the standard notations $M(r)$, $\mu(r)$ used in the papers [Ch74] for the orbital integrals of the functions (7.43) and (7.44):

$$\mu_{nk}(r) = \int_0^r \mu'_{nk}(r')\,dr'\,,\qquad M_{nk}(r) = \int_r^R \frac{\mu'_{nk}(r')}{r'^3}\,,$$

$$\mu'_{nk}(r') = \frac{2}{3}N^2(\omega_{nk})\,j_0\Big(\frac{\omega_{nk}}{R}r'\Big)\,j_1\Big(\frac{\omega_{nk}}{R}r'\Big)\,r'^3\,,\qquad \mu_{nk} = \mu_{nk}(R)\,. \tag{7.63}$$

A nonstatic part of the operator (7.62) which defines the non-diagonal transitions $s_{1/2} \longleftrightarrow p_{1/2}$, contains the following orbital part of the matrix element

$$J_{nk,\overline{nk}}(r) = \int_r^R J'_{nk,\overline{nk}}(r')\,dr'\,,$$

$$J'_{nk,\overline{nk}}(r') = N(\omega_{nk})N(\omega_{\overline{nk}})\Big[j_0\Big(\frac{\omega_{nk}}{R}r'\Big)\,j_0\Big(\frac{\omega_{\overline{nk}}}{R}r'\Big) - j_1\Big(\frac{\omega_{nk}}{R}r'\Big)\,j_1\Big(\frac{\omega_{\overline{nk}}}{R}r'\Big)\Big] \tag{7.64}$$

In the formula (7.62) the matrix elements of the matrices λ_a, σ are expressed through $\lambda_a^{\bar\alpha\alpha}$ and $\sigma^{\bar m m}$

$$\lambda_a^{\bar\alpha\alpha} = q_{\bar\alpha}^+ \lambda_a q_\alpha, \quad \sigma^{\bar m m} = U_{\bar m}^+ \sigma U_m \tag{7.65}$$

where q_α, U_m are the spinors of the corresponding groups SU_3^C and SU_2^S

$$q_1 = \begin{pmatrix}1\\0\\0\end{pmatrix}_C, \quad q_2 = \begin{pmatrix}0\\1\\0\end{pmatrix}_C, \quad q_3 = \begin{pmatrix}0\\0\\1\end{pmatrix}_C; \quad U_{1/2} = \begin{pmatrix}1\\0\end{pmatrix}_S, \quad U_{-1/2} = \begin{pmatrix}0\\1\end{pmatrix}_S. \tag{7.66}$$

The terms in the effective Hamiltonian of quark interaction in the bag which are obtained as a result of substitution of the operator (7.62) into the term of magnetic energy $H_M = -\frac{1}{2} \int_{V_{bag}} \mathbf{B}_a^2 \, d^3\mathbf{r}$ may be represented by diagrams. Let us restrict ourselves by the case $n = 1$, i.e. two orbital states $1\,s_{1/2}$ and $1\,p_{1/2}$ are allowed in the system (see diagrams of Fig. 25).

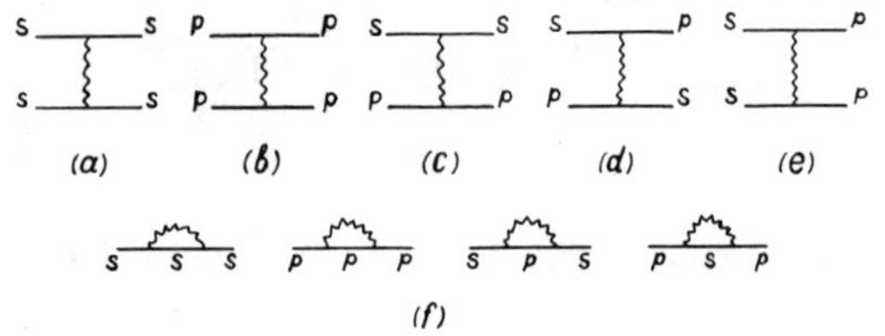

Fig. 25
Terms of the effective Hamiltonian H_{Bag}^{eff}.

Self-energy contributions (diagrams in Fig. 25 (f)) renormalize mass of a quark m_q which in the present model is a phenomenological constant ($m_p = 0$, $m_n = 0$, $m_s \neq 0$). The inclusion of self-energy contributions will change only the phenomenological constants of the model m_q and Z_0 and will not change substantially the hadron spectrum [Ch74]. The diagram in Fig. 25 (e) corresponds to a nonstatic contribution and should not be included in the calculations performed in the second order in g.

So, only the contributions of the diagrams in Figs. 25 (a)–(d) will enter the effective Hamiltonian of the colour magnetic quark interaction. In writing down the effective Hamiltonian it is convenient to pass from the operators $b^+_{\alpha T_z(nkjm)}$ to the ordinary quantum-mechanical operators operating on the wave functions. Let us write down the vector of the bag state as

$$|\mathrm{Bag}\rangle = \sum_{\alpha_1,\dots,m_N} \Psi(\alpha_1 T_{z_1} n_1 k_1 m_1, \dots, \alpha_N T_{z_N} n_N k_N m_N) \times$$

$$\times\, b^+_{\alpha_1 T_{z_1}(n_1 k_1 j m_1)} \cdots b^+_{\alpha_N T_{z_N} k_N j m_N)} |0\rangle \tag{7.67}$$

where N is the number of quarks in the bag (equal to the number of excitations of quark modes). Let i be the number of a quark ($i = 1, 2, \dots, N$). The operators entering (7.62)

$$\lambda^{\bar\alpha\alpha} \sigma^{\bar m m} b^+_{\bar\alpha T_z(\bar n \bar k j \bar m)} b_{\alpha T_z(nkjm)}$$

will operate on the state (7.67) in the same way as the matrices

$$\lambda_{a_i} \sigma_i$$

act on the wave functions

$$\Psi_{bag} = \sum_{\alpha_1,\ldots,m_N} \Psi(\alpha_1 T_{z_1} n_1 k_1 m_1,\ldots, \alpha_i T_{z_i} n_i k_i m_i,\ldots, \alpha_N T_{z_N} n_N k_N m_N) \times$$

$$\times\, \Phi_{n_1 k_1} q_{\alpha_1} t_{T_{z_1}} U_{m_1} \cdots \Phi_{n_i k_i} q_{\alpha_i} t_{T_{z_i}} U_{m_i} \cdots \Phi_{n_N k_N} q_{\alpha_N} t_{T_{z_N}} U_{m_N}, \quad (7.68)$$

where q_α, U_m are spinors (7.66),

$$t_{+1/2} = \begin{pmatrix}1\\0\end{pmatrix}_T, \qquad t_{-1/2} = \begin{pmatrix}0\\1\end{pmatrix}_T \qquad \text{are isospinors,}$$

$$\Phi_{1-1} = \begin{pmatrix}1\\0\end{pmatrix}_X, \qquad \Phi_{11} = \begin{pmatrix}0\\1\end{pmatrix}_X \qquad \text{are "spinors",}$$

which shall be used for designating the orbital states ($s_{1/2}$ and $p_{1/2}$, respectively). In what follows we treat the wave function $\Psi(\alpha_1 T_{z_1} n_1 k_1 m_1,\ldots, \alpha_N T_{z_N} n_N k_N m_N)$ as an ordinary non-relativistic function of the type (7.9) using the same set of quantum numbers for defining the basis vectors (only the orbital quantum numbers ωLM are not required since by the orbital state of a quark we now mean only the index which takes but two values: $s_{1/2}$ and $p_{1/2}$; in this case the quantum numbers $N_s N_p [f]^X (r_X)$ are sufficient for the orbital state; see Sec. 7.3 of this chapter).

In these terms the effective Hamiltonian of the quark bag for the $s_{1/2}^{N_s}\, p_{1/2}^{N_p}$ configuration

$$H_{bag}^{eff} = \frac{1}{R}\Bigg\{\sum_{i=1}^{N} \omega_i - Z_0 - \sum_{i>j=1}^{N} \lambda_{a_i}\lambda_{a_j}\sigma_i\sigma_j M(\omega_i,\omega_j) -$$

$$- \sum_{i>j=1}^{N} P_{ij}^X \lambda_{a_i}\lambda_{a_j}\sigma_i\sigma_j M^{ex}(\omega_s,\omega_p)\left[\delta_{\omega_i\omega_s}\delta_{\omega_j\omega_p} + \delta_{\omega_i\omega_p}\delta_{\omega_j\omega_s}\right]\Bigg\} + \frac{4\pi}{3} R^3 B. \quad (7.69)$$

The eigen frequencies of the states $s_{1/2}$ and $p_{1/2}$ are here denoted in the abbreviated form as $\omega_s = \omega_{1-1} = 2.043$, $\omega_p = \omega_{11} = 3.81$. P_{ij}^X is the permutation operator in the orbital space The exchange term in the second line of (7.69) corresponds to the diagram in Fig. 25 (d). It remains nonzero only for the quark pairs (ij) which are in the orbital state $s_{1/2}\, p_{1/2}$. The orbital integrals in the colour-magnetic term $-\frac{1}{2}\int_{V_{bag}} \mathbf{B}_a^2(\mathbf{r},t)\, d^3\mathbf{r}$, calculated for the functions occuring in (7.62), are expressed in terms of $M(\omega_i,\omega_j)$, $M^{ex}(\omega_s,\omega_p)$. Here, they are expressed following the MIT bag model [Ch74]

$$M(\omega_i,\omega_j) = 3\left(\frac{\alpha_s}{4}\right)\frac{\mu_i}{R}\frac{\mu_j}{R} I_{ij},$$

$$M^{ex}(\omega_s,\omega_p) = \frac{2}{3}\left(\frac{\alpha_s}{4}\right) I_{sp}^{ex}, \quad \alpha_s = \frac{g^2}{4\pi}; \quad (7.70)$$

$$I_{ij} \equiv 1 + \frac{2R^3}{\mu_i\mu_j}\int_0^R \frac{\mu_i(r)\,\mu_j(r)}{r^4}\, dr = 1 + \frac{2}{(4\,\omega_i + 3\,k_i)(4\,\omega_j + 3\,k_j)}$$

$$\left\{-3(2\,\omega_i + k_i)(2\,\omega_j + k_j) - 4\,\omega_i\omega_j + \frac{\omega_i\omega_j}{\sin^2\omega_i\,\sin^2\omega_j}\,[2\,\omega_i\,\mathrm{Si}(2\,\omega_i) + 2\,\omega_j\,\mathrm{Si}(2\,\omega_j) - \right.$$

$$\left. - (\omega_i + \omega_j)\,\mathrm{Si}\,(2\omega_i + 2\omega_j) - (\omega_i - \omega_j)\mathrm{Si}(2\omega_i - 2\omega_j)]\right\} ; \tag{7.71}$$

$$\mu_i = \frac{R}{12}\;\frac{4\,\omega_i + 3\,k_i}{\omega_i\,(\omega_i + k_i)} ; \tag{7.72}$$

$$I^{ex}_{sp} \equiv R \int_0^R J^2_{sp,ps}(r)\,r^2\,dr = \frac{\omega_s^2\,\omega_p^2}{\sqrt{(\omega_s^2 - \sin^2\omega_s)\,(\omega_p^2 - \sin^2\omega_p)}} \times$$

$$\times \frac{2}{3} \int_0^1 \left[\frac{\cos(\omega_s x - \omega_p x)}{\omega_s\,\omega_p\,x^2} + \frac{(\omega_s + \omega_p)\sin(\omega_s x + \omega_p x)}{2\,\omega_s^2\,\omega_p^2\,x^3} + \frac{(\omega_p - \omega_s)\sin(\omega_s x - \omega_p x)}{2\,\omega_s^2\,\omega_p^2\,x^3} + \right.$$

$$\left. + \frac{\cos(\omega_s x + \omega_p x) - \cos(\omega_s x - \omega_p x)}{2\,\omega_s^2\,\omega_p^2\,x^4}\right] \times \left[\frac{(\omega_s^2 + \omega_p^2)\cos(\omega_s x - \omega_p x)}{2(\omega_s^2 - \omega_p^2)\,\omega_s^2\,\omega_p^2} - \right.$$

$$- \frac{\cos(\omega_s x - \omega_p x)}{2\,\omega_s^2\,\omega_p^2} + \frac{x\sin(\omega_s x - \omega_p x)}{\omega_s\omega_p(\omega_s - \omega_p)} - \frac{\mathrm{Ci}(\omega_s x - \omega_p x) - \mathrm{Ci}(\omega_s x + \omega_p x)}{2\,\omega_s^2\,\omega_p^2} -$$

$$\left. - \frac{1}{\omega_s\,\omega_p(\omega_s - \omega_p)^2}\right] dx\,. \tag{7.73}$$

7.2.5 The mass formula for the $s^{N_s}_{1/2}\,p^{N_p}_{1/2}$ configuration

According to the considerations in section 7.2.4 in the quark bag the basis of states (7.68) may be constructed with the same set of quantum numbers as in the case of the shell model states (7.9)

$$\Psi^{bag}_n(q^N) = |s^{N_s}_{1/2}\,p^{N_p}_{1/2}\,[f]^X, [g_C]\,CS\,[h_{CS}]\,T\,[\tilde{f}\,]^{CST}\rangle . \tag{7.74}$$

Here energies of these states are calculated by averaging over these wave functions the effective Hamiltonian (7.69)

$$E^{bag}_n = \langle\Psi^{bag}_n\,|H^{eff}_{bag}\,|\Psi^{bag}_n\,\rangle . \tag{7.75}$$

In the quark bag, just as in the oscillatory shell model, we meet the problem of exclusion of spurious states of c.m. motion, which is not finally solved. According to the estimates [Wo77; Gr76] the c.m. motion provides a contribution to E^{bag}_n of the order of a hundred of MeV. Therefore, mass of the system $M_n(q^N)$ is not strictly equal to E^{bag}_n

$$M_n(q^N) \lesssim E^{bag}_n\,. \tag{7.76}$$

In what follows in a rough approximation we put the equality sign in (7.76).

Consider first a more simple case, when $N_p = 0$ (the configuration $s^N_{1/2}$). Since here $\omega_i = \omega_j = \omega_s$ the orbital part of the colour magnetic interaction $M(\omega_s, \omega_s)$ can be taken outside the summation sign.

As a result we obtain

$$E_n^{\text{bag } s^N}(R) = \frac{1}{R}\left[N\omega_s - Z_0 + M(\omega_s,\omega_s)\sum_{i>j}^{N}\Delta_{ij}\right] + \frac{4}{3}\pi R^3 B\,, \tag{7.77}$$

where

$$\Delta_{ij} = -\langle [g_C]\,CS\,[h_{CS}]\,T[\tilde{f}\,]^{CST}\,|\lambda_{a_i}\lambda_{a_j}\sigma_i\,\sigma_j|\,[g_C]\,CS[h_{CS}]\,T[\tilde{f}\,]^{CST}\rangle\,. \tag{7.78}$$

Since $[\tilde{f}\,]^{CST} = [1^N]$, it is sufficient to calculate Δ_{ij} only for one pair $N-1, N$: $\Delta_{ij} = \Delta_{N-1,N}$. Then,

$$\Delta \equiv \sum_{i>j}^{N}\Delta_{ij} = \frac{1}{2}N(N-1)\,\Delta_{N-1,N}\,. \tag{7.79}$$

Let us briefly describe the technique of Casimir operators developed in [Ja77;Ch74] for calculating Δ.

The Casimir operator $C_2^{(n)}$ of the SU_n group is determined in terms of the generators of this group $\frac{1}{2}\Lambda_A$ by the equality

$$C_2^{(n)} = \sum_{A=1}^{n^2-1}\left(\frac{1}{2}\Lambda_A\right)^2. \tag{7.80}$$

Normalization of the operators $\frac{1}{2}\Lambda_A$ corresponds to the standard form of the commutator

$$\left[\frac{1}{2}\Lambda_A,\ \frac{1}{2}\Lambda_B\right] = i\, f_{ABC}\,\frac{1}{2}\Lambda_C\,, \tag{7.81}$$

where f_{ABC} are the structural constants of the Lie algebra of the SU_n group which satisfy the antisymmetrization relations $f_{ABC} = -f_{BAC} = -f_{CAB} = -f_{ACB}$. It may be shown with the help of (7.81) that $C_2^{(n)}$ satisfies the relation:

$$[C_2^{(n)},\ \Lambda_A] = 0,\quad A = 1, 2, \dots, n^2-1\,.$$

For a spinor representation $(\chi_i^{(n)})$ of the SU_n group the operators $\Lambda_{Ai}^{(n)}$ are the $n \times n$ matrices normalized by the condition

$$\mathrm{Sp}\left\{\frac{1}{2}\Lambda_A^{(n)}\,\frac{1}{2}\Lambda_B^{(n)}\right\} = \frac{1}{2}\delta_{AB}\,. \tag{7.82}$$

In particular, for the SU_2- and SU_3-groups the following relations are valid:

$$\Lambda_\alpha^{(2)} = \sigma_\alpha,\ \alpha - 1,2,3\,,\qquad \mathrm{Sp}\left\{\frac{1}{2}\sigma_\alpha\,\frac{1}{2}\sigma_\beta\right\} = \frac{1}{2}\delta_{\alpha\beta}\,;$$

$$\Lambda_a^{(3)} = \lambda_a,\ a = 1,2,\dots,8\,,\quad \mathrm{Sp}\left\{\frac{1}{2}\lambda_a\,\frac{1}{2}\lambda_b\right\} = \frac{1}{2}\delta_{ab}\,.$$

An irreducible representation of the SU_n group corresponding to the Young scheme $[f] = [f_1 f_2,\dots, f_n\ ]$ $(f_1 + f_2 + \dots + f_n = N)$ can be constructed in the space of a direct production of spinors $\chi_1^{(n)} \times \chi_2^{(n)} \times \chi_3^{(n)} \times \dots \times \chi_N^{(n)}$. Then, the Casimir operator is equal to

$$C_2^{(n)} = \sum_{A=1}^{n^2-1}\left[\sum_{i=1}^{N}\frac{1}{2}\Lambda_{Ai}^{(n)}\right]^2 \tag{7.83}$$

and its eigenvalue can be expressed in terms of the length f_k of the Young scheme lines

$$C_2^{(n)}([f]) \equiv \langle [f] | C_2^{(n)} | [f] \rangle =$$

$$= \frac{1}{2} [f_1'(f_1' + n - 1) + f_2'(f_2' + n - 3) + \ldots + f_{n-1}'(f_{n-1}' - n + 3)] - \frac{1}{2n} (f_1' + f_2' + \ldots + f_n')^2 , \tag{7.84}$$

where $f_k' = f_k - f_n$, $k = 1, 2, \ldots, n-1$.

In particular,

$$C_2^{(2)}([f]) = S(S+1), \quad S = \frac{1}{2}(f_1 - f_2) ;$$

$$C_2^{(3)}([f]) = \frac{1}{3}(\lambda + \mu)(\lambda + 3) - \frac{1}{6}\mu^2 , \quad \lambda = f_1 - f_2 , \quad \mu = f_2 - f_3 ;$$

$$C_2^{(6)}([f]) = \frac{1}{2}[f_1'(f_1' + 5) + f_2'(f_2' + 3) + \ldots + f_5'(f_5' - 3)] - \frac{1}{12}(f_1' + f_2' + \ldots + f_5')^2 ;$$

$$C_2^{(2)}([1]) = \frac{3}{4} , \quad C_2^{(3)} = \frac{4}{3} , \quad C_2^{(6)}([1]) = \frac{35}{12} .$$

The relation (7.83) permits the operator $\sum_{i<j}^{N} \lambda_{ai} \lambda_{aj} \boldsymbol{\sigma}_i \boldsymbol{\sigma}_j$ to be also expressed in terms of the Casimir operators $C_2^{(6)}$, $C_2^{(3)}$ and $C_2^{(2)}$. The matrices $\Lambda_A^{(6)}$ of the spinor representation of SU_6^{CS} may be written as a direct product of the matrices I^λ, λ_a affecting the colour part (χ_i^C) and the matrices I^σ, σ_α affecting the spin part (χ_i^S) of a six-component spinor of the SU_6^{CS} group $(\chi_i^{CS} \equiv \chi_i^C \times \chi_i^S)$

$$\left\{ \frac{1}{2} \Lambda_{Ai}^{(6)} \right\} = \left\{ \frac{1}{2\sqrt{2}} \lambda_{ai} \times I_i^\sigma , \quad \frac{1}{2\sqrt{3}} I_i^\lambda \times \sigma_{\alpha i} , \quad \frac{1}{2\sqrt{2}} \lambda_{ai} \times \sigma_{\alpha i} \right\} ,$$

$$i = 1, 2, \ldots, N; \; A = 1, 2, \ldots, 35; \; a = 1, 2, \ldots, 8; \; \alpha = 1, 2, 3 . \tag{7.85}$$

Then, substituting (7.85) into (7.83) we obtain

$$C_2^{(6)} \equiv \sum_{A=1}^{35} \left[\sum_{i=1}^{N} \frac{1}{2} \Lambda_{Ai}^{(6)} \right]^2 = \sum_{i=1}^{N} \sum_{A=1}^{35} \left(\frac{1}{2} \Lambda_{Ai}^{(6)} \right)^2 + 2 \sum_{i<j}^{N} \sum_{A=1}^{35} \frac{1}{2} \Lambda_{Ai}^{(6)} \frac{1}{2} \Lambda_{Aj}^{(6)} =$$

$$= N C_2^{(6)}([1]) + 2 \sum_{i<j}^{N} \left\{ \frac{1}{2} \sum_{a=1}^{8} \left(\frac{1}{2} \lambda_{ai} \times I_i^\sigma \right) \left(\frac{1}{2} \lambda_{aj} \times I_j^\sigma \right) + \right.$$

$$\left. + \frac{1}{3} \sum_{\alpha=1}^{3} \left(I_i^\lambda \times \frac{1}{2} \sigma_{\alpha i} \right) \left(I_j^\lambda \times \frac{1}{2} \sigma_{\alpha j} \right) + 2 \sum_{a=1}^{8} \sum_{\alpha=1}^{3} \left(\frac{1}{2} \lambda_{ai} \times \frac{1}{2} \sigma_{\alpha i} \right) \left(\frac{1}{2} \lambda_{aj} \times \frac{1}{2} \sigma_{\alpha j} \right) \right. \tag{7.86}$$

From the above-relation it follows the Jaffe formula [Ja77]

$$\Delta \equiv - \langle [g_C] \, S \, [h_{CS}] \, | \sum_{i>j} \lambda_{ai} \boldsymbol{\sigma}_i \lambda_{aj} \boldsymbol{\sigma}_j \, | [g_C] \, S \, [h_{CS}] \rangle =$$

$$= -4 \, C_2^{(6)}([h_{CS}]) + 2 \, C_2^{(3)}([g_C]) + \frac{4}{3} S(S+1) + 8N \tag{7.87}$$

The mass formula for the $s_{1/2}^N$ configuration is obtained by a substitution of the value of the equilibrium radius of the bag into (7.77)

$$\bar{R} = \left[\frac{N\omega_s - Z_0 + M(\omega_s, \omega_s)\Delta}{4\pi B}\right]^{1/4},$$

$$M_n^{bag}(s^N) \cong E_n^{bag}(s^N) = \frac{4}{3}(4\pi B)^{1/4}\left[N\omega_s - Z_0 + M(\omega_s, \omega_s)\Delta\right]^{3/4} \quad (7.88)$$

It will be noted that the equilibrium radius **R** could be determined from the condition of minimum of the bag energy

$$\frac{dE_n^{bag}}{dR} = 0, \quad R = \bar{R}.$$

Three phenomenological parameters ($\alpha_s/4$, B and Z_0) in (7.88) are fitted to masses N(938), Δ(1236) and ω(783)

$$B^{1/4} = 146\ \text{MeV}, \quad Z_0 = 1.84, \quad \frac{\alpha_s}{4} = 0.55\,. \quad (7.89)$$

With the introduction of the fourth phenomenological parameter, mass of a strange quarks $m_s = 279$ MeV, the mass formula (7.88) describes satisfactory the spectrum of all hadrons (except for a π-meson which is not fitted into the scheme with the colour magnetic splitting of levels).

In the s^N configurations for nonstrange quarks the Young scheme $[h_{CS}]$ is unambigously associated with the isospin T and the isospin Young scheme ($[h_{CS}] = [\widetilde{\varphi}_T]$), since the Pauli principle is consistent only with complete antisymmetry $[1^N]^{CST}$ in the CST-space. The relationship between $[h_{CS}]$ and the total isospin T is:

$$[h_{CS}] = [2^{\frac{1}{2}N-T}\,1^{2T}], \quad C_2^{(6)}([h_{CS}]) = 4N - \frac{1}{3}N^2 - T(T+1)\,.$$

For nonstrange (ns) quarks the Jaffe formula (7.87) in the case of neutral-colour hadrons (nc) has the form

$$\Delta^{nc,ns} = \frac{4}{3}S(S+1) + 4T(T+1) - \frac{4}{3}N(6-N)\,. \quad (7.90)$$

It is seen that at $N \geqslant 6$ the colour magnetic forces ensure the *repulsion* of these systems.

Let us come back to the $s_{1/2}^{N_s}\,p_{1/2}^{N_p}$ configurations without an unambigous connection between $[h_{CS}]$ and the isospin T. The formula (7.90) does not hold here. The Jaffe formula (7.88) cannot be used too for $s_{1/2}^{N_s}\,p_{1/2}^{N_p}$ since in the $s_{1/2}^{N_s}\,p_{1/2}^{N_p}$ configurations the colour-magnetic interaction in the Hamiltonian (7.69) depends on the orbital states in the pairs (ij). For the matrix element of the colour magnetic interaction to be calculated by the formula (7.75), it is necessary to write down the fractional parentage expansion of $\Psi_n^{bag}(q^N)$ for separating a pair of quarks ($q^N \to q^{N-2} \times q^2$).

In Section 3 the technique of obtaining arbitrary fractional parentage expansions $q^N \to q^{N'} \times q^{N''}$ in the multiquark systems is described and the necessary fractional parentage coefficients for the expansions $q^6 \to q^4 \times q^2$ and $q^6 \to q^3 \times q^3$ are calculated. So all the necessary fractional parentage coefficients (FPC) are assumed to be known here. Let us divide FPC ($q^N \to q^{N-2} \times q^2$) into three groups:

i) coefficients $\Gamma_{s^2}(Q', Q'')$ for separating a pair $(N-1, N)$ in the orbital state $s_{1/2}^2$,

ii) coefficients $\Gamma_{p^2}(Q', Q'')$ for separating a pair $(N-1, N)$ in the orbital state $p^2_{1/2}$,
iii) coefficients $\Gamma_{sp}(Q', Q'')$ for separating a pair $(N-1, N)$ in the orbital state $s_{1/2}p_{1/2}$.
Here we denoted by Q' a set of quantum numbers defining the state vectors in the subsystem $(1, 2, \dots, N')$ and by Q'' a set of quantum numbers defining the state in the subsystem $(N-1, N)$; $N' = N'_s + N'_p = N-2$, $N'' = N''_s + N''_p = 2$; namaly,

$$\begin{aligned}\{s_{1/2}^{N'_s} p_{1/2}^{N'_p} Q'\} &= \{s_{1/2}^{N'_s} p_{1/2}^{N'_p}[f']^X, \quad [g'_C] C'S'[h'_{CS}] T'[\tilde{f}']^{CST}\},\\ \{s_{1/2}^{N''_s} p_{1/2}^{N''_p} Q''\} &= \{s_{1/2}^{N''_s} p_{1/2}^{N''_p}[f'']^X, \quad [g''_C] C''S''[h''_{CS}] T''[\tilde{f}'']^{CST}\}.\end{aligned} \tag{7.91}$$

The fractional parentage expansion of the vector $\Psi_n(q^N)$ appears as

$$\begin{aligned}\Psi_n(q^N) &\equiv |s^{N_s}p^{N_p}[f]^X, [g_C] CS[h_{CS}] T[\tilde{f}]^{CST}\rangle =\\ &= \sum_{Q', Q''} \Gamma_{s^2}(Q', Q'') \{|s^{N_s-2}p^{N_p}Q'\rangle |s^2 Q''\rangle\}_{CST} +\\ &+ \sum_{Q', Q''} \Gamma_{p^2}(Q', Q'') \{|s^{N_s}p^{N_p-2}Q'\rangle |p^2 Q''\rangle\}_{CST} +\\ &+ \sum_{Q', Q''} \Gamma_{sp}(Q', Q'') \{|s^{N_s-1}p^{N_p-1}Q'\rangle |spQ''\rangle\}_{CST}.\end{aligned} \tag{7.92}$$

From the unitarity of matrix of the transformation (7.92) it follows that FPCs are normalized by the relation

$$\sum_{Q', Q''} \Gamma^2_{s^2}(Q', Q'') + \sum_{Q', Q''} \Gamma^2_{p^2}(Q', Q'') + \sum_{Q', Q''} \Gamma^2_{sp}(Q', Q'') = 1. \tag{7.93}$$

Using the expansion (7.92) we obtain the following expression for the bag energy

$$\begin{aligned}E_n^{bag}(R) &\equiv \langle \Psi_n^{bag}(q^N) | H_{bag}^{eff} | \Psi_n^{bag}(q^N)\rangle =\\ &= \frac{1}{R}\Big\{N_s\omega_s - Z_0 + \frac{N(N-1)}{2} M(\omega_s, \omega_s) \sum_{\tilde{Q}''} \tilde{\Gamma}^2_{s^2}(\tilde{Q}'') \Delta_{N-1,N}(Q'') +\\ &+ \frac{N(N-1)}{2} M(\omega_p, \omega_p) \sum_{\tilde{Q}''} \tilde{\Gamma}^2_{p^2}(\tilde{Q}'') \Delta_{N-1,N}(Q'') +\\ &+ \frac{N(N-1)}{2} M(\omega_s, \omega_p) \sum_{\tilde{Q}''} \tilde{\Gamma}^2_{sp}(\tilde{Q}'') \Delta_{N-1,N}(Q'') +\\ &+ \frac{N(N-1)}{2} M^{ex}(\omega_s, \omega_p) \sum_{\tilde{Q}''} \Lambda_{f''} \tilde{\Gamma}^2_{sp}(\tilde{Q}'') \Delta_{N-1,N}(Q'')\Big\} + \frac{4}{3}\pi R^3 B.\end{aligned} \tag{7.94}$$

Here we denoted $Q'' = (C'', \tilde{Q}'')$,

$$\tilde{\Gamma}^2_{s^2}(\tilde{Q}'') = \sum_{Q'C''} \Gamma^2_{s^2}(Q', Q''),$$

$$\tilde{\Gamma}^2_{p^2}(\tilde{Q}'') = \sum_{Q'C''} \Gamma^2_{p^2}(Q', Q''), \quad \tilde{\Gamma}^2_{sp}(\tilde{Q}'') = \sum_{Q'C''} \Gamma^2_{sp}(Q', Q''),$$

$$\Lambda_{f''} = \begin{cases} +1, & [f'']^X = [2], \\ -1, & [f'']^X = [1^2]. \end{cases} \tag{7.95}$$

The matrix element of the colour magnetic operator in the subsystem (N − 1, N) is equal, according to the Jaffe formula (7.87), to:

$$\begin{aligned} \Delta_{N-1,N}(Q'') \equiv & -\langle [g''_C] C'' S'' [h''_{CS}] T'' [\tilde{f}'']^{CST} | \lambda_{aN-1} \sigma_{N-1} \lambda_{aN} \sigma_N | \times \\ & \times [g''_C] C'' S'' [h''_{CS}] T'' [\tilde{f}'']^{CST} \rangle = \\ & = -4 C_2^{(6)}([h''_{CS}]) + 2 C_2^{(3)}([g''_C]) + \frac{4}{3} S''(S''+1) + 8 \cdot 2 . \end{aligned} \tag{7.96}$$

In Table 7.1 are given the values of the coefficients (7.95) for the $[42]^{CS}$ state in the $s^4 p^2 [42]^X$ configuration. Here are also given the values of the sums occuring in the formula (7.94) and denoted by $\Delta_{xy}, \Delta_\pm$

$$\Delta_{xy} = \frac{N(N-1)}{2} \sum_{\tilde{Q}''} \tilde{\Gamma}^2_{xy}(\tilde{Q}'') \Delta_{N-1,N}(Q''), \quad xy = s^2, p^2, sp,$$

$$\Delta_+ = \frac{N(N-1)}{2} \sum_{\tilde{Q}'', [f'']^X = [2]} \tilde{\Gamma}^2_{sp}(\tilde{Q}'') \Delta_{N-1,N}(Q''),$$

$$\Delta_- = \frac{N(N-1)}{2} \sum_{\tilde{Q}'', [f'']^X = [1^2]} \tilde{\Gamma}^2_{sp}(\tilde{Q}'') \Delta_{N-1,N}(Q''). \tag{7.97}$$

In these terms the Jaffe mass formula generalized to the case of the $s_{1/2}^{N_s} p_{1/2}^{N_p}$ configuration appears as

$$M_n^{bag}(s^{N_s} p^{N_p}) \cong E_n^{bag}(s^{N_s} p^{N_p}) = \frac{4}{3}(4\pi B)^{1/4} \Big[N_s \omega_s + N_p \omega_p - Z_0 + \\ + \sum_{xy = s^2, p^2, sp} M(\omega_x, \omega_y) \Delta_{xy} + M^{ex}(\omega_s, \omega_p)(\Delta_+ - \Delta_-) \Big]^{3/4} . \tag{7.98}$$

It will be noted that the coefficients $\Delta_{xy}, \Delta_{\pm}$ are related with the coefficient Δ

$$\Delta = \Delta_{s^2} + \Delta_{p^2} + \Delta_{sp}, \quad \Delta_{sp} = \Delta_+ + \Delta_- . \tag{7.99}$$

The equalities (7.99) follow directly from the normalizing relation (7.93) for the FPCs and determinations of the values of Δs in equalities (7.79) and (7.97).

Table 7.1: Coefficients $\tilde{\Gamma}^2_{xy}(\tilde{Q}'')$, Δ_{xy}, $\Delta_{\pm}$ for the state $\Psi_1 = |s^4p^2[42]^X L=0,$ $[2^3]^C C=0\ S=1[42]^{CS} T=0[2^21^2]^{CST}\rangle$

	$\tilde{\Gamma}^2_{xy}(\tilde{Q}'')$								
Q''_{CST}	$[2]^C s''=1[2]^{CS}$		$[1^2]^C s''=0[2]^{CS}$		$[2]^C s''=0[1^2]^{CS}$		$[1^2]^C s''=1[1^2]^{CS}$		Δ_{xy}
$xy[f]^{X''}$	$T''=0$	$T''=1$	$T''=0$	$T''=1$	$T''=1$	$T''=0$	$T''=1$	$T''=0$	
$s^2[2]$	$\frac{3}{20}$	0	$\frac{1}{20}$	0	$\frac{1}{50}$	0	$\frac{9}{50}$	0	$-\frac{3}{5}$
$p^2[2]$	$\frac{11}{300}$	0	$\frac{3}{100}$	0	0	0	0	0	$-\frac{13}{3}$
$sp[2]$	$\frac{9}{200}$	0	$\frac{11}{200}$	0	$\frac{1}{100}$	0	$\frac{9}{100}$	0	$\Delta_+ = -\frac{33}{10}$
$sp[1^2]$	0	$\frac{27}{200}$	0	$\frac{33}{200}$	0	$\frac{1}{300}$	0	$\frac{3}{100}$	$\Delta_- = -\frac{211}{10}$
$\Delta_{N-1,N}$	$-\frac{4}{3}$		-8		4		$\frac{8}{3}$		

7.2.6 Level succession in the six-quark systems

The results obtained in sections 7.2.4 and 7.2.5 permit one to calculate the levels in the multiquark systems by means of ordinary quantum-mechanical methods, for example, by the method of diagonalization of the Hamiltonian. Consider the state $\Psi_d(q^6)$ with the deuteron quantum numbers $(S,T)=(1,0)$ (we take only one partial wave $L=0$). The solutions are written as an expansion in the set of states in the configurations s^6 and s^4p^2

$$\Psi_d(q^6) = C_0\Psi_0^{bag}(q^6) + \sum_{n=1}^{6} C_n\Psi_n^{bag}(q^6)\,, \tag{7.100}$$

where $\Psi_0(q^6)$ and $\Psi_1(q^6)$ are basis vector of the type (7.15) and (7.16); $n=2,3,4,5$ corresponds to the basis vectors (7.74) in the configuration $s^4_{1/2}\,p^2_{1/2}\,[42]^X$ with the Young schemes $[h_{CS}]=[321],[2^3],[31^3]$ and $[21^4]$; $n=6$ cooresponds to the state $s^4_{1/2}\,p^2_{1/2}\,[6]^X\,[2^3]^{CS}$. The searched levels E_d are determined from the solution of the secular equation

$$\langle\Psi_n^{bag}(q^6)\,|H^{eff}_{bag}|\,\Psi_{n'}^{bag}(q^6)\rangle - E_d\,\delta_{nn'} = 0\,, \tag{7.101}$$

where $n(n')=0,1,2,\ldots,6$.

It will be noted that the mass formula (7.98) gives only diagonal matrix elements of the matrix $\langle\Psi_n|H^{eff}|\Psi_{n'}\rangle$. Nondiagonal elements may be analogously calculated using the fractional parentage expansions of the vectors $\Psi_n(q^6)$ and $\Psi_{n'}(q^6)$. The mixing of the states Ψ_n with different Young schemes $[h_{CS}]$ is due to dependence of the colour magnetic interaction term upon the orbital quark states $M(\omega_s,\omega_s)\neq M(\omega_s,\omega_p)\neq M(\omega_p,\omega_p)$. At the values of the parameters (7.89) we obtain

$$M(\omega_s,\omega_s) = 0.0968,\qquad M(\omega_p,\omega_p) = 0.0651,$$

$$M(\omega_s,\omega_p) = 0.0615,\quad M^{ex}_{sp}(\omega_s,\omega_p) = 0.01185\,. \tag{7.102}$$

Thus, the SU_6^{CS}-symmetry is broken. However, the mixing will be significant only for those states Ψ_n which are degenerate or nearly degenerate.

Not applying to the fractional parentage coefficient and the mass formula (7.98) we may roughly estimate the energy splitting of the states in the s^4p^2 configuration by calculating the colour magnetic contribution by the Jaffe formula $E_M = \frac{1}{R} M(\omega_s, \omega_s)\Delta$.

From Table 7.2 it follows that the mixing of states $[2^3]^{CS}$ and $[31^3]^{CS}$ may be expected while the remaining states Ψ_n, $n = 0, 1, 2, 5, 6$ are separated by large energy gaps. The state with the Young scheme $[42]^{CS}$ is the lowest-lying level in the s^4p^2 configuration and, most likely, mixed with the S-shell levels. For a comparison, in Table 7.3 are presented the results of a calculation by the Jaffe formula for the S-shell [Ch74; Ae78; Ja77].

Table 7.2: Colour magnetic energy $E_M = \frac{1}{R} M(\omega_s, \omega_s)\,\Delta$ of the states $s^4p^2[42]^X[h_{CS}]$, where $M(\omega_s, \omega_s) = M(\omega_p, \omega_p) = M(\omega_s, \omega_p) = 0.0968$ is supposed. $R^{-1} = 147$ MeV.

$[h_{CS}]$	[42]	[321]	$[2^3]$	$[31^3]$	$[21^4]$
Δ	$-\frac{88}{3}$	$-\frac{28}{3}$	$\frac{8}{3}$	$\frac{8}{3}$	$\frac{80}{3}$
E_M (MeV)	−432	−137	39	39	393

Table 7.3: Masses of nonstrange barions and dibarions in MIT bag model. The $s^3_{1/2}$ and $s^6_{1/2}$ configurations. $M(\omega_s, \omega_s) = 0.0968$.

Configuration	s^3		s^6					
(S, T)	$(\frac{1}{2}, \frac{1}{2})$	$(\frac{3}{2}, \frac{3}{2})$	(1, 0)	(0, 1)	(2, 1)	(1, 2)	(3, 0)	(0,3)
$[h_{CS}]$	[21]	$[1^3]$	$[2^3]$	$[2^21^2]$	$[2^21^2]$	$[21^4]$	$[2^3]$	$[1^6]$
Δ	−8	8	$\frac{8}{3}$	8	16	$\frac{80}{3}$	16	48
R (fm)	0.99	1.08	1.295	1.31	1.33	1.36	1.33	1.41
E_M (MeV)	−155	141	39	116	229	374	229	648
M_{bag} (MeV)	938	1233	2163	2241	2356	2507	2356	2801
B or B_1B_2	N	Δ	NN	NN	NΔ	NΔ	ΔΔ	ΔΔ
Threshold (MeV)	938	1236	1876	1876	2174	2174	2472	2472

An accurate calculation by the mass formula (7.98) gives for the state $\Psi_1(s^4p^2[42]^X[42]^{CS})$:

i) neglecting the exchange term ($M^{ex}(\omega_s, \omega_p) = 0$) [Ob79a]
$E_1 = 2380$ MeV, $R = 1.34$ fm, $E_M = -271$ MeV;

ii) including the exchange term ($M^{ex}(\omega_s, \omega_p) = 0.01185$)
$E_1 = 2411$ MeV, $R = 1.34$ fm, $E_M = -240$ MeV.

These results must be compared with an analogous result for the state $\Psi_0(s^6[6]^X[2^3]^{CS}$:

$E_0 = 2163$ MeV, $R = 1.3$ fm, $E_M = +39$ MeV.

It is seen that the energy of the state $\Psi_1(s^4p^2[42]^X[42]^{CS})$ is located really in the area of the S-shell states. The mixing of the states Ψ_0 and Ψ_1 will seemingly, be the most essential. Unfortunately, the mixing of configurations s^6 and s^4p^2 requires exceeding the limits of the static bag model since the diagram in Fig. 25 (e) of the transition $s_{1/2}^2 \longleftrightarrow p_{1/2}^2$ provides a nonstatic contribution to the effective Hamiltonian H_{bag}^{eff} proportional to $e^{\pm 2i\frac{1}{R}(\omega_s - \omega_p)t}$ [see the expression (7.62)].

We can obtain only a rough estimate in solving eq. (7.101) for the two-dimensional basis $\{\Psi_0, \Psi_1\}$ (with a neglect of the nonstatic effects). The contribution of the diagram of Fig.25 (e) to the effective Hamiltonian differs from the exchange term in the expression (7.69) by the replacement of the operator $P_{ij}^X[\delta_{\omega_i\omega_s}\delta_{\omega_j\omega_p} + \delta_{\omega_i\omega_p}\delta_{\omega_j\omega_s}]$ by the operator converting the $s_{1/2}^2$ state of the pair ij to $p_{1/2}^2$ (or vice versa). Using the fractional parentage expansion of the vectors Ψ_0 and Ψ_1 we obtain the following expression for the nondiagonal matrix element

$$\langle \Psi_0 | H_{bag}^{eff} | \Psi_1 \rangle = \frac{1}{R} \frac{N(N-1)}{2} M^{ex}(\omega_s, \omega_p) \times$$

$$\times \sum_{Q'Q''} \Gamma_{s^2}^{(0)}(Q', Q'')\,\Gamma_{p^2}^{(1)}(Q', Q'')\,\Delta_{n-1,N}(Q'') . \qquad (7.103)$$

Here we differ the fractional parentage coefficients for the states Ψ_0 and Ψ_1 with the help of a superscript (0) or (1) (for the values $\Gamma^{(0)}$ and $\Gamma^{(1)}$, see the tables in Sect. 7.3). An elementary calculation yields the value

$$\langle \Psi_0 | H_{bag}^{eff} | \Psi_1 \rangle = -5\sqrt{2}\,\frac{1}{R} M^{ex}(\omega_s, \omega_p) = -12.3 \text{ MeV}.$$

The results are

$E_{d_0} = 2162.4$ MeV, $\Psi_{d_0} = 0.9985\,\Psi_0 + 0.05\,\Psi_1$;

$E_{d_1} = 2411.6$ MeV, $\Psi_{d_1} = -0.05\,\Psi_0 + 0.9985\,\Psi_1$.

An analogous calculation in the configuration $s_{1/2}^4 p_{3/2}^2$ should give even a lower position of the level E (or E_{d_1}) since the eigenfrequency in the state $p_{3/2}$ ($\omega_{1-2} = 3.204$) is lower than in the state $p_{1/2}$ ($\omega_{1-2} = 3.81$). The difference for two quanta is equal to

$$\frac{2}{R}(\omega_{11} - \omega_{1-2}) \simeq 230 \text{ MeV} .$$

However, the Dirac states $p_{3/2}$ do not satisfy the condition of spherical symmetry (7.42), otherwise, in this case a spherical form of the bag must be instable [Jo76; Mu78]. If this circumstance is neglected, the calculation in the configuration $s_{1/2}^4 p_{3/2}^2$ is in principle analogous to the calculation for $s_{1/2}^4 p_{1/2}^2$.

As a summary of the above section we can say that we have found the symmetry channels, showing the nucleon-nucleon colour magnetic attraction, i.e. the channels of a "free nucleon-nucleon penetration". We may expect that NN-resonances are connected with such channels which should exist in each nucleon-nucleon partial wave.

However, not all the levels obtained should be observable – many of them can have very large widths, so the microscopical investigation of decay properties is necessary.

In the next section we write down the quark wave functions with the prescribed symmetry in terms of the translationally – invariant shell model (TISM) and study their cluster composition.

7.3 Construction of the Fractional Parentage Expansions in the Multiquark Systems

The technique of calculating the fractional parentage coeffients in the nucleus shell model (SM) and in the translationally invariant shell model (TISM) is described in the monograph [Ne 69] and the survey [Ne 79]. The main point of this technique is that the many-particle fractional parentage coefficients (FPC) are reduced to FPC of separation of a smaller number of particles and, eventually, to the oneparticle FPC. Use is made of the Racah algebra for the groups SU_n, O_n and etc. and of the formalism of the permutation group S_N, where N is the number of particles in the system. This approach is applicable to the multiquark systems too. However, the calculations are cumbersome due to the recurrent character of the approach (especially in the case of a large number of inner symmetries and particle charges). So, in order to construct the multiquark system wave functions, it is convenient to use another method, which is based on the complementarity of the permutation and unitary groups. The basic idea of the method is the following [Ob 79b, Ob 81]. The wave function $\Psi(q^N)$ of the N-quark system may be composed of the spatial (X), color (C), spin (S) and isospin (T) parts are multiplied with the help of the Clebsch-Gordan coefficients (CGC) for the permutation group S_N [see formulas (7.10) and (7.11)]. So the problem of the fractional parentage expansion of the multiquark wave function is reduced to construction of the fractional parentage expansions for its factors. As for the latter, the fractional parentage expansions for the spin and isospin functions are known [Ja 51]. For the colour part of the function, it is possible to use the orbital FPC for the p-shell, since in the colour space use is made of the quantum numbers of the reduction chain $SU_3^C \supset O_3^C \supset O_2^C$ (the equivalence of this chain to the SU_3-scheme for classifying the p-shell levels [El 58] is obvious). FPC for the spatial part of the wave function are not difficult to obtain either. The technique of calculating the Clebsch-Gordon coefficients for the group S_N is described in the monograph [Ha 64] and the transformation matrices which we used equally with CGC are studied by Kaplan [Ka 61] and tabulated for $N \leqslant 6$ in the ref. [Ka 69], so all necessary components for constructing the fractional parentage expansion of the total wave function of the N-quark system are available.

Eventually, FPC are reduced to the CGC for the unitary groups. It is shown in the present section that the CGC for the unitary groups of high rank ($m \geqslant 4$) may be reduced to CGC for the groups of the lowest rank (SU_2 or SU_3) using a remarkable relation between representations of the unitary group U_m and the permutation group S_N (the so-called property of complementarity). We shall obtain an explicit expression for CGC of an arbitrary unitary group U_m which, apart from CGC for subgroups SU_2 (SU_3), will contain the quantities of the formalism of the permutation groups only: the Clebsch-Gordan coefficients [Ha 64] and the transformation matrices [Ka61]. Then, these formulas are used for calculating FPC in the six-quark system. The results for the configurations s^6 and s^4p^2 are presented in the tables of the necessary factors.

7.3.1 The Complementarity of the Unitary U_m and Permutation S_N groups

Let us recall the principle of constructing the basis vectors for the part of the wave function, which is characterized by the U_m symmetry (m = 2 for the spin and isospin parts of the function, m = 3 for the color part, m = 6 for the function in the CS-space and m = 12, for the CST-space). We have N particles every of which may be in one of the states $\chi_1, \chi_2, \ldots, \chi_m$. A set of these states forms the basis of an irreducible representation (IR) of the group U_m. This IR $D^{[1]}$, characterized by the Young scheme [1], is the spinor of the group U_m

$$|[1]\rangle = \begin{pmatrix} \chi_1 \\ \chi_2 \\ \vdots \\ \chi_m \end{pmatrix}. \tag{7.104}$$

The N particle system wave functions are the linear combinations of the terms of the form

$$\chi_i(1)\,\chi_j(2)\ldots\chi_k(N)\ (i, j, \ldots, k = 1, 2, \ldots, m) \tag{7.105}$$

The total number of linear-independent terms is equal to m^N. These terms form the tensor of rank N in the m-dimensional vector space. The functions (7.105) form the basis of a direct product of N irreducible representations (IR) $D^{[1]}$ of the group U_m. This representation is reducible and can be expanded in IR $D^{[f]}$ of the group U_m as follows

$$[1]\times[1]\times\ldots\times[1]\equiv[1]^N = \sum_f \nu_f[f]. \tag{7.106}$$

According to the results of Weyl [We46], this expansion may include any Young scheme $[f]\equiv[f_1 f_2 \ldots f_m]$, containing N squares and no more than m lines ($N = f_1 + f_2 + \ldots + f_m$). The multiplicity ν_f of each IR $D^{[f]}$ is equal, in this sum, to the dimension of IR of the permutation group S_N with the same Young scheme $\{f\}$.

$$\nu_f = \dim\{f\}. \tag{7.107}$$

The matter is that the space L, stretched over the components of the tensor of rank N in the m-dimensional space, may be considered as the space of representation of the direct product of the groups $U_m \times S_N$. If the representation $[1]^N$ is decomposed in irreducible representations ([f], $\{f'\}$) of these two groups (the square brackets mean the Young scheme for the group U_m and the curly brackets, for the group S_N, to avoid confusion in using simultaneously both types of the Young schemes) the result of Weyl (7.106) is rewritten as

$$[1]^N = \sum_f ([f], \{f\}). \tag{7.108}$$

This remarkable result shows that in the space L of the representation $[1]^N$ each IR $D^{[f]}$ of the group U_m is combined only with one definite IR of the group S_N which characterized the same Young scheme $\{f\}$. It will be noted that (7.108) does not include the multiplicity ν_f which occurred in (7.106). (The matter is that the representation $\{f\}$ has the dimension ν_f).

IR of the groups U_m and S_N are in unambigous correspondence and each pair of the representations ([f], $\{f\}$) encountered in the space L no more than once. Owing to these properties the groups U_m and S_N may be called *the complementary groups within the space L*, following the concept of complementarity set forth by Moshinsky *et al* [Mo 70]. The noted complementarity allows us to establish a strong interrelation between different quantities of the groups U_m and S_N. For example, let the N-particle wave function $|\chi^N [f] \alpha>$ be constructed by some method from the states χ_i, this function being transformed according to IR $D^{[f]}$ of the group U_m (α is the remaining necessary quantum numbers, defining this function construction). Then, this function is simultaneously characterized by the permutation Young scheme $\{f\}$, although in constructing this function we did not take care of providing a definite permutation symmetry. On the contrary, if the function $|\chi^N \{f\} (r) \beta>$ is constructed with the permutation Young scheme $\{f\}$, and for example, with the Yamanouchi symbol (r), defining the row of IR$\{f\}$, one can state that it belongs to IR $D^{[f]}$ of the group U_m as well. This permits various quantities for the group U_m to be expressed through some quantities for the group S_N and vice versa. For example, the standard theory of fractional parentage coefficients [Ra 49; Ra 51; Ja 51; Fl 52] in which the calculation of FPC is reduced to the calculation of the matrix elements of operators of the permutations $P_{N-1,N}$, $P_{N-1,N-2}$ etc. is actually the reduction of the Clebsch-Gordan coefficients of the unitary groups U_m to permutation quantities. Further we shall consider it at greater length.

7.3.2 The standard Young-Yamanouchi basis in terms of irreducible representations of the unitary group

Consider in more detail the correspondence between quantum numbers of the groups U_m and S_N. Multiply successively the representations $D^{[1]}$ (7.104) of the group U_m to produce, with the help of CGC of this group, the IR with a definite Young scheme $[f^{(i)}]$ at each ith stage of multiplication: $[f^{(1)}] = [1]$, $[f^{(2)}] = [2]$ or $[1^2]$ and etc. As a result, we get the set of functions of the form

$$|[1]\,[1]\,([f^{(2)}]), [1]\,([f^{(3)}]), \ldots [1] : [f]\, \alpha>, \tag{7.109}$$

where both the resultant Young scheme [f] and the intermediate Young schemes $[f^{(i)}]$ ($i = 2, 3 \ldots, N-1$) for IR of the group U_m are defined. By the property of complementarity (7.108) we get that the function (7.109) belongs also to the irreducible representations $D^{\{g\}}$ ($\{g\} = \{f^{(2)}\}, \{f^{(3)}\}, \ldots, \{f\}$) of the groups S_i permutating the particles $1, 2, \ldots, i$ ($i = 2, 3, \ldots, N$). So, the functions (7.109) may be characterized by the set of permutation Young schemes

$$\{f\}, \{f^{(N-1)}\}, \ldots, \{f^{(i)}\}, \ldots, \{f^{(2)}\} \tag{7.110}$$

which show the IR of the subgroup chain

$$S_N \supset S_{N-1} \supset \ldots \supset S_i \supset \ldots \supset S_2 \tag{7.111}$$

according to which they are transformed. The Yamanouchi symbol (r) is just the compact representation of the set of the Young schmes (7.110). Thus, the functions (7.109) form the basis of the standard Young-Yamanouchi irreducible representation for the group S_N [Ya37]. So, the function (7.109) may be written in short as $|\chi^N [f] (r) \alpha>$, where the Yamanouchi symbol (r) implies the information on the intermediate Young schemes $[f^{(i)}]$ (or, which is the same, $\{f^{(i)}\}$). It will be recalled that the Yamanouchi symbol is the set of N figures $(r) \equiv (\rho_1\, \rho_2 \ldots \rho_i \ldots \rho_N)$, ρ_i indicates the number of the row of the standard Young table including the square with the number of the ith particle ($1 < \rho_i < m$). It is clear that by the symbol (r) we may define the Young scheme $\{f\}$. Moreover, by rewriting the symbol (r) as $(r' \rho_N)$ where (r') is the Yamanouchi symbol in the subsystem of particles numbered $1, 2, \ldots, N-1$, we may define the Young scheme for the group S_{N-1} in the chain (7.111). In an analogous manner we can determine by (r) the entire chain (7.111). In the functions (7.109) α are the quantum numbers numerating the rows of IR $D^{[f]}$ of the group U_m. Therefore the function (7.109) can be written as

$$|\chi^N [f] (r) \alpha> = \\ = \sum_{\alpha', i} \begin{pmatrix} [f^{(N-1)}] & [1] & [f] \\ \alpha' & i & \alpha \end{pmatrix} |\chi^{N-1} [f^{(N-1)}] (r') \alpha'> \chi_i(N). \tag{7.112}$$

Here $\begin{pmatrix} [f'] & [f''] & [f] \\ \alpha' & \alpha'' & \alpha \end{pmatrix}$ are the CGC for the direct product of IR $[f'] \times [f''] \to [f]$ of the group U_m. In the case $m = 2$ (for example, for the spin group U_2) the Young scheme $[f] = [f_1 f_2]$ unambigously sets the spin value $S = \frac{1}{2}(f_1 - f_2)$, the rows of IR of the group U_2 are numbered by the spin projection M ($\alpha \equiv M$, $i \equiv s_z = \pm \frac{1}{2}$) and the equality (7.112) takes the form

$$|\chi^N [f] (r) \alpha> \equiv |\chi^N SM(r)> = \\ = \sum_{M', s_z} (S'M' \tfrac{1}{2} s_z | SM) |\chi^{N-1} S'M'(r')> \chi_{s_z}(N) \tag{7.113}$$

in the right-hand side use is made of CGC of three-dimensional rotation group. The formula (7.112) is a simple example of the fractional parentage expansion of the N-particle state vector possessing definite transformation properties relative to the elements of the group U_m. It is seen that the FPC for separation of one particle (N^{th} particle) coincide with the CGC of the group U_m.

7.3.3 The Kaplan transformation matrices for the transition to nonstandard basis

Expansion of the vector $|\chi^N [f] (r) \alpha>$ for the separaton of two and more particles may be realized by applying consequently the formula (7.112). Isolating the $(N-1)^{th}$ particle in the Yamanouchi symbol $(r') = (\bar{r}' \rho_{N-1})$ we get

$$|\chi^N [f] (r) \alpha> = \sum_{\alpha',i} \sum_{\bar{\alpha}',j} \left(\begin{matrix} [f^{(N-2)}] & [1] \\ \bar{\alpha}' & j \end{matrix} \middle| \begin{matrix} [f^{(N-1)}] \\ \alpha' \end{matrix} \right) \times$$

$$\times \left(\begin{matrix} [f^{(N-1)}] & [1] \\ \alpha' & i \end{matrix} \middle| \begin{matrix} [f] \\ \alpha \end{matrix} \right) |\chi^{N-2} [f^{(N-2)}] (\bar{r}') \bar{\alpha}' > \chi_j(N-1)\, \chi_i(N). \tag{7.114}$$

Here $(r) = (r' \rho_N) = (\bar{r}' \rho_{N-1} \rho_N)$ and the present basis vector corresponds, as usual, to a subsequent multiplication of IR [1] of the group U_m which is fixed in (7.109). However, we can construct now the linear combinations of the basis functions (7.114) which shall correspond to another manner of multiplication of IR of the group U_m. The sequence of multiplication of IR in the subspace of the first $N-2$ particles is not changed, but the two remaining IR $[1] \times [1]$ corresponding to the particles numbered $N-1$ and N, are multiplied separatley to produce the functions

$$|\chi^2 [f''] (r'') \alpha'' > \equiv |[1] [1] : [f''] (r'') \alpha'' > =$$

$$= \sum_{i,j} \left(\begin{matrix} [1] & [1] \\ j & i \end{matrix} \middle| \begin{matrix} [f''] \\ \alpha'' \end{matrix} \right) \chi_j(N-1)\, \chi_i(N). \tag{7.115}$$

Now multiply IR $[f^{(N-2)}] \times [f'']$ to produce the function with the Young scheme [f]:

$$|\chi^N [f] (\bar{r}') (r'') \alpha > = \sum_{\bar{\alpha}',\alpha''} \left(\begin{matrix} [\bar{f}'] & [f''] \\ \bar{\alpha}' & \alpha'' \end{matrix} \middle| \begin{matrix} [f] \\ \alpha \end{matrix} \right) |\chi^{N-2} [\bar{f}'] (\bar{r}') \bar{\alpha}' > |\chi^2 [f''] (r'') \alpha'' > \tag{7.116}$$

(We denoted $[f^{(N-2)}] = [\bar{f}']$). The function (7.116) corresponds to the nonstandard reduction

$$S_N \supset S_{N-2} \times S_2 \tag{7.117}$$

and produces the nonstandard basis of IR $D^{\{f\}}$ of the group S_N. In the ref. [Ka61; Ka69] use is made of the transformation matrices (TM) for the transition from the nonstandard basis $|[f] (\bar{r}') (r'')>$ of IR of the permutation group to the standard basis

$$|[f] (\bar{r}') (r'') > = \sum_{\rho_{N-1} \rho_N} < [f] (r) | [f] (\bar{r}') (r'') > |[f] (r) >. \tag{7.118}$$

In the right-hand side of (7.118) $(r) = (\bar{r}' \rho_{N-1} \rho_N) = (r' \rho_N)$. Using (7.114)–(7.116) we can write down the equation (7.118) in terms of the CGC for the group U_m. As a result, the TM coincide with the Racah coefficients (or 6f-symbols) for the group U_m.

$$< [f] (r) | [f] (\bar{r}') (r'') > = U([\bar{f}'] [1] [f] [1]; [f'] [f'']) \tag{7.119}$$

This is a very important property which means that the 6f-symbols for the unitary group U_m do not depend on the rank of the group U_m and depend only on the form of the involved Young schemes. It will be shown below that this property is also true for higher 3nf-symbols so far as they will coincide with a more complex transformation matrix for the permutation groups. (This statement can be proved without applying to the permutation group but only on the basis of the properties of IR of the unitary group [Gu71]).

7.3.4 Relation between the Kaplan transformation matrices and the 3nf-symbols for the unitary groups

Consider now a more general case of the nonstandard reduction (7.117)

$$S_N \supset S_{N'} \times S_{N''}, \; N = N' + N'' \qquad (7.120)$$

where N'' is an arbitrary number of particles numbered $N'+1, N'+2, \ldots, N$. Multiply the IR $D^{[f']}$ and $D^{[f'']}$ of the group U_m and go over to the functions

$$|\chi^N [f] (r') (r'') \alpha >_\gamma = \sum_{\alpha' \alpha''} \left(\begin{matrix} [f'] & [f''] \\ \alpha' & \alpha'' \end{matrix} \middle| \begin{matrix} [f] \\ \alpha \end{matrix} \right)_\gamma \; |\chi^{N'} [f'] (r') \alpha' > |\chi^{N''} [f''] (r'') \alpha'' > \qquad (7.121)$$

forming the nonstandard basis of IR $D^{\{f\}}$ of the group S_N for the reduction (7.120). This basis can contain several times the linear independent orthonormalized vectors with the same quantum numbers in the groups $S_{N'}$ and $S_{N''}$: $[f'] (r')$, $[f''] (r'')$. We distinguish them with the help of the index γ ($\gamma = 1, 2, \ldots, \nu_f$) (it is sometimes convenient to consider the index γ as a constituent of the total set of quantum numbers α). From (7.121) it is seen that the number of this linear independent vectors coincides with the multiplicity ν_f

$$[f'] \times [f''] = \sum_f \nu_f [f] \qquad (7.122)$$

The nonstandard basis (7.121) may be related with the Young-Yamanouchi standard basis with the help of the Kaplan transformation matrices of a more general form than in the equation (7.118)

$$|[f] (r') (r'') >_\gamma = \sum_\rho < [f] (r) | [f'] (r'), [f''] (r'') >_\gamma | [f] (r) > \qquad (7.123)$$

where $(r) = (r' \rho_{N'+1} \rho_{N'+2} \ldots \rho_N) = (r' \rho)$, $\rho = \rho_{N'+1} \rho_{N'+2} \ldots \rho_N$. It is clear that the elements of the matrix $< [f] (r) | [f'] (r'), [f''] (r'') >_\gamma$ are the coefficients of transition from one coupling scheme of N representations of the form [1] of the group U_m to another coupling scheme. Therefore, they are some 3nf-symbols for the group U_m. (A particular case of this relation is the equality (7.119)). The unitarity of the transformation (7.123) implies the orthogonality properties of this quantities

$$\sum_\rho < [f] (r'\rho) | [f'] (r'), [f''] (r'') >_\gamma < [f] (r'\rho) | [f'] (r''), [\bar{f}''] (\bar{r}'') >_{\bar{\gamma}} =$$
$$= \delta_{f'' \bar{f}''} \delta_{r'' \bar{r}''} \, \delta_{\gamma \bar{\gamma}},$$

$$\sum_{[f''] (r'') \gamma} < [f] (r'\rho) | [f'] (r'), [f''] (r'') >_\gamma < [f] (r'\bar{\rho}) | [f'] (r'), [f''] (r'') >_\gamma = \delta_{\rho \bar{\rho}}. \qquad (7.124)$$

The method of calculating the transformation matrices using the Young projection operators was proposed in the paper [Ka61]. In the absence of the multiplicities ($\nu_f = 1$) the corresponding formula was of the form

$$<[f](r'\rho)|[f'](r'),[f''](r'')> = \frac{<[f](r'\rho)|C^{[f'']}_{r''s''}|[f](r'\bar{\rho})>}{(<[f](r'\bar{\rho})|C^{[f'']}_{s''s''}|[f](r'\bar{\rho})>)^{1/2}}, \tag{7.125}$$

where $\bar{\rho}$ and S″ are fixed and chosen so that the denominator is nonzero. The Young operators are of a common form

$$C^{[f]}_{rs} = \frac{n_f}{N!}\sum_P <[f](r)|P|[f](s)> P,$$

where n_f is the dimension of IR $D^{\{f\}}$ of the group S_N, P is the element of the group S_N (an arbitrary permutation), $<[f](r)|P|[f](s)>$ is the matrix element of the operator P in the Young-Yamanouchi basis. The relations of the type (7.125) may be treated as the method of calculating the 3nf-symbols for the unitary groups with the help of the quantities in the formalism of the permutation group. In the case $N'' = 2$ the formula (7.125) determines the Racah coefficient. So, the currently available expressions for the transformation matrices at $N'' = 2$ enable one to calculate the Racah coefficients for any group U_m by the general formulas obtained by the substitution of (7.125) into (7.119):

$$U([\bar{f}'][1][f][1];[f'][2]) = \sqrt{\frac{f_i - f'_k - i + k + 1}{2(f_i - f'_k - i + k)}};$$

$$U([\bar{f}'][1][f][1];[f'][1^2]) = \epsilon_{ik}\sqrt{\frac{f_i - f'_k - i + k - 1}{2(f_i - f'_k - i + k)}}. \tag{7.126}$$

Here i is the number of the row of the Young scheme [f] for which the equality $f_i = f'_i + 1$ is valid; k is the number of the row satisfying the equality $f'_k = \bar{f}'_k + 1$; $\epsilon_{ik} = +1$ for $i > k$ and $\epsilon_{ik} = -1$ for $i < k$. If $i = k$, the first Racah coefficient (7.126) should be assumed as equal to + 1 and the second one, to be zero. Analogously, at $k = -1$ $f_i = f'_k$ the first coefficient (7.126) is zero and the second one, equal to + 1.

In the papers [Ka61; Ne79] a discussion is made of the wave-functions with reductions of the type $S_N \supset (S_{N'} \times S_{N''}) \times S_{N'''}$ and $S_N \supset S_{N'} \times (S_{N''} \times S_{N'''})$. The transition from the first type of reduction to the second one consists here merely in changing the scheme of multiplying three representations of the unitary group $([f'] \times [f'']) \times [f'''] \to [f'] \times ([f''] \times [f'''])$. Therefore, the transformation matrix for the transition between the above-methods of reduction S_N coincides with the Racah coefficients for the unitary group. The orthonormalization and symmetry properties of these and also more complex transformation matrices were studied by Kramer in the papers [Kr67] and monograph [Kr79]. Obviously, they are a direct consequence of ordinary properties of orthonormalization and symmetry of the 3nf-symbols for the unitary group. Turn back to the fractional parentage coefficients.

7.3.5 Factorization of the fractional parentage coefficients as the product of TM and CGC of the complementary groups S_N and U_m

It has been already noted that the formula (7.112) is the simplest example of fractional parentage expansion for separation of one particle in the function with a definite permutation and unitary symmetry $[f](r)\alpha$. Define now the fractional parentage expansion for separation of an arbitrary number of particles N'' (N = N' + N'') as a unitary transformation relating the basis $|\chi^N[f](r)\alpha>$ to the product
$|\chi^{N'}[f'](r')\alpha'>|\chi^{N''}[f''](r'')\alpha''>$:

$$|\chi^N[f](r)\alpha> = \sum_{[f''](r'')}\sum_{\alpha'\alpha''} <\chi^N[f](r)\alpha|\chi^{N'}[f'](r')\alpha',\chi^{N''}[f'']r''\alpha''> \times$$
$$\times\, |\chi^{N'}[f'](r')\alpha'>|\chi^{N''}[f''](r'')\alpha''>. \tag{7.127}$$

From the discussion made in section 3.3 it follows that the fractional parentage coefficients $<\chi^N|\chi^{N'},\chi^{N''}>$ at $N''\geqslant 2$ are no longer reduced only to CGC for the group U_m since the equality (7.127) implies the transition from the standard reduction $S_N\supset S_{N-1}\supset\ldots$ to the nonstandard reduction $S_N\supset S_{N'}\times S_{N''}$. Yet, if the transition to the nonstandard reduction has already been realized in the vector occuring in the left-hand side of (7.127) we immediately obtain, using CGC for the unitary group, the required expansion in the form (7.121). Comparing (7.123) and (7.127) and (7.121) we obtain the equality relating the fractional parentage coefficients $<\chi^N|\chi^{N'},\chi^{N''}>$ to the transformation matrices of the permutation group and to the Clebsch-Gordan coefficients of the unitary group:

$$<\chi^N[f](r)\alpha|\chi^{N'}[f'](r')\alpha',\chi^{N''}(r'')\alpha''> =$$
$$= <[f](r)|[f'](r'),[f''](r'')>_{\gamma(\alpha)}\left(\begin{matrix}[f'] & [f''] \\ \alpha' & \alpha''\end{matrix}\middle|\begin{matrix}[f] \\ \alpha\end{matrix}\right).. \tag{7.128}$$

In (7.128) it is implied that the additional quantum numbers γ, occuring in the equality (7.123) due to the multiplicity $\nu_f > 1$ are fully set by the system of "inner" quantum numbers α of the unitary group. This is the consequence of the complementarity (7.108): in the space *L* of the direct product of IR of the groups $S_N\times U_m$ a set of quantum numbers of these groups $[f](r)\alpha$ is complete, i.e. it unambigously defines the basis vector. Therefore it might be written in (7.121) $\gamma=\gamma(\alpha)$ (or $\alpha=\alpha(\gamma)$). Yet, the transformation matrices can be calculated in advance by some method which is not connected with the group U_m. In this case the choice of an additional quantum number γ will not be consistent with the system of quantum numbers α. Here, it is possible that the operators Γ and A corresponding to these quantum numbers will not commute $[\Gamma, A]\neq 0$. Then, an additional summation over γ will arise in the formulas (7.127) and (7.128) in accordance to the unitary transformations relating the two sets of quantum numbers

$$|\alpha> = \sum_\gamma U_{\alpha\gamma}|\gamma>,\ \sum_\gamma |U_{\alpha\gamma}|^2 = 1. \tag{7.129}$$

The formula for FPC is here of the form

$$<\chi^N|\chi^{N'},\chi^{N''}>\equiv<\chi^N[f](r)\alpha|\chi^{N'}[f'](r')\alpha',\chi''[f''](r'')\alpha''>=$$

$$=\sum_{\gamma} U_{\alpha\gamma}<[f](r)|[f'](r'),[f''](r'')>_\gamma \left(\begin{matrix}[f'] & [f''] \\ \alpha' & \alpha''\end{matrix}\middle|\begin{matrix}[f] \\ \alpha\end{matrix}\right). \quad (7.130)$$

7.3.6 Total fractional parentage coefficient

Our eventual aim here will be to calculate the fractional parentage coefficients $<q^N|q^{N'}, q^{N''}>$ for the expansion of the completely antisymmetric vector of the multiquark system state $\Psi_n(q^N)$ (7.9). Therefore, the expansions (7.127) and the FPC (7.130) play but a subsidiary role. It will be shown now that the eventual result contains only CGC of unitary groups. Let us have the fractional parentage expansions (7.127) for the orbital and CST-parts of the wave function (7.9):

$$\Phi^X=\sum_{[\bar f''](\bar r'')}\ \sum_{\omega'L'M'\omega''L''M''} <[f](r)|[f'](r'),[\bar f''](\bar r'')>X$$

$$\times\left(\begin{matrix}[f'] & [\bar f''] \\ \omega'L'M' & \omega''L''M''\end{matrix}\middle|\begin{matrix}[f\ ] \\ \omega LM\end{matrix}\right)|s^{N'_s}p^{N'_p}[f'](r')\omega'L'M'>|s^{N''_s}p^{N''_p}[\bar f''](\bar r'')\omega''L''M''>$$
(7.131)

$$\chi^{CST}=\sum_{[\tilde f''](\tilde r'')}\ \sum_{\alpha'\alpha''}<[\tilde f](\tilde r)|[\tilde f'](r'),[\tilde f''](r'')>\left(\begin{matrix}[\tilde f'] & [\tilde f''] \\ \alpha' & \alpha''\end{matrix}\middle|\begin{matrix}[\tilde f] \\ \alpha\end{matrix}\right)\times$$

$$\times|\chi^{N'}[\tilde f'](\tilde r')\alpha'>|\chi^{N''}[\tilde f''](\tilde r'')\alpha''>. \quad (7.132)$$

Here α is the set of the internal quantum number of the unitary group U_{12}^{CST} which were used to set the basis in the product of spaces $SU_3^C\times SU_2^S\times SU_2^T$; ωLM is the set of quantum numbers of the unitary group U_4^X which set the basis in four-dimensional space of orbital states s and p_m $m=0,\pm1$ (the disignations $\beta=L\omega M$, $\beta'=L'\omega'M'$, $\beta''=L''\omega''M''$ will be used henceforth for the sake of brevity). Substituting eqs. (7.131) and (7.132) in the expressions (7.9) for the completely antisymmetric state in the CSTX-space, we obtain the expansions

$$\Psi_n(q^N)\equiv|q^N[f]^X\beta,[\tilde f]^{CST}\alpha:[1^N]^{CSTX}>=$$

$$=\sum_{i=1}^{n_f}\frac{\Lambda_i}{\sqrt{n_f}}\sum_{[\bar f''](\bar r'')}\ \sum_{[\tilde f''](\tilde r'')}<[\tilde f](\tilde r_i)|[\tilde f'](\tilde r'),[\tilde f''](\tilde r'')><[f](r_i)|[f'](r'),[\bar f''](r'')>\times$$

$$\times\sum_{\alpha'\alpha''}\sum_{\beta'\beta''}\left(\begin{matrix}[\tilde f'] & [\tilde f''] \\ \alpha' & \alpha''\end{matrix}\middle|\begin{matrix}[\tilde f] \\ \alpha\end{matrix}\right)\left(\begin{matrix}[f'] & [\bar f''] \\ \beta' & \beta''\end{matrix}\middle|\begin{matrix}[f] \\ \beta\end{matrix}\right)\times$$

$$\times|s^{N'_s}p^{N'_p}[f'](r')\beta'>|s^{N''_s}p^{N''_p}[\bar f''](\bar r'')\beta''>|\chi^{N'}[\tilde f'](\tilde r')\alpha'>|\chi^{N''}[\tilde f''](\tilde r'')\alpha''>$$
(7.133)

Here, we have introduced the index $i = 1, 2, \ldots, n_f$ which makes the Yamanouchi symbols (r) ordered following the conventional rule [Ha64]:

$$r_1 < r_2 < \ldots < r_i < \ldots < r_{n_f} \tag{7.134}$$

where the inequality relations in (7.134) are written for the numerals of the symbol (r) which must be read as decimal number. For example, we have the following ordering of the Yamanouchi symbols for the Young schema [31]:

$$\begin{array}{ccc}
\begin{array}{|c|c|c|}\hline 1&2&3\\ \hline 4 \\ \cline{1-1}\end{array} &
\begin{array}{|c|c|c|}\hline 1&2&4\\ \hline 3 \\ \cline{1-1}\end{array} &
\begin{array}{|c|c|c|}\hline 1&3&4\\ \hline 2 \\ \cline{1-1}\end{array} \\
(r_1) = (1112) & (r_2) = (1121) & (r_3) = (1211) \\
\Lambda_1 = +1 & \Lambda_2 = -1 & \Lambda_3 = +1.
\end{array}$$

According to [Ha64], the phase factor $\Lambda_i = +1$ for $i = 1$ and for all the symbols (r_i) which were obtained from (r_1) by means of even permutation of the cells in the standard Young table; $\Lambda_i = -1$ for the symbols (r_i) obtained from (r_1) by odd permutation of the cells in the Young scheme. It can readily be seen that the ordering of (7.134) is reversed when using the conjugate Young schemes

$$\tilde{r}_1 > \tilde{r}_2 > \ldots > \tilde{r}_{n_f}. \tag{7.136}$$

For example, in case of (7.135), we get for $[\tilde{f}] = [21^2]$:

$$\begin{array}{ccc}
\begin{array}{|c|c|}\hline 1&2\\ \hline 3 \\ \cline{1-1} 4 \\ \cline{1-1}\end{array} &
\begin{array}{|c|c|}\hline 1&3\\ \hline 2 \\ \cline{1-1} 4 \\ \cline{1-1}\end{array} &
\begin{array}{|c|c|}\hline 1&4\\ \hline 2 \\ \cline{1-1} 3 \\ \cline{1-1}\end{array} \\
(r_1) = (1123) = (\tilde{r}_3) & (r_2) = (1213) = (\tilde{r}_2) & (r_3) = (1231) = (\tilde{r}_1) \\
\Lambda_1 = +1 & \Lambda_2 = -1 & \Lambda_3 = +1.
\end{array} \tag{7.137}$$

The phase factors Λ_i fix a certain agreement in the choice of the phases of the basis IR functions of the permutation group S_N [Ha64]. When the states with conjugate Young schemes $[f](r_i) \leftrightarrow [\tilde{f}](\tilde{r}_i)$ are introduced, many of the relations between the basis vectors remain invariable accurate to within the phase factors Λ_i. The following symmetry properties of CGC of the S_N group are presented in [Ha64]:

$$\left[\begin{array}{c|c} [\tilde{f}'][\tilde{f}''] & [f] \\ (\tilde{r}_i')(\tilde{r}_j'') & (r_k) \end{array}\right] = \Lambda_i' \Lambda_j'' \left[\begin{array}{c|c} [f'][f''] & [f] \\ (r_i')(r_j'') & (r_k) \end{array}\right]$$

$$\left[\begin{array}{c|c} [f'][\tilde{f}''] & [\tilde{f}] \\ (r_i')(\tilde{r}_j'') & (\tilde{r}_k) \end{array}\right] = \Lambda_j'' \Lambda_k \left[\begin{array}{c|c} [f'][f''] & [f] \\ (r_i')(r_j'') & (r_k) \end{array}\right] \tag{7.138}$$

The symmetry properties of the transformation matrices (for a given choice of phase factors), which do not contradict the relations (7.138) are quite obvious:

$$< [\tilde{f}](\tilde{r}_k) | [\tilde{f}'](\tilde{r}_i'), [\tilde{f}''](\tilde{r}_j'') > = \Lambda_i' \Lambda_j'' \Lambda_k < [f](r_k) | [f'](r_i'), [f''](r_j'') >. \tag{7.139}$$

Another interesting relation follows from eq. (7.139) and from the orthogonality relations (7.124)

$$\sum_\rho \Lambda_k < [f](r_k = r_i'\rho) | [f'](r_i'), [\tilde{f}''](\tilde{r}_j'') > < [\tilde{f}](\tilde{r}_k = \tilde{r}_i'\rho) | [\tilde{f}'](\tilde{r}_i'), [\tilde{f}''](\tilde{r}_j) > =$$

$$= \Lambda_i' \Lambda_j'' \delta_{\bar{f}''f''} \delta_{\bar{r}''r''}, \tag{7.140}$$

which permits the expansion (7.133) to be significantly simplified. After making identical transformations which introduce CGC $\frac{\Lambda'}{\sqrt{n_{f'}}}$ and $\frac{\Lambda''}{\sqrt{n_{f''}}}$, in the right part of eq. (7.133), the antisymmetric states $[1N']^{CSTX}$ and $[1N']^{CSTX}$ will be formed

$$\frac{\Lambda}{\sqrt{n_f}} \equiv \sqrt{\frac{n_{f'} n_{f''}}{n_f}} \frac{\Lambda'}{\sqrt{n_{f'}}} \frac{\Lambda''}{\sqrt{n_{f''}}} (\Lambda \Lambda' \Lambda''); \sum_{i=1}^{n_f} \equiv \sum_{[f'](r'\rho)} \equiv \sum_{[f'](r')} \sum_{\rho}. \tag{7.141}$$

The summation over ρ gives rise to the Kronecker delta symbols (7.140) which are then summed up over $[\tilde{f}''](\tilde{r}'')$. Obtained eventually in the right part of (7.133) is the following fractional parentage expansion in the completely antisymmetric states in the N' and N'' subsystems:

$$\Psi_n(q^N) \equiv |q^N [f]^X \beta, [\tilde{f}]^{CST} \alpha : [1^N]^{CSTX} > =$$

$$= \sum_{[f'],[f'']} \sqrt{\frac{n_{f'} n_{f''}}{n_f}} \sum_{\alpha'\alpha''} \sum_{\beta'\beta'} \left(\begin{matrix} [\tilde{f}'] & [\tilde{f}''] \\ \alpha' & \alpha'' \end{matrix} \middle| \begin{matrix} [\tilde{f}] \\ \alpha \end{matrix} \right) \left(\begin{matrix} [f'] & [f''] \\ \beta' & \beta'' \end{matrix} \middle| \begin{matrix} [f] \\ \beta \end{matrix} \right) \times$$

$$\times |q^{N'} [f']^X \beta', [\tilde{f}']^{CST} \alpha' : [1^{N'}]^{CSTX} > | q^{N''} [f'']^X \beta'', [\tilde{f}'']^{CST} : [1^{N''}]^{CSTX} >. \tag{7.142}$$

Thus, we have obtained that the total fractional parentage coefficient for the antisymmetric state in the product of the spaces $U^a_{m_1} \times U^b_{m_2}$ equals the product of the CGC for groups $U^a_{m_1}$ and $U^b_{m_2}$ by each other and by the the weight factor $\sqrt{\frac{n_{f'} n_{f''}}{n_f}}$

$$< q^N | q^{N'}; q^{N''} > \equiv$$

$$\equiv < q^N [1^N] ([\tilde{f}]^a \alpha, [f]^b \beta) | q^{N'} [1^{N'}] ([\tilde{f}']^a \alpha', [f']^b \beta'), q^{N''} [1^{N''}] ([\tilde{f}'']^a \alpha'', [f'']^b \beta'') > =$$

$$= \sqrt{\frac{n_{f'} n_{f''}}{n_f}} \left(\begin{matrix} [\tilde{f}']^a & [\tilde{f}'']^a \\ \alpha' & \alpha'' \end{matrix} \middle| \begin{matrix} [\tilde{f}]^a \\ \alpha \end{matrix} \right) \left(\begin{matrix} [f']^b & [f'']^b \\ \beta' & \beta'' \end{matrix} \middle| \begin{matrix} [f]^b \\ \beta \end{matrix} \right). \tag{7.143}$$

Considered below will be the problem of calculating CGC for U_m. We follow here the papers [Ob79b, Ob81]. (See also [So79] in the case of $SU_6 \supset SU_3 \times SU_2$)

7.3.7 Calculations of CGC for U_m using the formalism of the permutation group S_N

In the previous Section, we reduced the total fractional parentage coefficient to the product of CGC for unitary groups. Presented here will be the technique of calculations of CGC for the unitary group U_m [Ob79b, Ob81]. It is well known that CGC for the high-rank groups may be factorized by separating out the scalar factor (SF) and CGC for a lower rank subgroup (in our case, SU_2 or O_3 will be used as such a lower-rank subgroup).

7.3.7.1 U_3: the simple example. Examine first the simplest version of factorization using the colour part of the wave function with symmetry U^C_3 as an example (it will be noted that the sole colour Young scheme $[2^3]^C$ is admissible if the total colour of the six-quark system is zero). If the colour moment (C, C_z) and the "multiplicity index" ω_c (see, eq.

(7.122)) are used to number the rows of IR $D^{(\lambda\mu)}$ of the SU_3^C group (in accordance with the reduction chain $SU_3^C \supset O_3^C \supset O_2^C$), then CGC for the SU_3 group, which are necessary to make the fractional parentage expansion of the colour function

$$|c^N[f_C](r_C)\,\omega_C CC_z> = \sum_{[f_C''](r_C'')} \sum_{\omega_C' C' C_z'} \sum_{\omega_C'' C'' C_z''} \times$$

$$\times <c^N[f_C](r_C)\,\omega_C CC_z|c^{N'}[f_C'](r_C')C'C_z',\ c^{N''}[f_C''](r'')\,\omega_C'' C'' C_z''>_\gamma \times$$

$$\times |c^{N'}[f_C'](r_C' C' C_z')> |c^{N''}[f_C''](r_C'')\, C'' C_z''>$$

[where

$$<c^N|c^{N'};c^{N''}>_\gamma = <[f_C](r_C)|[f_C'](r_C'),[f_C'']r_C''>_\gamma \begin{pmatrix} [f_C'] & [f_C''] & [f_C] \\ \omega_C' C' C_z' & \omega_C'' C'' C_z'' & \omega_C CC_z \end{pmatrix}_\gamma$$

are of the form (7.144)

$$\begin{pmatrix} [f_C'] & [f_C''] & [f_C] \\ \omega_C' C' C_z' & \omega_C'' C'' C_z'' & \omega_C CC_z \end{pmatrix}_\gamma = \left\langle \begin{matrix} [f_C'] & [f_C''] \\ \omega_C' C' & \omega_C'' C'' \end{matrix} \middle\| \begin{matrix} [f_C] \\ \omega_C C \end{matrix} \right\rangle_\gamma (C'C_z'C''C_z''|CC_z). \tag{7.145}$$

In the right part of equality (7.145), we used the CGC factorization property [Ra 51] and factorized CGC for U_3 into CGC for three-dimensional rotational group O_3 and SF $\left\langle \begin{matrix} [f_C'] & [f_C''] \\ \omega_C' C' & \omega_C'' C'' \end{matrix} \middle\| \begin{matrix} [f_C] \\ \omega_C C \end{matrix} \right\rangle_\gamma$ which is independent of quantum numbers C_z', C_z'', C_z (i.e. projections of coloured momentum) and invariant with respect to the O_3^C subgroup. In their essence, these SF are the fractional parentage coefficients in the colour space when the expansion (7.144) is written in the form

$$|c^N[f_C](r_C)\,\omega_C C>_\gamma = \sum_{[f_C''](r_C'')} <[f_C](r_C)|[f_C'](r_C'),[f_C''](r_C'')>_\gamma \times$$

$$\times \sum_{\omega_C' C' \omega_C'' C''} \left\langle \begin{matrix} [f_C'] & [f_C''] \\ \omega_C' C' & \omega_C'' C'' \end{matrix} \middle\| \begin{matrix} [f_C] \\ \omega_C C \end{matrix} \right\rangle_\gamma \left\{ |c^{N'}[f_C'](r_C')\,\omega_C' C'> |c^{N''}[f_C''](r_C'')\,\omega_C'' C''> \right\}_C \tag{7.146}$$

SF of CGC for the U_3 group were tabulated in [Ja 51; El 53] for $N'' = 1, 2$, and in [Ch 64; Ro 65] for $N'' = 3, 4$. Table 7.4 presents SF for the Young scheme $[2^3]^C$ in which neither the index ω_c nor the additional quantum number γ are necessary because of the absence of multiplicity.

7.3.7.2 Scalar factors (SF) of CGC of U_m group for the reduction $U_m^{ab} \supset U_{m_1}^a \times U_{m_2}^b$, $m = m_1 m_2$. The reduction of the form

$$U_m^{ab} \supset U_{m_1}^a \times U_{m_2}^b,\ m = m_1 m_2 \tag{7.147}$$

is used in all links of the reduction chain of the subgroups

$$U_{48}^{CSTX} \supset U_{12}^{CST} \times U_4^X,$$

$$U_{12}^{CST} \supset U_6^{CS} \times U_2^T \supset U_3^C \times U_2^S \times U_2^T \supset SU_3^C \times SU_2^S \times SU_2^T \tag{7.148}$$

and corresponds to the "physical" set of quantum numbers (a, b) = (CST, X), (CS, T), (C, S).

Let CGC for the $U^a_{m_1}$ and $U^b_{m_2}$ subgroups be known:

$$|a^N[f_a](r_a)\alpha>_{\gamma_a} = \sum_{[f_a''](r_a'')} <[f_a](r_a)|[f_a'](r_a'),[f_a''](r_a'')>_{\gamma_a} \sum_{\alpha'\alpha''} \left(\begin{matrix}[f_a'] & [f_a''] \\ \alpha' & \alpha''\end{matrix}\middle|\begin{matrix}[f_a] \\ \alpha\end{matrix}\right)_{\gamma_a} \times$$
$$\times |a^{N'}[f_a'](r_a')\alpha'>|a^{N''}[f_a''](r_a'')\alpha''>,$$

$$|b^N[f_b](r_b)\beta>_{\gamma_b} = \sum_{[f_b''](r_b'')} <[f_b](r_b)|[f_b'](r_b'),[f_b''](r_b'')>_{\gamma_b} \sum_{\beta'\beta''} \left(\begin{matrix}[f_b'] & [f_b''] \\ \beta' & \beta''\end{matrix}\middle|\begin{matrix}[f_b] \\ \beta\end{matrix}\right)_{\gamma_b} \times$$
$$\times |b^{N'}[f_b'](r_b')\beta'>|b^{N''}[f_b''](r_b'')\beta''>, \tag{7.149}$$

where

$$(r_a) = (r_a'\rho_a), \quad (r_b) = (r_b'\rho_b). \tag{7.150}$$

Now we can calculate CGC for the unitary group $U^{ab}_m \supset U^a_{m_1} \times U^b_{m_2}$ by expanding the space of the direct composition of the vectors (7.149) $|a^N>|b^N>$ in the IR of the group U^{ab}_m. The basis vectors exhibiting certain permutational (S_N) and unitary (U_m) symmetry $[f_{ab}](r_{ab})$ will be obtained by multiplying the states $[f_a](r_a)$ and $[f_b](r_b)$ by CGC of the S_N group

$$|(ab)^N[f_{ab}](r_{ab})([f_a]\alpha,[f_b]\beta)\,\omega_{ab}>_{\gamma_a\gamma_b} =$$
$$= \sum_{(r_a)(r_b)} \left(\begin{matrix}[f_a] & [f_b] \\ (r_a) & (r_b)\end{matrix}\middle|\begin{matrix}[f_{ab}] \\ (r_{ab})\end{matrix}\right)_{\omega_{ab}} |a^N[f_a](r_a)\alpha>_{\gamma_a}|b^N[f_b](r_b)\beta>_{\gamma_b}. \tag{7.151}$$

In such a way, we set the basis vectors of IR $D^{[f_{ab}]}$ of the U^{ab}_m group using the quantum numbers of the $U^a_{m_1} \times U^b_{m_2}$ subgroups and the multiplicity index ω_{ab} which numbers the multiple representations ($\nu_f > 1$) in the Clebsch-Gordon series for the inner product of the Young schemes [Ha64]

$$[f_a]\cdot[f_b] = \sum_{f_{ab}} \nu_f[f_{ab}]. \tag{7.152}$$

It should be noted that the procedure for calculating CGC of the permutation group S_N has been described in [Ha64]. The method is used in which the additional quantum numbers ω_{ab} may be introduced in any arbitrary manner; the only requirement is that the orthogonality condition for the indices ω_{ab} be satisfied:

$$\sum_{(r_a)(r_b)} \left[\begin{matrix}[f_a] & [f_b] \\ (r_a) & (r_b)\end{matrix}\middle|\begin{matrix}[f_{ab}] \\ (r_{ab})\end{matrix}\right]_{\omega_{ab}} \left[\begin{matrix}[f_a] & [f_b] \\ (r_a) & (r_b)\end{matrix}\middle|\begin{matrix}[\bar f_{ab}] \\ (\bar r_{ab})\end{matrix}\right]_{\bar\omega_{ab}} = \delta_{f_{ab}\bar f_{ab}}\,\delta_{r_{ab}\bar r_{ab}}\,\delta_{\omega_{ab}\bar\omega_{ab}};$$

$$\sum_{[f_{ab}](r_{ab})\omega_{ab}} \left[\begin{matrix}[f_a] & [f_b] \\ (r_a) & (r_b)\end{matrix}\middle|\begin{matrix}[f_{ab}] \\ (r_{ab})\end{matrix}\right]_{\omega_{ab}} \left[\begin{matrix}[f_a] & [f_b] \\ (\bar r_a) & (\bar r_b)\end{matrix}\middle|\begin{matrix}[f_{ab}] \\ (r_{ab})\end{matrix}\right]_{\omega_{ab}} = \delta_{r_a\bar r_a}\delta_{r_b\bar r_b}. \tag{7.153}$$

Let now CGC for the U_m^{ab} group be introduced. With this purpose, we shall use the complementarity of representations of the U_m^{ab} and S_N groups in the space L $(\underbrace{|ab>|ab>\ldots|ab>}_{N})$ thereby permitting the basis vectors (7.151) to be constructed again (using in this case, however, the vectors $|(ab)^{N'}>|(ab)^{N''}>$) as the states with definite unitary symmetry. Using CGC of the U_m^{ab} group

$$\begin{pmatrix} [f'_{ab}]\,[f''_{ab}] & [f_{ab}] \\ Q'_{ab}\;\; Q''_{ab} & Q_{ab} \end{pmatrix}_{\gamma_{ab}},$$

$$|(ab)^N[f_{ab}](r_{ab})\tilde{Q}_{ab}>_{\gamma_{ab}} = \sum_{[\bar{f}''_{ab}](\bar{r}''_{ab})} <[f_{ab}](r_{ab})|[f'_{ab}](r'_{ab}),[\bar{f}''_{ab}](\bar{r}''_{ab})>_{\gamma_{ab}} \times$$

$$\times \sum_{\omega_{ab}} U_{\omega_{ab}\gamma_{ab}} \sum_{Q'_{ab}\bar{Q}''_{ab}} \begin{pmatrix} [f'_{ab}]\,[\bar{f}''_{ab}] & [f_{ab}] \\ Q'_{ab}\;\; \bar{Q}''_{ab} & Q_{ab} \end{pmatrix}_{\gamma_{ab}} |(ab)^{N'}[f'_{ab}](r'_{ab})Q'_{ab}>|(ab)^{N''}[\bar{f}''_{ab}](\bar{r}''_{ab})\bar{Q}''_{ab}>, \tag{7.154}$$

where $(r_{ab}) = (r'_{ab}\rho_{ab})$. Here, Q_{ab}, Q'_{ab}, Q''_{ab}, $\tilde{Q}_{ab}$ are the quantum numbers which determine the IR rows for the $U_m^{ab} \supset U_{m_1}^a \times U_{m_2}^b$ group:

$$Q_{ab} = ([f_a]\alpha,[f_b]\beta)\,\omega_{ab},\; Q'_{ab} = ([f'_a]\alpha',[f'_b]\beta')\,\omega'_{ab},\; \bar{Q}''_{ab} = ([f''_a]\alpha'',[f''_b]\beta'')\,\bar{\omega}''_{ab},$$

In other word, we have assumed that the states $|(ab)^{N'}>$ and $|(ab)^{N''}>$ in (7.154) are constructed out of the vectors $|a^{N'}>|b^{N'}>$ and $|a^{N''}>|b^{N''}>$ using the formulas similar to (7.151):

$$|(ab)^{N'}[f'_{ab}](r'_{ab})Q'_{ab}> \equiv |(ab)^{N'}[f'_{ab}](r'_{ab})([f'_a]\alpha',[f'_b]\beta')\,\omega'_{ab}> =$$

$$= \sum_{(r'_a)(r'_b)} \begin{bmatrix} [f'_a]\,[f'_b] & [f'_{ab}] \\ (r'_a)\,(r'_b) & (r'_{ab}) \end{bmatrix}_{\omega'_{ab}} |a^{N'}[f'_a](r'_a)\alpha'>|b^{N'}[f'_b](r'_b)\beta'>,$$

$$|(ab)^{N''}[\bar{f}''_{ab}](\bar{r}''_{ab})\bar{Q}''_{ab}> \equiv |(ab)^{N''}[\bar{f}''_{ab}](\bar{r}''_{ab})([f''_a]\alpha'',[f''_a]\beta'')\,\bar{\omega}''_{ab}> =$$

$$= \sum_{(r''_a)(r''_b)} \begin{bmatrix} [f''_a]\,[f''_b] & [\bar{f}''_{ab}t] \\ (r''_a)\,(r''_b) & (\bar{r}''_{ab}) \end{bmatrix} \sum_{\bar{\omega}''_{ab}} |a^{N''}[f''_a](r''_a)\alpha''>|b^{N''}[f''_b](r''_b)\beta''>. \tag{7.155}$$

The basises (7.151) and (7.154) are complete in the examined space and coincide with each other to within the multiplicity indices γ_{ab}, ω_{ab}. By expanding the vector (7.154) in the states (7.151), we obtain the equality:

$$|(ab)^N[f_{ab}](r_{ab})\tilde{Q}_{ab}>_{\gamma_{ab}} = \sum_{\omega_{ab}} U_{\omega_{ab}\gamma_{ab}}|(ab)^N[f_{ab}](r_{ab})Q_{ab}>_{\gamma_a\gamma_b} \tag{7.156}$$

which may be used as equation for determining CGC for the unitary group $U_m^{ab} \supset U_{m_1}^a \times U_{m_2}^b$ if the matrix elements $U_{\omega_{ab}\gamma_{ab}}$ of unitary transformation from the index ω_{ab} to the index γ_{ab} are known:

$$U_{\omega_{ab}\gamma_{ab}} = <\omega_{ab}|\gamma_{ab}>,\; \sum_{\omega_{ab}} |U_{\omega_{ab}\gamma_{ab}}|^2 = 1 \tag{7.157}$$

The sought equation for CGC for the U_m^{ab} group will be obtained by substituting the expansions (7.154) and (7.151) in the left and right sides of the equality (7.156) respectively. After that, the expressions (7.155) for the vectors $|(ab)^{N'}>$ and $|(ab)^{N''}>$ and (7.149) and (7.150) for the vectors $|a^N>$ and $|b^N>$ will be used. In such a way, we shall obtain in the left and right parts of the equality (7.156) the expansions in the elementary basis of the form

$$|a^{N'}>|a^{N''}>|b^{N'}>|b^{N''}>\equiv|a^{N'}[f_a'](r_a')\alpha'>|a^{N''}[f_a''](r_a'')\alpha''>|b^{N'}[f_b'](r_b')\beta'> |b^{N''}[f_b''](r_b'')\beta''>. \tag{7.158}$$

By equating the coefficients at the indentical basis vectors (7.158) in the right and left parts of the equality (7.156), we obtain the following set of algebraic equations for CGC of U_m^{ab}

$$\sum_{[\bar f_{ab}''](\bar r_{ab}'')\bar\omega_{ab}''} <[f_{ab}](r_{ab})|[f_{ab}'](r_{ab}'),[\bar f_{ab}''](\bar r_{ab}'')>_{\gamma_{ab}} \sum_{\omega_{ab}} U_{\omega_{ab}\gamma_{ab}}$$

$$\left(\begin{matrix}[f_{ab}'] & [f_{ab}''] \\ ([f_a']\alpha',[f_b']\beta')\omega_{ab}' & ([f_a'']\alpha'',[f_b'']\beta'')\bar\omega_{ab}''\end{matrix}\middle|\begin{matrix}[f_{ab}] \\ ([f_a]\alpha,[f_b]\beta)\omega_{ab}\end{matrix}\right)_{\gamma_{ab}} \times$$

$$\times \begin{bmatrix}[f_a'] & [f_b'] & [f_{ab}'] \\ (r_a') & (r_b') & (r_{ab}')\end{bmatrix}_{\omega_{ab}'} \begin{bmatrix}[f_a''] & [f_b''] & [\bar f_{ab}''] \\ (r_a'') & (r_b'') & (\bar r_{ab}'')\end{bmatrix}_{\bar\omega_{ab}''} =$$

$$= \sum_{\rho_a\rho_b} \begin{bmatrix}[f_a] & [f_b] & [f_{ab}] \\ (r_a) & (r_b) & (r_{ab})\end{bmatrix}_{\omega_{ab}} \left(\begin{matrix}[f_a'] & [f_a''] \\ \alpha' & \alpha''\end{matrix}\middle|\begin{matrix}[f_a] \\ \alpha\end{matrix}\right)_{\gamma_a} \left(\begin{matrix}[f_b'] & [f_b''] \\ \beta' & \beta''\end{matrix}\middle|\begin{matrix}[f_b] \\ \beta\end{matrix}\right)_{\gamma_b} \times$$

$$\times <[f_a](r_a)|[f_a'](r_a'),[f_a'']r_a''>_{\gamma_a} <[f_b](r_b)|[f_b'](r_b'),[f_b''](r_b'')>_{\gamma_b}. \tag{7.159}$$

It will be noted that the sum over $\rho_a\rho_b$ at the right is what is left after the summation over (r_a) and (r_b) in (7.151). It is quite obvious that the solution for the equations (7.159) may be sought in factorized form be determining SF of CGC from the relation

$$U_{\omega_{ab}\gamma_{ab}}\left(\begin{matrix}[f_{ab}'] & [f_{ab}''] \\ Q_{ab}' & Q_{ab}''\end{matrix}\middle|\begin{matrix}[f_{ab}] \\ Q_{ab}\end{matrix}\right) \equiv$$

$$\equiv \left(\begin{matrix}[f_{ab}'] & [f_{ab}''] \\ ([f_a']\alpha',[f_b']\beta')\omega_{ab}' & ([f_a'']\alpha'',[f_b'']\beta'')\omega_{ab}''\end{matrix}\middle|\begin{matrix}[f_{ab}] \\ ([f_a]\alpha,[f_b]\beta)\omega_{ab}\end{matrix}\right)^{\gamma_{ab}}_{\gamma_a\gamma_b} =$$

$$= \left\langle\begin{matrix}[f_{ab}'] & [f_{ab}''] \\ ([f_a'],[f_b'])\omega_{ab}' & ([f_a''],[f_b''])\omega_{ab}''\end{matrix}\middle\|\begin{matrix}[f_{ab}] \\ ([f_a],[f_b])\omega_{ab}\end{matrix}\right\rangle^{\gamma_{ab}}_{\gamma_a\gamma_b} \times \tag{7.160}$$

$$\times \left(\begin{matrix}[f_a'] & [f_a''] \\ \alpha' & \alpha''\end{matrix}\middle|\begin{matrix}[f_a] \\ \alpha\end{matrix}\right)_{\gamma_a} \left(\begin{matrix}[f_b'] & [f_b''] \\ \beta' & \beta''\end{matrix}\middle|\begin{matrix}[f_b] \\ \beta\end{matrix}\right)_{\gamma_b}$$

After substituting the relation (7.160) in eq. (7.159), CGC for $U_{m_1}^a$ and $U_{m_2}^b$ may be cancelled in the right and left sides of the equation. As a result, we get the set of SF equations where the coefficients of the equations are expressed through only the values from the formalism of the permutation group, namely through the tranformational matrices and CGC. Let another simplification be made in eq. (7.159). The left and right parts of the equation will be multiplied by CGC for $S_{N''}$

$$\begin{pmatrix} [f''_a]\,[f''_b] & [f''_{ab}] \\ (r''_a)\,(r''_b) & (r''_{ab}) \end{pmatrix}_{\omega''_{ab}}$$

and then summarized over the Yamanouchi symbols (r''_a) and (r''_b)). As a result of this, and becuase of the orthogonality of the expressions (7.153), the left part of the equation will contain the Kronecker delta symbols $\delta_{f''_{ab}\bar{f}''_{ab}}\,\delta_{r''_{ab}\bar{r}''_{ab}}\,\delta_{\omega''_{ab}\bar{\omega}''_{ab}}$ which can trivially be summarized over $[\bar{f}''_{ab}](\bar{r}''_{ab})\,\bar{\omega}''_{ab}$. In such a way, the summation in the right part of eq. (7.159) disappears and we get the explicit form of the solution:

$$\left\langle \begin{matrix} [f'_{ab}] & [f''_{ab}] \\ ([f'_a],[f'_b])\,\omega'_{ab} & ([f''_a],[f''_b])\,\omega''_{ab} \end{matrix} \middle\| \begin{matrix} [f_{ab}] \\ ([f_q],[f_b])\,\omega_{ab} \end{matrix} \right\rangle^{\gamma_{ab}}_{\gamma_a\gamma_b} =$$

$$= \sum_{(r''_a),(r''_b)} \begin{bmatrix} [f''_a]\,[f''_b] & [f''_{ab}] \\ (r''_a)\,(r''_b) & (r''_{ab}) \end{bmatrix} \sum_{\rho_a\rho_b} \frac{\begin{bmatrix} [f_a]\,[f_b] & [f_{ab}] \\ (r_a)\,(r_{b'}) & (r_{ab}) \end{bmatrix}_{\omega_{ab}}}{\begin{bmatrix} [f'_a]\,[f'_b] & [f'_{ab}] \\ (r'_a)\,(r'_b) & (r'_{ab}) \end{bmatrix}_{\omega'_{ab}}} \times$$

$$\times \frac{<[f_a](r_a)|[f'_a](r'_a),[f''_a](r''_a)>_{\gamma_a} <[f_b](r_b)|[f'_b](r'_b),[f''_b](r''_b)>_{\gamma_b}}{<[f_{ab}](r_{ab})|[f'_{ab}](r'_{ab}),[f''_{ab}](r''_{ab})>_{\gamma_{ab}}}, \tag{7.161}$$

where $(r_a) = (r'_a\rho_a)$, $(r_b) = (r'_b\rho_b)$, $(r_{ab}) = (r'_{ab}\rho_{ab})$. It can easily be seen that the right hand part of (7.161) is independent of particular selection of the Yamanouchi symbols (r'_a), (r'_b), (r'_{ab}). The only requirement ist that CGC $\begin{bmatrix} [f'_a]\,[f'_b] & [f'_{ab}] \\ (r'_a)\,(r'_b) & (r'_{ab}) \end{bmatrix}_{\omega'_{ab}}$ be not vanishing at the selected values of (r'_a), (r'_b), (r'_{ab}).

It will be noted that the formula (7.161) (in combination with the representation (7.160)) is the main result of the present Section. This formula can be used to calculate CGC of any unitary groups from the reduction series (7.148).

7.3.7.3 Some of the results. Fractional parentage expansion of six-quark states. It follows from the formulas (7.160) and (7.161) that the factor of CGC of the group U^{ab}_m $(m = m_1 m_2)$ scalar with respect to $U^a_{m_1} \times U^b_{m_2}$ can completely be expressed through the permutational values (transformational Kaplan matrices and CGC for the S_N group). Therefore, these SF are universal for all the unitary groups and prove to be independent of the ranks of the groups and determined exclusively by the form of the Young scheme comprised in them. This fact was known earlier for only the special case of completely symmetric representations [Ma67]. At present, this fact haş been proved in the general form and makes it possible, for example, to use the spin-isospin FPC corresponding to the reduction $U^{ST}_4 \supset U^S_2 \times U^T_2$ as SF of CGC for the unitary groups of any rank, $m \geqslant 2$ (if the Young schmes $[f_a]$, $[f_b]$ comprise not more than tworows, while the scheme $[f_{ab}]$ comprises not more than fourrows). After that, using the symmetry properties (7.138) and (7.139) of the transformational matrices and CGC of the S_N group when going to the conjugate Young schemes $[f](r) \to [\widetilde{f}](\widetilde{r})$, we can obtain all the values for the two-column and four-column Young schemes of interest to us proceeding from the two-row and four-row Young schemes.

It is clear that this procedure exhaust all the possible versions which may be found in the six-quark system. It will be noted that the matrices $U_{\omega_{ab}\gamma_{ab}}$ will not be practically required in the six-quark system. In fact, the indices γ meant the multiplicity indices in the outer products (7.122) of the Young schemes (on the contrary, the indices ω are the multiplicity indices in the inner products (7.152) of the Young schemes). In the six-quark system, the multiple representation appears in but a single outer product $[21]\times[21]\to 2\,[321]$, i.e. the matrices $U_{\omega_{ab}\gamma_{ab}}$ may have been required only when $[f_a]=[321]$, $[f_b]=[321]$ and $[f_{ab}]=[321]$ respectively. However, we shall deal with the Young scheme [321] in only the CS-space; at the same time, it is impossible for the Young scheme [321] to be also in the C- and S-space.

Summarizing all the results obtained in this Section, we shall write the following eventual expression for the total fractional parentage coefficient (7.143) in the multi-quark system whose wave function is defined by the set of quantum numbers from the reduction group series (7.148) (in accordance with the above, we omit the multiplicity indices γ as applied to the 6q-system):

$$< q^N [1^N]^{CSTX} ([\tilde{f}]^{CST} Q_{CST}, [f]^X \omega_X LM) | q^{N'} [1^{N'}]^{CSTX} ([\tilde{f}']^{CST} Q'_{CST}, \times$$

$$\times [f']^X \omega'_X L'M'), q^{N''} [1^{N''}]^{CSTX} ([\tilde{f}'']^{CST} Q''_{CST}, [f'']^X \omega''_X L''M'') > =$$

$$= \sqrt{\frac{n_{f'} n_{f''}}{n_f}} \left\langle \begin{matrix} [f']^X & [f'']^X \\ L'\omega'_X & L''\omega''_X \end{matrix} \middle\| \begin{matrix} [f]^X \\ L\omega_X \end{matrix} \right\rangle (L'M'L''M''|LM) \times$$

$$\times \left\langle \begin{matrix} [\tilde{f}']^{CST} & [\tilde{f}'']^{CST} \\ ([h'_{CS}], T)\,\omega'_{CST} & ([h''_{CS}], T'')\,\omega''_{CST} \end{matrix} \middle\| \begin{matrix} [\tilde{f}]^{CST} \\ ([h_{CS}], T)\,\omega_{CST} \end{matrix} \right\rangle \times$$

$$\times \left\langle \begin{matrix} [h'_{CS}] & [h''_{CS}] \\ ([g'_C], S)\,\omega'_{CS} & ([g''_C], S'')\,\omega''_{CS} \end{matrix} \middle\| \begin{matrix} [h_{CS}] \\ ([g_C], S)\,\omega_{CS} \end{matrix} \right\rangle \times$$

$$\times \left\langle \begin{matrix} [g'_C] & [g''_C] \\ C'\omega'_C & C''\omega''_C \end{matrix} \middle\| \begin{matrix} [g_C] \\ C\omega_C \end{matrix} \right\rangle (C'C'_z\, C''C''_z | C\, C_z) \times$$

$$\times (S'S'_z S''S''_z | S\, S_z)(T'T'_z T''T''_z | T\, T_z), \tag{7.162}$$

where $Q_{CST} = ([h_{CS}]([g_C](C\,C_z)\,\omega_C, (S\,S_z))\,\omega_{CS}, (T\,T_z))\,\omega_{CST}$ are the quantum numbers in CST-space.

We used the formula (7.161) to calculate SF in the CS- and CST-spaces (SF_{CS}, SF_{CST}) for two lower six-quark states in the channel (S, T) = (1.0)

$$|s^6 [6]^X L=0, [2^3]^C\ C=0\ S=1\ [2^3]^{CS}\ T=1\ [1^6]^{CST} > \tag{7.163}$$

$$|s^4 p^2\ [42]^X L=0, [2^3]^C\ C=0\ S=1\ [42]^{CS}\ T=1\ [2^2 1^2]^{CST} >. \tag{7.164}$$

The results obtained are presented in Tables 7.5, 7.6, 7.7 and 7.8. It will be noted that the multiplicity indices ω were not used at all, so we refrained from writing their explicit form.

7.3.8 Orbital states

It was emphasized in the Introduction to the present Chapter that the wave functions of multiquark systems for the various confinement models differ from each other by only the orbital part, whereas the CST-part of the wave function was universal for various models. Considered here will be the orbital parts of the wave functions (and fractional parentage coefficients) in the following two models:

(a) relativistic quark states in the MIT bag model;

(b) nonrelativistic oscillator shell model (in terms of TISM basis).

7.3.8.1 Orbital fractional parentage coefficients for the bag. It was shown in Section 7.2 above that, to calculate the levels in the $s_{1/2}^{N_s}\, p_{1/2}^{N_p}$ configurations of quark bag, it was necessary to know the fractional parentage coefficients for separating two quarks $q^N \to q^{N-2} \times q^2$ in all the possible orbital states:

$$s_{1/2}^2[2]^X,\ p_{1/2}^2[2]^X,\ s_{1/2}\, p_{1/2}[2]^X,\ s_{1/2}\, p_{1/2}[1^2]^X.$$

It will be reminded that, in the effective Hamiltonian of quark bag (7.69) written in terms of the Pauli spinors, the "orbital state" is understood to be only its eigenfrequency ($\omega_s = 2.04$ for $s_{1/2}$ and $\omega_p = 3.81$ for $p_{1/2}$), whereas the projections of the total angular momentum $j = \frac{1}{2}$ of the Dirac states $s_{1/2}$ and $p_{1/2}$ are treated as the particle "spin". Therefore, we can apply the representation of the total fractional parentage coefficient in the form (7.162) where the orbital and spin parts are separated. In this case, the "obital part" of FPC is of notably simple form, since it is deprived now of the orbital moments proper (L, L′, L″) and of their projections (M, M′ M″) and has but a single index $\omega = \omega_s$ or ω_p. The natural procedure is to combine the two orbital states into the spinor $\xi_{+1/2} = \omega_s$, $\xi_{-1/2} = \omega_p$ of the unitary group U_2^X and to set the "orbital states" by the Young schemes $[f]^X$, Yamanouchi symbols (r_X), and numbers N_s, N_p ($N_s + N_p = N$) which designate the total number of the $s_{1/2}$ and $p_{1/2}$ states:

$$\phi^X = |s_{1/2}^{N_s}\, p_{1/2}^{N_p}\, [f_X](r_X)>. \tag{7.165}$$

The fractional parentage expansion of the orbital state (7.165) will be written according to the general form (7.126)–(7.127)

$$|s_{1/2}^{N_s} p_{1/2}^{N_p} [f_X](r_X)> = \sum_{[f''_X](r''_X)} \cdot <[f_X](r_X)|[f'_X](r'_X), [f''_X](r''_X)> \times$$

$$\times \sum_{\substack{N'_s+N'_p=N' \\ N''_s+N''_p=N''}} \left(\begin{matrix} [f'_X] & [f''_X] \\ s_{1/2}^{N'_s}\, p_{1/2}^{N'_p} & s_{1/2}^{N''_s}\, p_{1/2}^{N''_p} \end{matrix} \middle| \begin{matrix} [f_X] \\ s_{1/2}^{N_s}\, p_{1/2}^{N_p} \end{matrix} \right) |s_{1/2}^{N'_s} p_{1/2}^{N'_p} [f'_X](r'_X)> |s_{1/2}^{N''_s} p_{1/2}^{N''_p} [f''_X](r''_X)> \tag{7.166}$$

The last line of (7.166) includes CGC for the unitary group U_2^X which coincide with the conventional Clebsch-Gordan coefficients of the rotational group $(j'\, m'\, j''\, m''|jm)$ for the values of the momenta $j'\, j''\, j$ determined by the lengths of the Young scheme

$$j' = \tfrac{1}{2}(f'_{1X} - f'_{2X}),\ j'' = \tfrac{1}{2}(f''_{1X} - f''_{2X}),\ j = \tfrac{1}{2}(f_{1X} - f_{2X}),$$

and for the values of the projections m′, m″, m determined by the differences in the numbers of the states N_s and N_p:

$$m' = \frac{1}{2}(N_s' - N_p'),\ m'' = \frac{1}{2}(N_s'' - N_p''),\ m = \frac{1}{2}(N_s - N_p).$$

As a result, we obtain the following Table 7.9 of orbital FPC which will be substituted in the formula (7.162) for the $s_{1/2}^4\ p_{1/2}^2\ [42]^X$ configurations in quark bag.

7.3.8.2 The oscillator basis states. TISM basis. We shall construct here the explicit form of the oscillator shell states (7.163) and (7.164) in the s^6 and s^4p^2 configurations. After that, by separating the c.m. motion of the 6q system

$$\mathbf{X} = \frac{1}{6}(\mathbf{r}_1 + \mathbf{r}_2 + \ldots + \mathbf{r}_6) \tag{7.167}$$

we shall proceed to the TISM basis and present the fractional parentage expansion of the states (7.163) and (7.164) $q^N \to q^{N'} + q^{N''}$, $N' = N'' = 3$ in these terms.

The orbital part of FPC (7.162) has been written in terms of nonrelativistic LS-coupling thereby permitting the orbital states to be treated separately from the spin states. However, the formulation (7.162) fails to reflect the fact that, in the general case, we deal with the occupation of two shells, s and p. The shell composition of the subsystems $q^{N'}$ and $q^{N''}$ will be reflected in the form of the following designations in the orbital-scalar factor:

$$SF_X = \left(\begin{matrix} [f_X'] & [f_X''] \\ (s^{N_s'}, p^{N_p'}[f_p']\,L')\ & (s^{N_s''}, p^{N_p''}[f_p'']\,L'') \end{matrix} \middle\| \begin{matrix} [f_X] \\ (s^{N_s}, p^{N_p}[f_p]\,L) \end{matrix} \right) \tag{7.168}$$

In the case of mere S-shell, this factor turns out to be unity:

$$\left(\begin{matrix} [N'] & [N''] \\ (s^{N'}L'=0) & (s^{N''}L''=0) \end{matrix} \middle\| \begin{matrix} [N] \\ (s^N L=0) \end{matrix} \right) = 1,$$

and the shell functions prove to be of very simple form. For example, the shell states for the s^3 and s^6 configurations are

$$|s^3[3]^X\rangle_{SM} = \left(\frac{\Omega}{\pi}\right)^{9/4} \exp\left[-\frac{\Omega}{2}(r_1^2 + r_2^2 + r_3^2)\right] =$$

$$= \Phi_{00}(R_1)|0\,(00)\,0\,[3]^X\rangle_{TISM} \tag{7.169}$$

$$|s^6[6]^X\rangle_{SM} = \left(\frac{\Omega}{\pi}\right)^{9/2} \exp\left[-\frac{\Omega}{2}(r_1^2 + r_2^2 + \ldots + r_6^2)\right] =$$

$$= \Phi_{00}(X)|0\,(00)\,0\,[6]^X\rangle_{TISM} =$$

$$= \Phi_{00}(X)\,\Phi_{00}(R)|0\,(00)\,0\,[3]^{X'}\rangle_{TISM}|0\,(00)\,0\,[3]^{X''}\rangle_{TISM}. \tag{7.170}$$

Here, Φ_{nl} are the oscillator nl-states; $|N_p(\lambda\mu)\,L\,[f]^X(r_X)\rangle_{TISM}$ are the TISM states, N_p is the number of the p-excitation quanta; $(\lambda\mu)$ are the quantum numbers of the SU_3-scheme of classification of the p-shell levels in TISM: $\lambda = f_{p1} - f_{p2}$, $\mu = f_{p1} - f_{p2}$ (where $[f_{p1} f_{p2} f_{p3}]$ is the Young scheme for p-quanta); L is the total orbital momentum of the state.

We make use of the following set of Jacobi coordinates:

$$x_1 = r_1 - r_2,\ y_1 = \frac{1}{2}(r_1 + r_2) - r_3,\ x_2 = r_4 - r_5,\ y_2 = \frac{1}{2}(r_4 + r_5) - r_6$$

$$R_1 = \frac{1}{3}(r_1 + r_2 + r_3),\ R_2 = \frac{1}{3}(r_4 + r_5 + r_6),\ R = R_1 - R_2, \qquad (7.171)$$

satisfying the relations

$$r_1^2 + r_2^2 + \ldots + r_6^2 = \frac{1}{2}x_1^2 + \frac{2}{3}y_1^2 + \frac{1}{2}x_2^2 + \frac{2}{3}y_2^2 + \frac{3}{2}R^2 + 6X^2,$$

$$d^3r_1\, d^3r_2 \ldots d^3r_6 = d^3x_1\, d^3y_1\, d^3x_2\, d^3y_2\, d^3R\, d^3X. \qquad (7.172)$$

The functions written in (7.169) and (7.170) at the right are of the form

$$|0(00)0\,[3]^{X'}>_{TISM} = 3^{-3/4}\left(\frac{\Omega}{\pi}\right)^{3/2} \exp\left[-\frac{\Omega}{2}\left(\frac{1}{2}x_1^2 + \frac{2}{3}y_1^2\right)\right],$$

$$|0(00)0\,[6]^{X}>_{TISM} = 6^{-3/4}\left(\frac{\Omega}{\pi}\right)^{15/4} \exp\left[-\frac{\Omega}{2}\left(\frac{1}{2}x_1^2 + \frac{2}{3}y_1^2 + \frac{1}{2}x_2^2 + \frac{2}{3}y_2^2 + \frac{3}{2}R^2\right)\right],$$

$$\phi_{00}(R_1) = \left(\frac{3\Omega}{\pi}\right)^{3/4} \exp\left[-\frac{\Omega}{2}3R_1^2\right];\ \phi_{00}(R) = \left(\frac{3\Omega}{2\pi}\right)^{3/4} \exp\left[-\frac{\Omega}{2}\cdot\frac{3}{2}R^2\right],$$

$$\phi_{00}(X) = \left(\frac{6\Omega}{\pi}\right)^{3/4} \exp\left[-\frac{\Omega}{2}6X^2\right].$$

In the case of the s^4p^2 configuration, we deal with the two-shell system, and FPC (7.168) are not trivial. Consider briefly the problem of their calculations. Some general propositions will be formulated at first. The basis vectors in the sp-shell

$$|s^{N_s}p^{N_p}[f_p]L : [f]^X L(r_X)>_{SM}$$

may be treated, from the viewpoint of the group theory, as the basis vectors of IR $D^{[f]}$ of the U_4^X group for the reduction $U_4^X \supset U_3^p \times U_1^s$. Here, the U_3^p group acts in the space of three one-particle states in the p-shell: $|\ell = 1, m>$, $m = 0, \pm 1$; IR of this group are set by the Young schemes $[f_p]$ for the p-states. The U_4^X group acts in the unified four-dimensional space of the one-particle s- and p-states. The two-shell FPC (7.168) is SF of CGC of the U_4^X with respect to the $U_3^p \times U_1^s$ subgroups. It will naturally be assumed that, similarly to the case of the reduction $U_m^{ab} \supset U_{m_1}^a \times U_{m_2}^b$ $(m = m_1 m_2)$, the given SF (7.168) is determined by only the permutational values (Young schemes) and is in no way related to the rank of the U_3^p, U_4^X groups. It seems reasonable to formulate this proposition in the general form.

Thus, let us have the system of two shells, $\ell_1^{N_1}$ and $\ell_2^{N_2}$. The basis inside either shell is constructed with the quantum numbers of the unitary groups U_{m_1}, U_{m_2}, $(m_1 = 2\ell_1 + 1,$ $m_2 = 2\ell_2 + 1)$

$$|\ell_1^{N_1}[f_1](r_1)\alpha_1>|\ell_2^{N_2}[f_2](r_2)\alpha_2>$$

while the basis of the entire system is described by the quantum numbers of the U_m group $(m = m_1 + m_2)$ for the reduction chain $U_m \supset U_{m_1} \times U_{m_2}$.

The fractional parentage expansion of such basis vector is written in the form

$$|\ell_1^{N_1}[f_1]\alpha_1, \ell_2^{N_2}[f_2]\alpha_2 : [f](r)\epsilon>_{\gamma(\gamma_1\gamma_2)} =$$

$$= \sum_{[f''](r'')} <[f](r)|[f'](r'), [f''](r'')>_\gamma \sum_{Q'Q''} \times$$

$$\times \left\langle \begin{matrix} [f'] & [f''] \\ ([f_1'],[f_2'])\,\epsilon' & ([f_1''],[f_2''])\,\epsilon'' \end{matrix} \middle\| \begin{matrix} [f] \\ ([f_1],[f_2])\,\epsilon \end{matrix} \right\rangle^{\gamma}_{\gamma_1\gamma_2} \times$$

$$\times |\ell_1^{N_1'}[f_1']\ell_2^{N_2'}[f_2'][f']\epsilon', \ell_1^{N_1''}[f_1'']\ell_2^{N_2''}[f_2''][f'']\epsilon'' : \ell_1^{N_1}[f_1](r_1)\alpha_1 \times$$

$$\times \ell_2^{N_2}[f_2](r_2)\alpha_2\,[f]\epsilon>_{\gamma_1\gamma_2}. \tag{7.173}$$

Summation in (7.173) is made over the sets of quantum numbers Q' and Q'' in the N' and N'' subsystems

$$Q' = N_1'N_2'[f_1'][f_2']\epsilon',$$

$$Q'' = N_1''N_2''[f_1''][f_2'']\epsilon''.$$

It is assumed in this case that the IR $[f_1']$ and $[f_1'']$ of U_{m_1} group in the right part of equality (7.173) are interrelated through CGC of this group

$$\begin{pmatrix} [f_1'] & [f_1''] & [f_1] \\ \alpha_1' & \alpha_1'' & \alpha_1 \end{pmatrix}_{\gamma_1}$$

while IR $[f_2']$ and $[f_2'']$ of the U_{m_2} group are interrelated through their CGC

$$\begin{pmatrix} [f_2'] & [f_2''] & [f_2] \\ \alpha_2' & \alpha_2'' & \alpha_2 \end{pmatrix}_{\gamma_2}.$$

The following designations are used here: ϵ, ϵ' and ϵ'' are the multiplicity indices in the outer products of the Young schemes $[f_1]\cdot[f_2]$, $[f_1']\cdot[f_2']$ and $[f_1'']\cdot[f_2'']$ respectively; γ_1, γ_2 and γ are the multiplicity indices in the outer products of the Young schmes $[f_1']\cdot[f_1'']$, $[f_2']\cdot[f_2'']$ and $[f']\cdot[f'']$ respectively.

As it was shown in [Kr67, Ne79], the two-shell FPC may be expressed through the permutational values in the following way:

$$<\ell_1^{N_1}[f_1]\ell_2^{N_2}[f_2]:[f]\epsilon|\ell_1^{N_1'}[f_1']\ell_2^{N_2'}[f_2'][f']\epsilon', \times$$

$$\times \ell_1^{N_1''}[f_1'']\ell_2^{N_2''}[f_2''][f'']\epsilon''>_{\gamma(\gamma_1\gamma_2)} =$$

$$= \binom{N_1}{N_1'}^{1/2}\binom{N_2}{N_2'}^{1/2}\binom{N_1+N_2}{N_1'+N_2'}^{-1/2}(-1)^{N_1'+N_2'}\sqrt{\frac{n_{f_1'}n_{f_1''}n_{f_2'}n_{f_2''}n_f}{n_{f_1}n_{f_2}n_{f'}n_{f''}}} \times$$

$$\times \begin{pmatrix} f_1' & f_1'' & f_1 \\ f_2' & f_2'' & f_2 \\ f' & f'' & f \end{pmatrix}^{\gamma(\gamma_1\gamma_2)}_{\epsilon(\epsilon'\epsilon'')}. \tag{7.174}$$

Again, this result means that the scalar (with respect to the $U_{m_1}\times U_{m_2}$ groups) factors of CGC for the $U_{m_1+m_2}$ group are universal and depend on only the form of the

Young schemes comprised in them, and not on the ranks (m_1 and m_2) of the U_{m_1}, and U_{m_2} groups.

In particular, if the Young schemes [f′] [f″] [f] are of two-row form

$$[f] = [\varphi_1 \varphi_2], [f'] = [\varphi_1' \varphi_2'] \text{ and } [f''] = [\varphi_1'' \varphi_2''],$$

then SF (7.174) coincides with the conventional CGC for the SU_2 group. For example, in the $s^{N_s} p^{N_p} [f]^X$ configuration for the completely symmetric states in the s and p shells $[f_s] = [N_s]$, $[f_p] = [N_p]$ the multiplicities are absent in the products $[f_s'] \times [f_p']$, $[f_s''] \times [f_p'']$ and $[f_s] \times [f_p]$ so we get

$$\left\langle \begin{matrix} [f'] & [f''] \\ ([f_s'],[f_p']) & ([f_s''],[f_p'']) \end{matrix} \middle\| \begin{matrix} [f] \\ ([f_s],[f_p]) \end{matrix} \right\rangle \equiv$$

$$\equiv \left\langle \begin{matrix} [\varphi_1'\varphi_2'] & [\varphi_1''\varphi_2''] \\ ([N_s'],[N_p']) & ([N_s''],[N_p'']) \end{matrix} \middle\| \begin{matrix} [\varphi_1\varphi_2] \\ ([N_s],[N_p]) \end{matrix} \right\rangle =$$

$$= \left(\frac{\varphi_1'-\varphi_2'}{2} \; \frac{N_s'-N_s''}{2} \; \frac{\varphi_1''-\varphi_2''}{2} \; \frac{N_s''-N_p''}{2} \middle| \frac{\varphi_1-\varphi_2}{2} \; \frac{N_s-N_p}{2} \right). \tag{7.175}$$

In the $s^4 p^2$ $[42]^X$ configurations, therefore, the table of SF_x (7.168) of nonrelativistic orbital states at $L = L' + L'' = 0$ (the symmetric state in p-shell) coincides with Table 7.9.

This table of coefficients

$$\left\langle \begin{matrix} [f']^X & [f'']^X \\ (s^{N_s'}, p^{N_p'})L' & (s^{N_s''}, p^{N_p''})L'' \end{matrix} \middle\| \begin{matrix} [42]^X \\ (s^4 p^2) L=0 \end{matrix} \right\rangle$$

will be used to carry out the partial summation of the terms of the fractional parentage expansion of the orbital shell function $|s^4 p^2 [42]^X (r) L=0>_{SM}$. Thus, we summarize the terms of the fractional parentage expansion of the form

$$\sum_{\substack{N_s'+N_s''=N_s=4,\\ N_p'+N_p''=N_p=2, L'L''}} \left\langle \begin{matrix} [f']^X & [f'']^X \\ (s^{N_s'}, p^{N_p'})L' & (s^{N_s''}, p^{N_p''})L'' \end{matrix} \middle\| \begin{matrix} [42]^X \\ (s^4, p^2) L=0 \end{matrix} \right\rangle \times$$

$$\times \left\{ |s^{N_s'} p^{N_p'} [f']^X (r_X') L' >_{SM} |s^{N_s''} p^{N_p''} [f'']^X (r_X'') L'' >_{SM} \right\}_{L=0} \equiv$$

$$\equiv \left\{ |[f']^X (r_X') >_{SM} |[f'']^X (r_X'') >_{SM} \right\}^{s^4 p^2}_{L=0} \tag{7.176}$$

over the quantum numbers N_s', N_s'', N_p', N_p'', L', L'' (the summation, however, does not include the Young schemes), and eventually obtain the new form of fractional parentage expansion, namely in terms of the TISM orbital states. In particular, after summing in the form (7.176) and proceeding to the Jacobi coordinates (7.167) and (7.171), the terms of the fractional parentage expansion $q^6 \to q^3 \times q^3$ make it possible to separate both the motion of the common c.m. X and the motion along the relative coordinate R of two three-quark clusters. In this case, we have only four terms of the form (7.176) which are all proportional to the non-excited state function $\Phi_{00}(X)$ (in accordance with the result, which is well known in nuclear physics [Ne 69], that the $s^4 p^2 [42]^X$ configuration is devoid of spurious excitations of c.m.):

i) $$\left\{ |[3]^{X'}>_{SM}|[3]^{X''}>_{SM}\right\}^{s^4p^2}_{L=0} = \Phi_{00}(X)\Big[-\sqrt{\tfrac{4}{5}}\Phi_{20}(R)|0(00)0[3]^{X'}>_{TISM}\times$$
$$\times|0(00)0[3]^{X''}>_{TISM}+\sqrt{\tfrac{1}{5}}\Phi_{00}(R)\Big(\sqrt{\tfrac{1}{2}}|2(20)0[3]^{X'}>_{TISM}|0(00)0[3]^{X''}>_{TISM}+$$
$$+\sqrt{\tfrac{1}{2}}|0(00)0[3]^{X'}>_{TISM}|2(20)0[3]^{X''}>_{TISM}\Big)\Big];$$

ii) $$\left\{|[3]^{X'}>_{SM}|[21]^{X''}(r''_X)>_{SM}\right\}^{s^4p^2}_{L=0} = \Phi_{00}(X)\Big[\sqrt{\tfrac{1}{2}}\Phi_{00}(R)|0(00)0[3]^{X'}>_{TISM}\times$$
$$\times|2(20)0[21]^{X''}(r''_X)>_{TISM}+\sqrt{\tfrac{1}{2}}\Big\{\Phi_{11}(R)|0(00)0[3]^{X'}>_{TISM}$$
$$|1(10)1[21]^{X''}(r''_X)>_{TISM}\Big\}_{L=0}\Big];$$

iii) $$\left\{|[21]^{X'}(r'_X)>_{SM}|[3]^{X''}>_{SM}\right\}^{s^4p^2}_{L=0} =$$
$$=\Phi_{00}(X)\Big[\sqrt{\tfrac{1}{2}}\Big\{\Phi_{11}(R)|1(10)1[21]^{X'}(r'_X)>_{TISM}\Big\}_{L=0}|0(00)0[3]^{X''}>_{TISM}-$$
$$-\sqrt{\tfrac{1}{2}}\Phi_{00}(R)|2(20)0[21]^{X'}(r'_X)>_{TISM}|0(00)0[3]^{X''}>_{TISM}\Big];$$

iv) $$\left\{|[21]^{X'}(r'_X)>_{SM}|[21]^{X''}(r''_X)>_{SM}\right\}^{s^4p^2}_{L=0} = \Phi_{00}(X)\,\Phi_{00}(R)\times$$
$$\left\{|1(10)1[21]^{X'}(r'_X)>_{TISM}|1(10)1[21]^{X''}(r''_X)>_{TISM}\right\}_{L=0}.$$

The TISM functions $|N'_p(\lambda'\mu')L'[f]^{X'}(r'_X)>$ are related to the oscillator functions $\Phi_{n\ell}$, depending on the Jacobi coordinates x_1 and y_1 as

$$|0(00)0[3]^{X'}> = \Phi_{00}(x_1)\,\Phi_{00}(y_1);$$
$$|1(10)1[21]^{X'}r_X^{(1)'}> = \Phi_{00}(x_1)\,\Phi_{11}(y_1);$$
$$|1(10)1[21]^{X'}r_X^{(2)'}> = \Phi_{11}(x_1)\,\Phi_{00}(y_1);$$
$$|2(20)0[3]^{X'}> = \sqrt{\tfrac{1}{2}}\Phi_{20}(x_1)\,\Phi_{00}(y_1)+\sqrt{\tfrac{1}{2}}\Phi_{00}(x_1)\,\Phi_{20}(y_1);$$
$$|2(20)0[21]^{X'}r_X^{(1)'}> = \sqrt{\tfrac{1}{2}}\Phi_{20}(x_1)\,\Phi_{00}(y_1)-\sqrt{\tfrac{1}{2}}\Phi_{00}(x_1)\,\Phi_{20}(y_1);$$
$$|2(20)0[21]^{X'}r_X^{(2)'}> = \left\{\Phi_{11}(x_1)\,\Phi_{11}(y_1)\right\}_{L=0}.$$

It can be seen now that the examined oscillator shell states (7.163) and (7.164) are very simply related to the basis functions in TISM, namely the TISM states will be obtained by dividing the shell functions (7.163) and (7.164) by the wave function of ground state $\Phi_{00}(X)$ in c.m.

Concluding this Section, we shall write the explicit form of the fractional parentage expansions $q^6 \to q^3 \times q^3$ of given wave functions of TISM in terms of the barion-barion states

i) $s^6 : |0(00)0[6]^X L=0, [2^3]^C S=1[2^3]^{CS} T=0[1^6]^{CST}>_{TISM} \equiv$

$$\equiv \frac{1}{\Phi_{00}(X)} |s^6[6]^X L=0, [2^3]^C S=1[2^3]^{CS} T=0[1^6]^{CST}>_{SM} =$$

$$= \sqrt{\frac{1}{5}} \Phi_{00}(R) \Big\{ |0(00)0[3]^{X'}, [1^3]^{CS'} T'=\frac{3}{2}[1^3]^{CST'}> |0(00)0[3]^{X''}, [1^3]^{CS''} T''= =\frac{3}{2}[1^3]^{CST''}> \Big\}^{(-)}_{CST} +$$

$$+ \sqrt{\frac{4}{5}} \Phi_{00}(R) \Big\{ |0(00)0[3]^{X'}, [21]^{CS'} T'=\frac{1}{2}[1^3]^{CST'}> |0(00)0[3]^{X''}, [21]^{CS''} T''= =\frac{1}{2}[1^3]^{CST''}> \Big\}^{(-)}_{CST}; \tag{7.177}$$

ii) $s^4 p^2 : |2(20)0[42]^X L=0, [2^3]^C S=1[42]^{CS} T=0[2^2 1^2]^{CST}>_{TISM} \equiv$

$$\equiv \frac{1}{\Phi_{00}(X)} |s^4 p^2 [42]^X L=0, [2^3]^C S=1[42]^{CS} T=0[2^2 1^2]^{CST}>_{SM} =$$

$$= -\sqrt{\frac{4}{45}} \Phi_{20}(R) \Big\{ |0(00)0[3]^{X'}, [21]^{CS'} T'=\frac{1}{2}[1^3]^{CST'}> |0(00)0[3]^{X''}, [21]^{CS''} T''= =\frac{1}{2}[1^3]^{CST''}> \Big\}^{(-)}_{CST} +$$

$$+ \sqrt{\frac{10}{45}} \Big\{ \Big\{ \Phi_{11}(R) \Big[\sqrt{\frac{4}{5}} |1(10)1[21]^{X'}, [3]^{CS'} T'=\frac{1}{2}[21]^{CST'}> +$$

$$+ \sqrt{\frac{1}{5}} |1(10)1[21]^{X'}, [21]^{CS'} T'=\frac{1}{2}[21]^{CST'}> \Big] \Big\}_{L=0} |0(00)0[3]^{X''}, [21]^{CS''} T''= =\frac{1}{2}[1^3]^{CST''}> \Big\}^{(-)}_{CST} +$$

$$+ \sqrt{\frac{11}{45}} \Phi_{00}(R) \Big\{ \Big[\sqrt{\frac{1}{11}} |2(20)0[3]^{X'}, [21]^{CS'} T'=\frac{1}{2}[1^3]^{CST'}> +$$

$$+ \sqrt{\frac{8}{11}} |2(20)0[21]^{X'}, [3]^{CS'} T'=\frac{1}{2}[1^3]^{CST'}> + \sqrt{\frac{2}{11}} |2(20)0[21]^{X'}, [21]^{CS'} T'= =\frac{1}{2}[21]^{CST'}> \Big] \times$$

$$\times |0(00)0[3]^{X''}, [21]^{CS''}T''=\tfrac{1}{2}[1^3]^{CST''}>\Big\}^{(-)}_{CST} +$$

$$+\sqrt{\frac{20}{45}}\,\Phi_{00}(R)\Big\{\frac{1}{2}|1(10)1[21]^{N'}, [3]^{CS'}T'=\frac{1}{2}[21]^{CST'}>|1(10)1[21]^{X''}, [3]^{CS''}T''=$$

$$=\frac{1}{2}[21]^{CST''}>+$$

$$+\sqrt{\frac{9}{20}}|1(10)1[21]^{X'}, [21]^{CS'}T'=\frac{3}{2}[21]^{CST'}>|1(10)1[21]^{X''}, [21]^{CS''}T''=$$

$$=\frac{3}{2}[21]^{CST''}>+$$

$$+\sqrt{\frac{3}{10}}|1(10)1[21]^{X'}, [21]^{CS'}T'=\frac{1}{2}[21]^{CST'}>\Big[\sqrt{\frac{2}{3}}|1(10)1[21]^{X''}, [3]^{CS''}T''=$$

$$=\frac{1}{2}[21]^{CST''}>+$$

$$+\sqrt{\frac{1}{3}}|1(10)1[21]^{X''}, [21]^{CS''}T''=\frac{1}{2}[21]^{CST''}>\Big]\Big\}^{(-)}_{CST, L=0}. \qquad (7.178)$$

Here, the bracket $\Big\{\ldots\Big\}^{(-)}_{CST}$ means the sum of the barion-barion wave functions composed of barions in various coloured and spin states $([g'_C]C'S')\times([g''_C]C''S'')\times SF_C\times SF_{CS}$ with the coefficients SF_C and SF_{CS} from Tables 7.4–7.6. Besides that, each term of this sum is antisymmetrized in the barion permutations are the whole. The wave functions of barions $|N'_p(\lambda'\mu')L'[f']^{X'}, [g'_C]C'S'[h'_{CS}]T'[\tilde{f}']^{CST'}>$ and $|N''_p(\lambda''\mu'')L''[f'']^{X''}$, $[g''_C]C''S''[h''_{CS}]T''[\tilde{f}]^{CST''}>$ are written in the TISM basis; for example,

$$|N^{C=0}_{1/2}> = |0(00)0[3]^X, [1^3]^C C=0\ S=\frac{1}{2}[21]^{CS}\ T=\frac{1}{2}[1^3]^{CST}>,$$

$$|\Delta^{C=0}_{3/2}> = |0(00)0[3]^X, [1^3]^C C=0\ S=\frac{3}{2}[1^3]^{CS}\ T=\frac{3}{2}[1^3]^{CST}>,$$

$$|N^{C=1;2}_{1/2}> = |0(00)0[3]^X, [21]^C C\ S=\frac{1}{2}[21]^{CS}\ T-\frac{1}{2}[1^3]^{CST}>,$$

Table 7.4: Scalar factors $SF_C = \left\langle \begin{matrix}[f'_C]\,[f''_C] \\ C'\ \ C''\end{matrix} \middle\| \begin{matrix}[2^3]^C\\ C=0\end{matrix}\right\rangle$ for the reduction chain $SU^C_3 \supset O^C_3$

A. $q^6 \to q^3 \times q^3$.

$[g'_C]\times[g''_C]$	$[1^3]\times[1^3]$	$[21]\times[21]$	
(C', C'')	(0,0)	(1,1)	(2.2)
SF_C	1	$-\sqrt{\frac{3}{8}}$	$\sqrt{\frac{5}{8}}$

B. $q^6 \to q^4 \times q^2$.

$[g'_C]\times[g''_C]$	$[21^2]\times[1^2]$	$[2^2]\times[2]$	
(C', C'')	(1,1)	(0,0)	(2,2)
SF_C	1	$-\sqrt{\frac{1}{6}}$	$-\sqrt{\frac{5}{6}}$

Table 7.5: The $s^6[6]^X$ configuration. Scalar factors

$$SF_{CS} = \left\langle \begin{matrix} [h'_{CS}] & [h''_{CS}] \\ ([g'_C], S') & ([g''_C], S'') \end{matrix} \middle\| \begin{matrix} [2^3]^{CS} \\ ([2^3]^C, S=1) \end{matrix} \right\rangle$$ for the reduction chain $SU_6^{CS} \supset SU_3^C \times SU_2^S$.

A. $q^6 \to q^3 \times q^3$.

$[h'_{CS}] \times [h''_{CS}]$	$[1^3] \times [1^3]$		$[21] \times [21]$				
$([g'_C], S')$ $\times ([g''_C], S'')$	$([1^3], \frac{3}{2})$ $\times ([1^3], \frac{3}{2})$	$([21], \frac{1}{2})$ $\times ([21], \frac{1}{2})$	$([1^3], \frac{1}{2})$ $\times ([1^3], \frac{1}{2})$	$([21], \frac{3}{2})$ $\times ([21], \frac{3}{2})$	$([21], \frac{3}{2})$ $\times ([21], \frac{1}{2})$	$([21], \frac{1}{2})$ $\times ([21], \frac{3}{2})$	$([21], \frac{1}{2})$ $\times ([21], \frac{1}{2})$
SF_{CS}	$-\sqrt{\frac{4}{9}}$	$\sqrt{\frac{5}{9}}$	$\sqrt{\frac{5}{36}}$	$\sqrt{\frac{1}{36}}$	$\sqrt{\frac{5}{18}}$	$\sqrt{\frac{5}{18}}$	$\sqrt{\frac{5}{18}}$

B. $q^6 \to q^4 \times q^2$.

$[h'_{CS}] \times [h''_{CS}]$	$[2^2] \times [2]$			$[21^2] \times [1^2]$			
$([g'_C], S')$ $\times ([g''_C], S'')$	$([21^2], 1)$ $\times ([1^2], 0)$	$([2^2], 2)$ $\times ([2], 1)$	$([2^2], 0)$ $\times ([2], 1)$	$([21^2], 2)$ $\times ([1^2], 1)$	$([21^2], 1)$ $\times ([1^2], 1)$	$([21^2], 0)$ $\times ([1^2], 1)$	$([2^2], 1)$ $\times ([2], 0)$
SF_{CS}	$\sqrt{\frac{5}{12}}$	$\sqrt{\frac{1}{6}}$	$-\sqrt{\frac{5}{12}}$	$-\sqrt{\frac{2}{27}}$	$\sqrt{\frac{5}{9}}$	$-\sqrt{\frac{5}{54}}$	$-\sqrt{\frac{5}{18}}$

$$|N_{3/2}^{C=1;2}\rangle = |0(00)0[3]^X, [21]^C\, C\, S=\tfrac{3}{2}\, [21]^{CS}\, T=\tfrac{1}{2}\, [1^3]^{CST}\rangle,$$

$$|\Delta_{1/2}^{C=1;2}\rangle = |0(00)0[3]^X, [21]^C\, C\, S=\tfrac{1}{2}\, [1^3]^{CS}\, T=\tfrac{3}{2}\, [1^3]^{CST}\rangle,$$

$$|N_{1/2}^{*C=0}\rangle = |1(10)1[21]^X, [1^3]^C\, C=0\, S=\tfrac{1}{2}\, [21]^{CS}\, T=\tfrac{1}{2}\, [21]^{CST}\rangle,$$

$$|N_{1/2}^{**C=0}\rangle = |2(20)0[3]^X, [1^3]^C\, C=0\, S=\tfrac{1}{2}\, [21]^{CS}\, T=\tfrac{1}{2}\, [1^3]^{CST}\rangle,$$

$$|\widetilde{N}_{1/2}^{**C=0}\rangle = |2(20)0[21]^X, [1^3]^C\, C=0\, S=\tfrac{1}{2}\, [21]^{CS}\, T=\tfrac{1}{2}\, [21]^{CST}\rangle, \ldots \text{etc.} \tag{7.179}$$

Or else, by rewritting the first addent of the expansion (7.178) concise designations (7.179) we obtain the barion-barion composition of such component of the wave function of 6q-system $s^4p^2\,[42]^X\,[42]^{CS}$ that has two quanta of excitation along the relative coordinate R and, in accordance with this, a node of the wave function

$$\Phi_{20}(R) = \left(\frac{2}{3}\right)^{1/2} \left(\frac{3\Omega}{2\pi}\right)^{3/4} \left(\frac{3}{2} - \frac{3}{2}\Omega R^2\right) \exp\left[-\frac{1}{2}\left(\frac{3}{2}\Omega\right) R^2\right] \tag{7.180}$$

Table 7.6: The s^4p^2 $[42]^X$ configuration. Scalar factors $SF_{CS} = \left\langle \begin{matrix}[h'_{CS}] & [h''_{CS}] \\ ([g'_C],S') & ([g''_C],S'')\end{matrix} \middle\| \begin{matrix}[42]^{CS} \\ ([2^3]^C, S=1)\end{matrix} \right\rangle$ for the reduction chain $SU_6^{CS} \supset SU_3^C \times SU_2^S$

A. $q^6 \to q^3 \times q^3$.

$[h'_{CS}] \times [h''_{CS}]$	$[3] \times [3]$	$[3] \times [21]$		$[21] \times [3]$		$[21] \times [21]$				
$([g'_C],S') \times ([g''_C],S'')$	$([21],\frac{1}{2}) \times ([21],\frac{1}{2})$	$([21],\frac{1}{2}) \times ([21],\frac{3}{2})$	$([21],\frac{1}{2}) \times ([21],\frac{1}{2})$	$([21],\frac{3}{2}) \times ([21],\frac{1}{2})$	$([21],\frac{1}{2}) \times ([21],\frac{1}{2})$	$([21],\frac{3}{2}) \times ([21],\frac{3}{2})$	$([21],\frac{3}{2}) \times ([21],\frac{1}{2})$	$([21],\frac{1}{2}) \times ([21],\frac{3}{2})$	$([21],\frac{1}{2}) \times ([21],\frac{1}{2})$	$([1^3],\frac{1}{2}) \times ([1^3],\frac{1}{2})$
SF_{CS}	1	$\sqrt{\frac{4}{5}}$	$-\sqrt{\frac{1}{5}}$	$\sqrt{\frac{4}{5}}$	$\sqrt{\frac{1}{5}}$	$\frac{1}{2}$	$-\sqrt{\frac{1}{10}}$	$\sqrt{\frac{1}{10}}$	$\sqrt{\frac{1}{10}}$	$\sqrt{\frac{9}{20}}$

B. $q^6 \to q^4 \times q^2$.

$[h'_{CS}] \times [h''_{CS}]$	$[4] \times [2]$	$[31] \times [2]$		$[31] \times [1^2]$			$[2^2] \times [2]$		
$([g'_C],S') \times ([g''_C],S'')$	$([2^2],0) \times ([2],1)$	$([2^2],1) \times ([2],1)$	$([21^2],1) \times ([1^2],0)$	$([2^2],1) \times ([2],0)$	$([21^2],1) \times ([1^2],1)$	$([21^2],0) \times ([1^2],1)$	$([2^2],0) \times ([2],1)$	$([2^2],2) \times ([2],1)$	$([21^2],1) \times ([1^2],0)$
SF_{CS}	1	$\sqrt{\frac{2}{5}}$	$-\sqrt{\frac{3}{5}}$	$-\sqrt{\frac{1}{10}}$	$\sqrt{\frac{3}{5}}$	$\sqrt{\frac{3}{10}}$	$\sqrt{\frac{1}{20}}$	$-\sqrt{\frac{1}{2}}$	$\sqrt{\frac{9}{20}}$

at the point $R_{node} = \Omega^{-1/2}$.

$$|2(20)0[42]^X L=0, [2^3]^C S=1[42]^{CS} T=0[2^2 1^2]^{CST} >_{TISM} =$$

$$= -\sqrt{\frac{1}{25}}\Phi_{20}(R)\left\{|N_{1/2}^{C'=0}>|N_{1/2}^{C''=0}>\right\}_{S=1,T=0} +$$

$$+\sqrt{\frac{1}{300}}\Phi_{20}(R)\left\{|N_{1/2}^{C'=1}>|N_{1/2}^{C''=1}>\right\}_{C=0,S=1,T=0} -$$

$$-\sqrt{\frac{1}{180}}\Phi_{20}(R)\left\{|N_{1/2}^{C'=2}>|N_{1/2}^{C''=2}>\right\}_{C=0,S=1,T=0} -$$

$$-\sqrt{\frac{1}{150}}\Phi_{20}(R)\left\{\sqrt{\frac{1}{2}}|N_{3/2}^{C'=1}>|N_{1/2}^{C''=1}>-\sqrt{\frac{1}{2}}|N_{1/2}^{C'=1}>|N_{3/2}^{C''=1}>\right\}_{C=0,S=1,T=0} +$$

$$+\sqrt{\frac{1}{90}}\Phi_{20}(R)\left\{\sqrt{\frac{1}{2}}|N_{3/2}^{C'=2}>|N_{1/2}^{C''=2}>-\sqrt{\frac{1}{2}}|N_{1/2}^{C'=2}>|N_{3/2}^{C''=2}>\right\}_{C=0,S=1,T=0} +$$

$$+\sqrt{\frac{1}{120}}\Phi_{20}(R)\left\{|N_{3/2}^{C'=1}>|N_{3/2}^{C''=1}>\right\}_{C=0,S=1,T=0} -$$

$$-\sqrt{\frac{1}{72}}\Phi_{20}(R)\left\{|N_{3/2}^{C'=2}>|N_{3/2}^{C''=2}>|N_{3/2}^{C''=2}>\right\}_{C=0,S=1,T=0} +$$

+ the terms comprising the orbital-excited barions (N*, N**, Δ*, Δ**). (7.181)

The part of the wave function (7.178), which describes the barion-barion components, without hidden colour, is also of interest:

$$|2(20)0[42]^X L=0, [2^3]^C S=1\ [42]^{CS} T=0\ [2^2 1^2]^{CST} >_{TISM} =$$

$$= -\sqrt{\frac{1}{25}}\Phi_{20}(R)\left\{|N_{1/2}^{C'=0}>|N_{1/2}^{C''=0}>\right\}_{S=1,T=0} -$$

$$-\sqrt{\frac{1}{50}}\left\{\Phi_{11}(\mathbf{R})\left[\sqrt{\frac{1}{2}}|N_{1/2}^{C'=0}>|N_{1/2;3/2}^{*C''=0}>-\sqrt{\frac{1}{2}}|N_{1/2;3/2}^{*C'=0}>|N_{1/2}^{C''=0}>\right]\right\}_{S=1,T=0,L=0} +$$

$$+\sqrt{\frac{1}{100}}\Phi_{00}(R)\left\{\sqrt{\frac{1}{2}}|N_{1/2}^{C'=0}>|N_{1/2}^{**C''=0}>-\sqrt{\frac{1}{2}}|N_{1/2}^{**C'=0}>|N_{1/2}^{C''=0}>\right\}_{S=1,T=0,L=0} -$$

$$-\sqrt{\frac{1}{50}}\Phi_{00}(R)\left\{\sqrt{\frac{1}{2}}|N_{1/2}^{C'=0}>|\widetilde{N}_{1/2}^{**C''=0}>-\sqrt{\frac{1}{2}}|\widetilde{N}_{1/2}^{**C'=0}>|N_{1/2}^{C''=0}>\right\}_{S=1,T=0,L=0} +$$

$$+\sqrt{\frac{9}{100}}\Phi_{00}(R)\left\{|\Delta_{1/2;3/2}^{*C'=0}>|\Delta_{1/2;3/2}^{*C''=0}>\right\}_{S=1,T=0,L=0} +$$

(7.182)

+ the terms with "hidden" colour.

Let the expansion (7.177) of the TISM wave function for the $s^6[6]^X[2^3]^{CS}$ state be rewritten in the same denominations:

$$|0(00)0[6]^X L=0, [2^3]^C\ S=1[2^3]^{CS}\ T=0[1^6]^{CST}>_{TISM} =$$

$$=\sqrt{\frac{1}{9}}\Phi_{00}(R)\left\{|N_{1/2}^{C'=0}>|N_{1/2}^{C''=0}>\right\}_{S=1,T=0} -$$

$$-\sqrt{\frac{4}{45}}\Phi_{00}(R)\left\{|\Delta_{3/2}^{C'=0}>|\Delta_{3/2}^{C''=0}>\right\}_{S=1,T=0} +$$

$$+\sqrt{\frac{1}{9}}\Phi_{00}(R)\left\{|\Delta_{1/2}^{C'=1;2}>|\Delta_{1/2}^{C''=1;2}>\right\}_{C=0,S=1,T=0} +$$

$$+\sqrt{\frac{2}{9}}\Phi_{00}(R)\left\{|N_{1/2}^{C'=1;2}>|N_{1/2}^{C''=1;2}>\right\}_{C=0,S=1,T=0} +$$

$$+\sqrt{\frac{4}{9}}\Phi_{00}(R)\left\{\sqrt{\frac{1}{2}}|N_{3/2}^{C'=1;2}>|N_{1/2}^{C''=1;2}>-\sqrt{\frac{1}{2}}|N_{1/2}^{C'=1;2}>|N_{3/2}^{C''=1;2}>\right\}_{C=0,S=1,T=0} +$$

$$+\sqrt{\frac{1}{45}}\Phi_{00}(R)\left\{|N_{3/2}^{C'=1;2}>|N_{3/2}^{C''=1;2}>\right\}_{C=0,S=1,T=0} \quad (7.183)$$

It can be seen that the barion components with (or without) hidden colour are of the same weights:

$$W(C'=0, C''=0)=\frac{1}{5},\ W(C'\neq 0, C''\neq 0)=\frac{4}{5}$$

in both configurations (s^6 and s^4p^2), which is accounted for by the identical symmetry of $[2^3]^C$ of the wave function in colour space. However, the barionic composition of the

Table 7.7: The $s^6[6]^X$ configuration. Scalar factors
$$SF_{CST}=\left\langle \begin{matrix}[f'_{CST}] & [f''_{CST}] \\ ([h'_{CS}],T') & ([h''_{CS}],T'')\end{matrix} \middle\| \begin{matrix}[1^6]^{CST} \\ ([2^3]^{CS},T=0)\end{matrix}\right\rangle$$ for the reduction chain $SU_{12}^{CST}\supset SU_6^{CS}\times SU_2^T$ and the factors $\sqrt{\frac{n_{f'}n_{f''}}{n_f}}$.

A. $q^6\to q^3\times q^3$.

$[f'_{CST}]\times[f''_{CST}]$	$[1^3]\times[1^3]$	
$([h'_{CS}],T')$ $\times([h''_{CS}],T'')$	$([1^3],\frac{3}{2})$ $\times([1^3],\frac{3}{2})$	$([21],\frac{1}{2})$ $\times([21],\frac{1}{2})$
SF_{CST}	$\sqrt{\frac{1}{5}}$	$\sqrt{\frac{4}{5}}$
$\sqrt{\frac{n_{f'}n_{f''}}{n_f}}$	1	

B. $q^6\to q^4\times q^2$.

$[f'_{CST}]\times[f''_{CST}]$	$[1^4]\times[1^2]$	
$([h'_{CS}],T')$ $\times([h''_{CS}],T'')$	$([21^2],1)$ $\times([1^2],1)$	$([2^2],0)$ $\times([2],0)$
SF_{CST}	$\sqrt{\frac{3}{5}}$	$\sqrt{\frac{2}{5}}$
$\sqrt{\frac{n_{f'}n_{f''}}{n_f}}$	1	

Table 7.8: The $s^4p^2[42]^X$ configuration. Scalar factors $SF_{CST} = \left\langle \begin{matrix} [f'_{CST}] & [f''_{CST}] \\ ([h'_{CS}],T') & ([h''_{CS}],T'') \end{matrix} \middle\| \begin{matrix} [2^2 1^2]^{CST} \\ ([42]^{CS},T) \end{matrix} \right\rangle$ for the reduction chain $SU_{12}^{CST} \supset SU_6^{CS} \times SU_2^T$ and the factors $\sqrt{\frac{n_{f'} n_{f''}}{n_f}}$.

A. $q^6 \to q^3 \times q^3$.

$[f'_{CST}] \times [f''_{CST}]$	$[1^3] \times [1^3]$	$[1^3] \times [21]$		$[21] \times [1^3]$		$[21] \times [21]$				
$([h'_{CS}],T') \times ([h''_{CS}],T'')$	$([21],\frac{1}{2}) \times ([21],\frac{1}{2})$	$([21],\frac{1}{2}) \times ([3],\frac{1}{2})$	$([21],\frac{1}{2}) \times ([21],\frac{1}{2})$	$([3],\frac{1}{2}) \times ([21],\frac{1}{2})$	$([21],\frac{1}{2}) \times ([21],\frac{1}{2})$	$([3],\frac{1}{2}) \times ([3],\frac{1}{2})$	$([3],\frac{1}{2}) \times ([21],\frac{1}{2})$	$([21],\frac{1}{2}) \times ([3],\frac{1}{2})$	$([21],\frac{1}{2}) \times ([21],\frac{1}{2})$	$([21],\frac{3}{2}) \times ([21],\frac{3}{2})$
SF_{CST}	1	$-\sqrt{\frac{4}{5}}$	$-\sqrt{\frac{1}{5}}$	$-\sqrt{\frac{4}{5}}$	$\sqrt{\frac{1}{5}}$	$\frac{1}{2}$	$\sqrt{\frac{1}{10}}$	$-\sqrt{\frac{1}{10}}$	$\sqrt{\frac{1}{10}}$	$\sqrt{\frac{9}{20}}$
$\sqrt{\frac{n_f' n_f''}{n_f}}$	$\sqrt{\frac{1}{9}}$	$\sqrt{\frac{2}{9}}$		$\sqrt{\frac{2}{9}}$		$\sqrt{\frac{4}{9}}$				

B. $q^6 \to q^4 \times q^2$.

$[f'_{CST}] \times [f''_{CST}]$	$[1^4] \times [1^2]$	$[21^2] \times [1^2]$		$[21^2] \times [2]$			$[2^2] \times [1^2]$		
$([h'_{CS}],T') \times ([h''_{CS}],T'')$	$([2^2],0) \times ([2],0)$	$([31],0) \times ([2],0)$	$([31],1) \times ([1^2],1)$	$([31],1) \times ([2],1)$	$([2^2],1) \times ([2],1)$	$([31],0) \times ([1^2],0)$	$([2^2],0) \times ([2],0)$	$([31],1) \times ([1^2],1)$	$([4],0) \times ([2],0)$
SF_{CST}	1	$-\sqrt{\frac{2}{5}}$	$-\sqrt{\frac{3}{5}}$	$\sqrt{\frac{3}{5}}$	$-\sqrt{\frac{3}{10}}$	$\sqrt{\frac{1}{10}}$	$\sqrt{\frac{1}{20}}$	$\sqrt{\frac{9}{20}}$	$\sqrt{\frac{1}{2}}$
$\sqrt{\frac{n_f' n_f''}{n_f}}$	$\sqrt{\frac{1}{9}}$	$\sqrt{\frac{3}{9}}$		$\sqrt{\frac{3}{9}}$			$\sqrt{\frac{2}{9}}$		

Table 7.9: The $s^4p^2[42]^X$ configuration. Orbital FPC $\Gamma_X = \left\langle \begin{matrix} [f'_X] & [f''_X] \\ s^{N'_s}p^{N'_p} & s^{N''_s}p^{N''_p} \end{matrix} \middle| \begin{matrix} [42]^X \\ s^4p^2 \end{matrix} \right\rangle$.

A. $q^6 \to q^3 \times q^3$

$[f'_X] \times [f''_X]$	$[3] \times [3]$			$[3] \times [21]$		$[21] \times [3]$		$[21] \times [21]$
$(s^{N'_s}p^{N'_p}) \times (s^{N''_s}p^{N''_p})$	$s^3 \times sp^2$	$s^2p \times s^2p$	$sp^2 \times s^3$	$s^3 \times sp^2$	$s^2p \times s^2p$	$s^2p \times s^2p$	$sp^2 \times s^3$	$s^2p \times s^2p$
Γ_X	$\sqrt{\frac{3}{10}}$	$-\sqrt{\frac{2}{5}}$	$\sqrt{\frac{3}{10}}$	$\sqrt{\frac{3}{4}}$	$-\sqrt{\frac{1}{4}}$	$\sqrt{\frac{1}{4}}$	$-\sqrt{\frac{3}{4}}$	1

B. $q^6 \to q^4 \times q^2$.

$[f'_X] \times [f''_X]$	$[4] \times [2]$			$[31] \times [2]$		$[31] \times [1^2]$	$[2^2] \times [2]$
$(s^{N'_s}p^{N'_p}) \times (s^{N''_s}p^{N''_p})$	$s^4 \times p^2$	$s^3p \times sp$	$s^2p^2 \times s^2$	$s^3p \times sp$	$s^2p^2 \times s^2$	$s^3p \times sp$	$s^2p^2 \times s^2$
Γ_X	$\sqrt{\frac{3}{5}}$	$-\sqrt{\frac{3}{10}}$	$\sqrt{\frac{1}{10}}$	$\sqrt{\frac{1}{2}}$	$-\sqrt{\frac{1}{2}}$	1	1

colourless components is different in the two cases, and this fact is already accounted for by the constraints due to the Pauli principle. In particular, the s^6 configuration is coupled with two open channels, NN and $\Delta\Delta$ (the $\Delta\Delta$ component of such type in deuteron wave function was analyzed in detail in [SM78]). On the contrary, the s^4p^2 $[42]^X$ $[42]^{CS}$ configuration is not coupled with the $\Delta\Delta$ channel and coupled with five open channels: $N_{1/2}N_{1/2}$, $N_{1/2}N^*_{1/2;3/2}$, $N_{1/2}N^{**}_{1/2}$, $N_{1/2}\tilde{N}^{**}_{1/2}$ and $\Delta^*_{1/2;3/2}\Delta^*_{1/2;3/2}$.

7.4 Results and Discussion; Some Physical Applications

The considerations presented above have shown that two types of symmetry (s^6 $[6]^X$ $[2^3]^{CS}$ and s^4p^2 $[42]^X$ $[42]^{CS}$) of the quark wave function in the region of complete nucleon overlapping are to be treated as most probable. However, in answering finally the question as to what of the two types of symmetry is actually realizable, we depend completely on a particular quark model. It is known [Wo77] that the non-relativistic potential models (oscillator, linear potential) are in a better agreement with the spectra of orbital excitations of hadrons than the MIT bag model. For example, the oscillator parameter which sets the sequence of orbital excitations of ρ-meson (0s-2s-4s- ...), $2\hbar\omega \simeq 0.67$ GeV, whereas the MIT bag model (in the case of massless quarks and at a phenomenological constant $Z_0 = 1.32$) gives two times as large an interval between the levels, namely $\sim \frac{2\pi}{R} \simeq 1.3$ GeV. At the same time, the quark magnetic moments in the non-relativistic models are independent of orbital state, thereby giving probably an overestimated value of the colour-magnetic contribution to the energy of the orbital-exicted state. Therefore, the results of relativistic and nonrelativistic calculations of splitting of the s^6 and s^4p^2 levels may be treated as only a rough estimate of the upper and lower boundaries of the splitting. The calculations in terms of nonrelativistic oscillator model [Li77] have given [Ob79a] $M_{ho}(s^4p^2) - M_{ho}(s^6) \simeq -200$ MeV, whereas we have obtained in the bag model that $M_{bag}(s^4_{1/2}p^2_{1/2}) - M_{bag}(s^6_{1/2}) \simeq +250$ MeV. It will be noted that the latter value may be reduced substantially bearing in mind that the spin-orbital splitting in the relativistic model is significant, namely $M_{bag}(s^4_{1/2}p^2_{3/2}) - M_{bag}(s^4_{1/2}p^2_{1/2}) \simeq -200$ MeV (the accurate calculations in the $s^4_{1/2}p^2_{3/2}$ configuration implies the use of nonspherical form of the bag [Jo76; Mu78] and have not been carried out yet). It may be expected that the relativistic model will give a strong mixing of the $s^6_{1/2}$ and $s^4_{1/2}p^2_{3/2}$ configurations and that, because of the nonsphericity of the bag for $s^4_{1/2}p^2_{3/2}$ the quadrupole moment of the given state will be different from zero. In this case, the small-distance region in deuteron ($R \lesssim 0.5$ fm) will make a non-zero contribution to the total quadrupole moment of deuteron. The calculations of the contribution of the region of small NN distances to the nonadditive part of deuteron magnetic moment in terms of the quark model is also of interest to nuclear physics as an alternative to the meson exchange currents in the region $R \lesssim 0.5$ fm [Du73; Mi78].

However, the static characteristics of deuteron (quadrupole moment, magnetic moment, charge radius), which depend little on the small-distance region, cannot be critical with respect to the form of configurations s^6 or s^4p^2. The deuteron electromagnetic formfactors are much more sensitive. At large transfer momenta, $-q^2 \gtrsim 1$ GeV^2/c^2, such

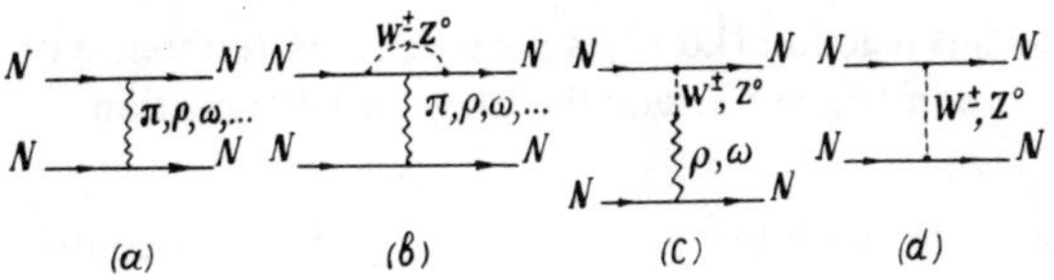

Fig. 26 Terms of the effective Hamiltonian of weak interaction H_{NN}^{W}. (a) The one bozon exchange term of strong interaction H_{NN}^{OBE}.

formfactors are essentially dependent on the form of the wave function at distances $R \lesssim 0.5$ fm. The set of the data on the electric monopole and quadrupole (separated) and magnetic dipole formfactors at large $-q^2$ may be a good test of the quark structure of deuteron in the region of distances $R \lesssim 0.5$ fm. The only thing known to us at present is that the summed data on the deuteron electric formfactors in the range $-q^2 \lesssim 6\ \mathrm{GeV}^2/c^2$ do not contradict the existence of a deuteron quark structure [Ar75; Br76].

The experiments in which the effects of spatial parity nonconservation in nuclei are observed [Cr76] may also be a sensitive test of the quark structure in the region of small internucleonic distances [Ne78]. The effective weak Hamiltonian H_{NN}^{W} describing the parity violation in NN-system is essentially dependent on the model of strong interaction. In the OBEP model H_{NN}^{W} is the sum of weak corrections to the diagram of one-meson exchange. Some of these corrections are shown in Fig. 26 in the form of the diagrams with meson vertex modified by the virtual emission of a weak boson ($W^{\pm}$, Z^0). In the conventional approach [Cr76; Du78b] the diagram of the contact nucleon interaction associated with direct exchange by the $W^{\pm}$ and Z^0 bosons (see Fig. 26d) is disregarded since the contribution from this diagram is proportional to $\Phi_{NN}(R=0)$ and should be vanishing in the NN interaction models with repulsive core. In the quark approach, however, the contact term of the weak interaction of ij-quark pair proportional to the δ-function $\delta^{(3)}(\mathbf{r}_i - \mathbf{r}_j)$ should be averaged over a certain finite volume in which the 6q-system is localized. The corresponding orbital matrix element

$$I_{ij}^{W} = \langle \phi_{NN}^{X}(\mathbf{r}_1 \dots \mathbf{r}_i \dots \mathbf{r}_6)\, \chi_{NN}^{CST}\, |W_{ij}\,\delta^{(3)}(\mathbf{r}_i - \mathbf{r}_j)|\, \Phi_{NN}^{X}(\mathbf{r}_1 \dots \mathbf{r}_j \dots \mathbf{r}_6)\, \chi_{NN}^{CST} \rangle\, d^3\mathbf{r}_1 \dots d^3\mathbf{r}_6 \tag{7.184}$$

(here W_{ij} is the spin-isospin part of the effective weak Hamiltonian from the unified gauge theory [We67]; see for example the model in ref. [Gl70]) is different from zero in any case. The quark diagram (Fig. 27), which makes a contribution proportional to the integral (7.184), should be included together with the diagrams of the type shown in Fig. 26b, c when calculating H_{NN}^{W}. This opinion was repeatedly expressed [Fr76, Ne78]. The contribution from the diagram shown in Fig. 27 to the parity nonconservation effect for the reaction of radiative capture of thermal neutrons by hadrogen $n + p \to d + \gamma$ was estimated in [Du78a]. The circular polarization of γ-quanta

$$P_{\gamma} = \frac{d(\sigma^{(+)} - \sigma^{(-)})/d\Omega}{d(\sigma^{(+)} + \sigma^{(-)})/d\Omega} \tag{7.185}$$

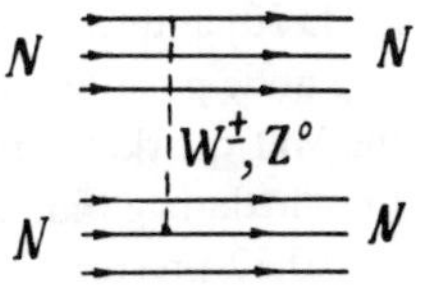

Fig. 27
The contact term of the weak interaction H_{qq}^{W}.

which is experimentally observable in this reaction [Lo 72] is the result of interference of the amplitudes of the regular M1-transition $^1S_0 \to {}^3S_1$ and the irregualr E1-transition associated with parity violation.

The long-range part of H_{NN}^{w} associated with the π-meson exchange does not contribute to the given E1-transition [Da 72] and, therefore, the data on P_γ may be a proper test of small distances in deuteron. The contribution to P_γ from the diagram shown in Fig. 27 was calculated in [Du 80]. Consideration was given to two phenomenological models of NN-interaction at small distances, namely the model with core [Re 68] and the model with nodal wave function [Ne 75]. The wave functions Φ_{NN}^{core} (R) and Φ_{NN}^{node} (R) were used in the representation (7.1) for the quark wave function Ψ_{NN} and, after that, the function Ψ_{NN} (q^6) was expanded in the basis of the quark shell model (see the expansion (7.17)). P_γ was calculated to the approximation (7.18), i.e. only the component $C_0 \Psi_0$ corresponding to the configuration s^6 was used in the case of Φ_{NN}^{core}, and the component $C_1 \Psi_1$ corresponding to the configuration $s^4 p^2$ was used in the case of Φ_{NN}^{node}. The quark wave functions of nucleons $\Psi_N(q^3)$ in the representation (7.1) were taken in the form of the TISM oscillator functions (7.179); the value of the oscillator parameter $\Omega = 7.56\ \mathrm{fm}^{-2}$ characterizing the functions (7.169) was taken from the ref. [Li 70] where the nucleon formfactor calculated with the functions (7.169) was fitted to the experimental data in the $0 < -q^2 \lesssim 25\ \mathrm{GeV}^2/c^2$ range of transferred momenta. It will be noted that, in order to satisfy the data in such a broad range of momenta, the authors of [Li 70] had to introduce the quark formfactors $F_q(q^2)$ which can effectively include the contribution from the ρ-meson pole to the nucleon formfactor:

$$F_q(-q^2) = \frac{1}{1 - \dfrac{q^2}{m_\rho^2}}, \quad m_\rho = 0.78\ \mathrm{GeV}. \tag{7.186}$$

On the contrary, when calculating the contribution from the diagram shown in Fig. 27 in [Du 79], the quark formfactors were not introduced $(F_q(-q^2) = 1)$ because the application of the form (7.186) would mean practically a double inclusion of the ρ-meson exchange which was already included in the diagrams shown in Fig. 26b, c.

It is of interest to note that, according to the fractional parentage expansion (7.178), the shell state Ψ_1 of the TISM basis (the $s^4 p^2$ configuration) has in its NN-component the node function (7.180) in which the node is located at the point $R_{node} = \Omega^{-1/2} = 0.365$ fm. This value should be compared with the position of the wave function node in the phenomenological model [Ne 75] whose parameters were fitted to the NN-scattering data at energies $0 < E < 400$ MeV. According to [Ne 75], $R_{node} = 0.36$–0.4 fm.

Such a coincidence between the values of R_{node} obtained as a result of pure phenomenological fitting of the absolutely different experimental data (for nucleon formfactor in [Li 70] and for phase shifts of NN-scattering in [Ne 75]) does not seem to be fortuitous. The node position at the point $R_{node} \simeq 0.36$ fm is in a bad agreement with the radius of the MIT quark bag $\bar{R} = 1.34$ fm for the $s^4 p^2$ state but, in turn, is in good agreement with the "little bag" model [Br 79]. It will be noted that the small radius in little bag $R \simeq 0.32$ fm (for s^3) makes it possible to obtain the necessary colour-magnetic splitting of hadron levels at the correct value of the constant of QCD $\alpha_S = 0.55$, the fact that is an undoubted advantage of this model over the MIT bag where the value $\alpha_S = 2.2$ is used.

Return now to the quark calculations of P_γ in [Du80]. The value (7.185) may be written through the amplitudes of the E1 and M1 transitions $T_{E1}^{M\lambda}$ and $T_{M1}^{M\lambda}$ (λ is the helicity of the emitted photon; M is the spin projection of the deuteron produced in the final state) as

$$P_\gamma = -2\,\mathrm{Re}\,\frac{T_{E1}^{+1-1}}{T_{M1}^{+1-1}}\,. \tag{7.187}$$

Here, the amplitude $T_{E1}^{M\lambda}$ is proportional to the orbital integrals (7.184) of the contact term $W_{ij}\,\delta^{(3)}(\mathbf{r}_i - \mathbf{r}_j)$ and to the weak constant $G = 10^{-5}/m_N^2$. The formula for P_γ obtained in [Du79] is of the form

$$P_\gamma^{(n)} = -2C_n\,\frac{G}{\sqrt{2}}\,\frac{\dfrac{\hbar}{2m_q}\sum\limits_{i<j}^{6} I_{ij}^{W}(n, {}^1S_0)}{\dfrac{\hbar}{2m_N}(\mu_P - \mu_n)\,E_d\,I(d, {}^1S_0)}\,, \tag{7.188}$$

where $I(d, {}^1S_0) = \int\limits_0^\infty \Phi_d(R)\,\Phi_{^1S_0}^{(+)}(R)\,4\pi R^2\,dR$ is the integral of overlapping of the deuteron wave function Φ_d and the scattering function in the ${}^1S_0$ state at $E_{NN} \to 0$;

$$I_{ij}^{W}(n, {}^1S_0) = \left\langle \Psi_n(q^6)\,|\,W_{ij}\,\delta^{(3)}(\mathbf{r}_i - \mathbf{r}_j)\,|\,\frac{1}{N_A}\,A\;\;\Phi_{^1S_0}\,\Psi_N(q^3)\,\Psi_N(q^3)\right\rangle$$

is the matrix element of th e contact term for the quark functions of the initial and final states; $m_q \simeq \frac{1}{3}\,m_N$ is the effective mass of quark used in the Feynman propagators which are relevant to the quark lines of the diagram shown in Fig. 27; $E_d = 2.2$ MeV is the deuteron binding energy; $\mu_P - \mu_n = 4.7$ is the difference in the anomalous magnetic moments of proton and neutron. The results of the calculations [Du79] are as follows.

(i) The model with core:

$C_0 = 0.06,\ P_\gamma^{(0)} = -38.4 \times 10^{-8}$;

(ii) the model with node:

$C_1 = 0.14,\ P_\gamma^{(1)} = -8.4 \times 10^{-8}$.

It can be seen that a significant difference is observed in individual factors in the expression (7.188) for P_γ. The wave function integral of overlapping with the s^4p^2 shell configuration ($C_1 = 0.14$) in the model with node is almost in 5 times as high as that of overlapping with the s^6 shell configuration ($C_0 = 0.06$) in the model with core. (See also Figures 28 and 29.)

The comparison between the results of the quark calculations of P_γ and conventional calculations in terms of ρ-meson exhange [Du78b] shows that the absolute values of the quark contributions are in excess of that of the ρ-meson contributions. This means that the P_γ data may actually be a proper test for the region of small NN distances in the

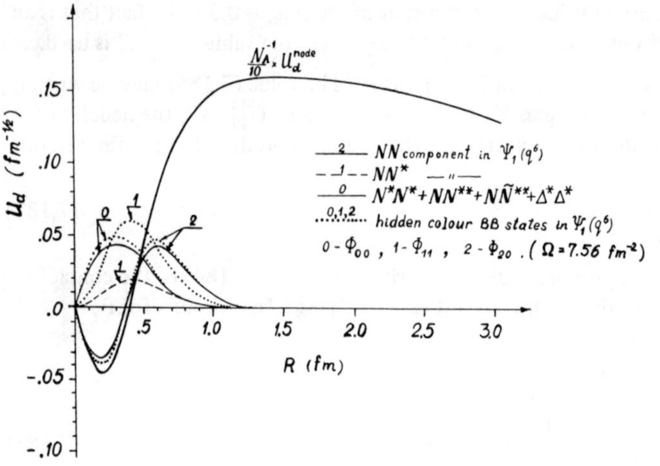

Fig. 28 The deuteron wave function, which is antisymmetrized with respect to quark permutations (see eq. (7.1)). In the overlap region a decomposition interms of barion-barion states is performed (see fractional parentage expantion (7.178)). Here $u_d^{mode}(R)$ is the nodal deuteron wave function of the NN interaction model with one forbidden state [Ne75].

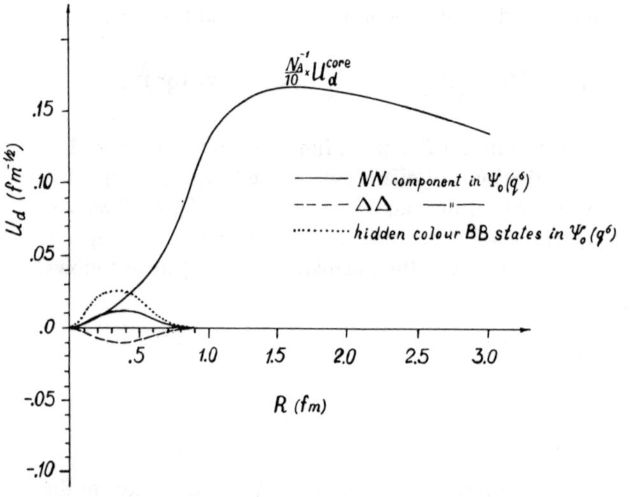

Fig. 29 The same as Fig. 28, but here $u_d^{core}(R)$ is the deuteron wave function of the NN interaction model with repulsive core [Re68].

deuteron. The experimental data on P_γ [Lo72] should have been obtained within a higher accuracy. More accurate data are also necessary for the effect of parity non-conservation in the polarized-proton scattering to be used in the calculations of the above presented type.

References

[Aa 66] *Aaron, R., P.E. Shanley* and *Y. Y. Yam*:Phys. Rev. **142**, 608 (1966).

[Ac 72] *Achmadhodjaev, B., V.B. Beliaev* and *A. L. Zubarev*: Preprint JINR P4-5318 (1972).

[Ae 78] *Aert, A. Th.M., P.J. G. Mulders* and *J.J. de Swart*: Phys. Rev. **D 17**, 260 (1978).

[Ag 77] *Aguilera-Navarro, V.C.* and *O. Portilho*: Ann. Phys. **107** (1977) 126.

[Al 70] *Allrow, M.* et al.: Nucl. Phys. **B 23**, 445 (1970).

[Al 65t *De Alfaro, V.* and *T. Regge*: Potential Scattering. North-Holland Publ. Comp., 1965. Amsterdam.

[Al 75] *Aljadir, M.N.*: Nucl. Phys. **A 251**, 156 (1975); **A 318**, 441 (1979).

[Am 67] *Amos, K.A.: V.A. Madsen and I.E. McCarthy*, Nucl. Phys., **A 34**, 103 (1967).

[Am 70] *Amusja, M.A.* and *B.E. Starodubsky*: Izv. AN SSSR, ser. Fiz., **34**, 1699 (1970).

[An 79] *Ando, T., K. Ikeda* and *A. Tohsaki-Suzuki*: Progr. Theor. Phys., **61**, 342 (1979).

[Ar 75] *Arnold, R., B.T. Chertok* et al.: Phys. Rev. Lett. **35**, 776 (1975).

[Au 72] *Auerbach, N., J. Hüfner, A.K. Kerman* and *C.H. Shakin*: Rev. Mod. Phys., **44**, 48 (1972)

[Au 73] *Austern, N.*: Phys. Letters **46 B**, 590 (1973).

[Au 77] *Auer, I.P.* et al.: Phys. Lett. **67 B**, 113 (1977).

[Au 78] *Austern, N.* et al.: Phys. Rev. **C 18**, 1577 (1978).

[Ba 59] *Balashov, V.V., V.G. Neudatchin, Yu.F. Smirnov* and *N.P. Yudin*: ZhETF, **37**, 1385 (1959).

[Ba 66] *Baz, A.I., V.F. Dyomin* and *I.I. Kuz'min*: Yad. Fiz., (Sov. Journ.) **4**, 737 (1966).

[Ba 68] *Barsella, B., L. Lovitch* and *S. Rosati*: Nucl. Phys. **A 117**, 638 (1969).

[Ba 71a] *Banerjee, M.K., C. Levinson, M. Shister* and *D. Zallman*: Phys. Rev. **C 3**, 509 (1971);

[Ba 71b] *Baz, A.I., Ya.B. Zeldovich* and *A.M. Perelomov*: "Scattering, Reactions and Decays in Non-relativistic quantum mechnics" (in Russian), Nauka, 1971.

[Ba 73] *Baldin, A.M.* et al.: Yad. Fiz. **18**, 79 (1973).

[Ba 78a] *Bang, J.* et al.: Nucl. Phys. **A 309**, 381 (1978).

[Ba 78b] *Barbour, I.M.* and *D.K. Ponting*: Preprint "The baryon and the three-body problem", Glasgow Univ., 1978, Glasgow.

[Ba 79a] *Bang, J.* and *G. Gignoux*: Nucl. Phys. **A 313**, 119 (1979).

[Ba 79b] *Barbour, I.M.* and *D.K. Ponting*: Nucl. Phys. **B 149**, 534 (1979).

[Be 70a] *Beljaev, V.B.* et al.: Preprint JINR, P 4–5000, Dubna 1970.

[Be 70b] *Beregi, P., I. Lovas* and *J. Revai*: Ann. Phys. (N.Y.) **61**, 57 (1970).

[Be 71] *Bertero, M.* and *G. Passatore*: Nuovo Cim. **2 A**, 579 (1971).

[Be 73] *Bencze, Gy*: Report TET 12–73, Univ. Helsinki; Lecture at the Finnish Summer School on Nucl. Phys., Liperi 1973.

[Be 75b] *Beregi, P., I. Lovas* and *J. Revai*: Ann. Phys. (N.Y.) **61**, 57 (1975).

[Be 79a] *Bencze, Gy* and *C. Chandler*: Preprint Centr. Research Inst. Phys., KFKI-1979-14, Budapest, 1979.

[Be 79b] *Bencze, Gy*: 1979 private communication

[Bj 64] *Bjorken, J.D.* and *S.D. Drell*: Relativistic Quantum Mechanics. McGraw-Hill Book Company, N.Y. (1964).

[Bo 79] *Bohr, A.* and *B.R. Mottelson*: "Nuclear Structure", V.I., 1969, W.A. Benjamin N.Y.-Am.

[Br 74] *Brayshaw, D.D.*: Phys. Rev. **D 10**, 2827 (1974).

[Br 76] *Brodsky, S.J.* and *B.T. Chertok*: Phys. Rev. **D 14**, 3003 (1976).

[Br 78] *Brinati, J.R.* and *G.W. Bund*: Nucl. Phys. **A 306**, 139 (1978).

[Br 79] *Brown, G.E.* and *M. Rho*: Phys. Rev. Lett. **82 B**, 177 (1979).

[Bu 70] *Burke, P.G.* and *A.L. Sin Failam*: J. Phys. B.; Atom. Molec. Phys. **3**, 641 (1970).

[Bu 72] *Burke, P. G.* and *N. Chandra*: J. Phys. B: Atom. Molec. Phys. **5**, 1696 (1972).

[Bu 75] *Buck, B., C.B. Dover* and *J.P. Vary*: Phys. Rev. **C 11**, 1803 (1975).

[Bu 77a] *Buck, B.* and *A.A. Pilt*: Nucl. Phys. **A 280**, 133 (1977);

[Bu 77b] *Buck, B., H. Friedrich* and *A.A. Pilt*: Nucl. Phys., **A 290**, 205 (1977).

[Bu 77c] *Buckley, B.D.* and *P. G. Burke*: J. Phys. B: Atom, Molec. Phys. **10**, 725 (1977).

[Ca 78] *Cattapan, G.* and *V. Vanzani*: Proceed. Int. Conf. on Few-Body Systems and Nuclear Forces (Lecture Notes in Physics) 82, Conf. No. 9.5., 1978.

[Ca 79] *Callan, C. G., Jr., R.F. Dashen* and *D.J. Gross*: Phys. Rev. **D 19**, 1826 (1979); ibid **D 17**, 2717 (1978).

[Ce 77] *Celmaster, W.*: Phys. Rev. **D 15**, 1391 (1977).

[Ch 64] *Chlebowska, D.*: Acta. Phys. Polonica **25**, 313 (1964).

[Ch 74] *Chodos, A., R.L. Jaffe, K. Johnson, C.B. Thorn* and *V.F. Weiskopf*: Phys. Rev. **D 9**, 3471 (1974). In addition, see, for example, Chodos, A., Jaffe, R.L. and Thorn, C.B., Phys. Rev. **D 10**, 2599 (1974), Jaffe, R.L., Phys. Rev. **D 11**, 1953 (1975); DeGrand, T., Jaffe, R.L., Johnson, K. and Kiskis, J., Phys. Rev. **D 12**, 2060 (1975).

[Ch 75] *Chodos, A.* and *C.B. Thorn*: Phys. Rev. **D 12**, 2733 (1975).

[Ch 76] *Chandra, N.* and *A. Temkin*: Phys. Rev. **A 13**, 188 (1976).

[Ch 77] *Charnomordic, B., C. Fayard* and *G.H. Lamot*: Phys. Rev. **C 15**, 864 (1977).

[Cl 74] *Clement, D., E. W. Schmid* and *A. G. Teufel*: Phys. Letters **49 B**, 308 (1974).

[Co 43] *Courant, R.*: Bull. Am. Math. Soc. **49**, 1 (1943).

[Cr 76] *Craver, B.A., E. Fishbach, Y.E. Kim* and *A. Tubis*: Phys. Rev. **D 13**, 1376 (1976).

[Da 72] *Danilow, G.S.*: Yad. Fiz. **14**, 443 (1972).

[Da 74] *Davison, N.E., P. Fintz* and *A. Gallmann*: Nucl. Phys. **A 220**, 166 (1974).

[De 75] *Demkov, Yu.N.* and *V.N. Ostrovsky*: "The zero-range potential method in atomic physics", Leningrad Univ. 1975, p. 20–21.

[Dr 72] *Drell, S.D.* and *K. Johnson*: Phys. Rev. **D 6**, 3248 (1972).

[Du 73] *Dushenko, V.F.* and *A.P. Kobushkin*: Nuovo Cimento Lett. 7, 529 (1973).

[Du 78a] *Dubovik, V.M.* and *A.P. Kobushkin*: Preprint ITP-78-85E Kiev (1978).

[Du 78b] *Dubovik, V.M.* and *V.S. Zamiralov*: Lett. Nuovo Cimento, 22, 21 (1978); Dubovik, V.M., Zamiralov, V.S. and Zenkin, S.V., JINR Preprint E 2-12381, Dubna (1979).

[Du 80] *Dubovik, V.M.* and *I. T. Obukhovsky*: JINR Preprint E 1–80–555 Dubna (1980)

[El 53] *Elliott, J.P., J. Hope* and *H.A. Jahn*: Trans. Roy. Soc., **A 246**, 241 (1953).

[El 58] *Elliott, J.P.*: Proc. Roy. Soc. **A 245**, 128, 562 (1958).

[En 74] *Englefield, M.J.* and *H.M.S. Shoukry*: Progr. Theor. Phys. **52**, 1554 (1974).

[Fa 63] *Faddeev, L.D.*: Mathematical Aspects of the Three-Body Problem in the Quantum Scattering Theory, (Proc. Steklov Math. Inst. v. **61**, 1963) (Israel Program for Scientific Translations, Jer. 1965).

[Fe 58] *Feshbach, H.*: Ann. Phys. (N.Y.) **5**, 357 (1958).

[Fe 62] *Feshbach, H.*: Ann. Phys. **19**, 287 (1962).

[Fe 71] *Feyrman, R.P., M. Kisslinger* and *F. Ravndal*: Phys. Rev. **D 3**, 2706 (1971).

[Fe 73] *Feshbach, H.*: Topics in the Theory of Nucl. Reactions, in "Reaction Dynamics". Part II (Gordon & Breach, N.Y. 1973).

[Fl 52] *Flowers, B.H.*: Proc. Roy. Soc., **A 212**, 248 (1952).

[Fo 76] *Fong, J.C., G. Igo* and *V. Perez-Mendez*: Nucl. Phys. **A 262**, 365 (1976).

[Fr 73] *Fritzsch, H., M. Gell-Mann* and *H. Leutwyller*: Phys. Lett. **47 B**, 365 (1973).

[Fr 76] *Frankfurt, L.L.* and *M.I. Strikman*: Preprint LNPI No. 238 Leningrad (1976).

[Fu 71] *Fuller, R.C.*: Phys. Rev. **C 3**, 1042 (1971).

[Fu 77] *Fukushima, Y.* and *M. Kamimura*: Fizika, **9**, Suppl. 2 (1977) 36.

[Fu 79] *Fujiwara, Y.*: Preprints Kyoto Univ., RIEP-351, RIEP-352, RIEP-353, 1979.

[Ga 74] *Gambhir, B.L.* and *J.J. Griffin*: Phys. Letters **50 B**, 407 (1974).

[Ga 76] *Garsia-Calderon, G.* and *R. Peierls*: Nucl. Phys., **A 265**, 443 (1976).

[Gi 72] *Gignoux, C.* and *A. Laverne*: Phys. Rev. Lett. **29**, 436 (1972).

[Gi 77] *Giraud, B.G.*: Fizika **9**, Suppl 4, 345 (1977) (Proceed. Int. Symp. Nucl. Collisions and Their Micr. Description).

[Gl 70] *Glashow, S.L., J. Illiopoulos* and *L. Maiani*: Phys. Rev. **D 2**, 185 (1970).

[Gl 76a] *Glöckle, W.* and *D. LeTurneux*: Nucl. Phys. **A 269**, 16 (1976).

[Gl 76b] *Glowinski, R., J.L. Lions* and *R. Tremolieres*: "Analyze numerique des inequations variationnelles", v. 1,2., Dunod, Paris, 1976.

[Go 64] *Goldberger, M.L.* and *K.M. Watson*: "Collision Theory". John Wiley and Sons, Inc., N.Y. 1964.

[Go 76] *Golovanova, N.F., I.M. Il'in, V.G. Neudatchin, Yu.F. Smirnov* and *Yu.M. Tchuvil'sky*: Nucl. Phys., **A 262**, 444 (1976); **A 285**, 531 (1977) Err.

[Gr 76] *DeGrand, T.A.* and *R.L. Jaffe*: Ann. Phys. (N.Y.), **100**, 425 (1976); ibid **101**, 395 (1976).

[Gu 71] *Gurbanovich, I.S., Yu.F. Smirnov, V.N. Tolstoy*: Czech. J. Phys. **B 21** (1971)

[Gu 75] *Gunion, J.F.* and *R.S. Willey*: Phys. Rev. **D 12**, 174 (1975).

[Ha 62] *Hamada, T.* and *I.D. Johnston*: Nucl. Phys. **34**, 382 (1962). See also, for example, Breit G. et al. Phys. Rev. **126**, 881 (1962); Vinh Mau R. in "Mesons in Nuclei", ed. by Rho M. and Wilkinson, D., North Holland, Amsterdam (1978).

[Ha 64] *Hamermesh, M.*: Group Theory and its Application to Physical Problems. Addison-Wesly Publishing Company, Inc., N.Y. (1964).

[Ha 65] *Han, M.Y.* and *J. Numbu*: Phys. Rev. **139**, B 1006 (1965).

[Ha 69] *Hamza, K.A.* and *S. Edwards*: Phys. Rev. **181**, 1494 (1965).

[Ha 78] *Hasenfratz, P.* and *J. Kuti*: Phys. Reports **40 C**, 75 (1973).

[Hi 77] *Hidaka, K.* et al.: Phys. Lett. **70 B**, 479 (1977).

[Ho 73] *Horgan, R.* and *R.H. Dalitz*: Nucl. Phys. **B 66**, 135 (1973).

[Ho 74] *Horiuchi, H.*: Progr. Theor. Phys. **51** (1974) 1266.

[Ho 75] *Horiuchi, H.*: Progr. Theor. Phys. **53** (1975) 447.

[Ho 76] *Horiuchi, H.*: Progr. Theor. Phys. **55** (1976) 1448.

[Ho 77a] *Horiuchi, H.*: Proc. Intern. Symposium on Nuclear Collisions and Their Microscopic Description, Bled, Yugoslavia, 1977, Fizika, **9**, Suppl. 3 (1977) 251.

[Ho 77b] *Horiuchi, H.*: Progr. Theor. Phys. **58**, 204 (1977).

[Ho 77c] *Horiuchi, H.*: Proc. Intern. Symp. on Nucl. Collisions, Bled (Yugosl.), Fizika, **9**, Suppl. 3, 251 (1977).

[Ho 77d] *Horiuchi, H.*: Progr. Theor. Phys. Suppl. No. 62, pp. 90-190 (1977).

[Ho 77e] *Hoshizaki, N.*: Progr. Theor. Phys. **58**, 716 (1977) ibid; **60**, 1796 (1978); **61**, 129 (1979).

[Ho 78] *Horiuchi, H.*: in "Clustering Aspects of Nuclear Structure and Nuclear Reactions". (Proc. III Intern. Conf., Winnipeg, 1978) p. 144. W.T.H. van Oers, Y.P. Svenne, J.S.C. McCee, W.R. Falk Editors, AIP Conference Procedings Nr. 47, AIP, N.Y., 1978.

[Hφ 79] *Hφgaasen, H.* and *P. Sorba*: Nucl. Phys. **B 150**, 427 (1979).

[Hu 27] *Hund, F.*: Zs. Phys. **43**, 788 (1927).

[Hu 37] *Hund, F.*: Zs. Phys. **105**, 202 (1937).

[Il 74] *Illiopoulos, J.I.*: Proceedings of the XVII Intern. Conf. on High Energy Physics, London (1974); Illiopoulos, J.I., An Introduction to Gauge Theories, CERN Preprint 76-11, Geneva (1976).

[Ja 51] *Jahn, H.A.* and *H. Van Wieringen*: Proc. Roy. Soc. **A 209**, 502 (1951).

[Ja 58] *Jauch, J.M.*: Helv. Phys. Acta **31**, 127 (1958).

[Ja 70] *Jackson, A.D., A. Lande* and *P. U. Sauer*: Nucl. Phys. **A 156**, 1 (1970).

[Ja 71] *Jackson, A.D., A. Lande* and *P. U. Sauer*: Phys. Lett. **35 B**, 365 (1971)

[Ja 77] *Jaffe, R.L.*: Phys. Rev. Lett. **38**, 195 (1977); Erratum **38**, 617 (1977).

[Ji 74] *Jigunov, V.P.* and *B.N. Zakharjev*: "The methods of strong coupling channels in scattering theory" (in russian) M. Atomizdat, 1974.

[Jo 70] *Johnson, R.C.* and *P.J.R. Soper*: Phys. Rev. **C 1**, 976 (1970).

[Jo 72] *Johnson, R.C.* and *P.J.R. Soper*: Nucl. Phys. **A 182**, 613 (1972).

[Jo 76] *Johnson, K.* and *C.B. Thorn*: Phys. Rev. **D 13**, 1934 (1976).

[Ka 35] *Kaczmarz, A.* and *H. Steinhaus*: "Theorie der Orthohonalreihen" (Warszawa-Lwow) (1935).

[Ka 61] *Kaplan, I.G.*: Zh. Eksp. Theor. Fiz., **41**, 560 (1961); ibid, **41**, 790 (1961).

[Ka 69] *Kaplan, I.G.*: Symmetry of Many-Electron Systems, Nauka, Moscow (1969) (in Russian).

[Ka 75] *Kanada, H., T. Kaneko* and *S. Saito*: Progr. Theor. Phys., **54**, 747 (1975).

[Ka 77a] *Kamae, T.* et al.: Phys. Rev. Lett. **38**, 468 (1977).

[Ka 77b] *Kamae, T.* and *T. Fujita*: Phys. Rev. Lett. **38**, 471 (1977).

[Ka 77c] *Kamimura, M., Y. Fukushima* and *A. Tohsaki-Suzuki*: Fizika, **9**, Suppl. 2 (1977) 33.

[Ke 74] *Kellet, B.H.*: Ann. Phys. (N.Y.) **87**, 60 (1974).

[Ko 65] *Kowalski, K.L.*: Phys. Rev. Lett. **15**, 798 (1965).

[Ko 69] *Kokkedee, J.J.J.*: The Quark Model. W.A. Benjamin, Inc., N.Y.–Amsterdam (1969).

[Ko 72] *Korennoy, V.P., I. V. Kurdyumov, V.G. Neudatchin* and *Yu.F. Smirnov*: Phys. Lett. **40 B**, 607 (1972); Yad. Fiz., **17**, 750 (1973).

[Ko 74] *Kogut, J.* and *L. Susskind*: Phys. Rev. **D 9**, 3501 (1974).

[Kr 66] *Kreps, R.E.* and *P. Nath*: Phys. Rev. **152**, 1475 (1966).

[Kr 67] *Kramer, P.*: Zs. Phys. **205**, 181 (1967); ibid **216**, 68 (1969).

[Kr 69] *Kramer, P.* and *M. Moshinsky*: Nucl. Phys. **A 125**, 321 (1969).

[Kr 74] *Krasnopol'sky, V.M.* and *V.I. Kukulin*: Yad. Fiz. **20**, 883 (1974).

[Kr 75] *Krasnopol'sky, V.M.* and *V.I. Kukulin*: Proced. II Inter. Conf. on Clustering Phenomena in Nuclei, Maryland, 1975.

[Kr 77] *Krasnopol'sky, V.M.* and *V.I. Kukulin*: J. Phys. G: Nucl. Phys. **3**, 795 (1977).

[Kr 79] *Kramer, P., G. John* and *D. Schenzle*: Group Theory and the Interaction of Composite Nucleon Systems. Vieweg, Braunschweig 1980.

[Ku 70] *Kurdyumov, I. V., Yu.F. Smirnov, K. V. Shitikova* and *S.H. El-Samarae*: Nucl. Phys. **A 145**, 593 (1970).

[Ku 75a] *Kukulin, V.I.*: Isv. Akad. Nauk (ser. fiz.) **39**, 535 (1975).

[Ku 75b] *Kukulin, V.I., V.G. Neudatchin* and *Yu.F. Smirnov*: Nucl. Phys. **A 245**, 429 (1975).

[Ku 76a] *Kukulin, V.I.* and *V.N. Pomerantsev*: Theor. Math. Fiz. (Soviet Journal of Theoretical and Mathematical Physics) **27**, 373 (1976).

[Ku 76b] *Kukulin, V.I., V.G. Neudatchin* and *V.N. Pomerantsev*: Yad. Fiz., **24**, 298 (1976); J. Phys. G: Nucl. Phys., **4**, 1409 (1978).

[Ku 77a] *Kukulin, V.I.*: Fizika **9**, Suppl. **4**, 395 (1977) (Proceed. Int. Symp. on Nucl Collisions and Their Microscopic Description Bled, Sept. 1977).

[Ku 77b] *Kukulin, V.I.* and *V.M. Krasnopol'sky*: Czech. Phys. Journ. **B 27**, 290 (1977).

[Ku 78a] *Kukulin, V.I.* and *N. V. Pomerantsev*: Yad. Fiz., **27**, 1668 (1978).

[Ku 78b] *Kukulin, V.I.* and *V.N. Pomerantsev*: Ann. Phys., (N.Y.) **111**, 330 (1978).

[Ku 78c] *Kukulin, V.I.*: Appendix I. to the russian translation (1978) of the book by E. Schmidt [Sc 74a].

[Ku 78d] *Kukulin, V.I.* and *V.N. Pomerantsev*: Proceed. Inter. Conf. Few-Body Problem at Graz, 1978, Contributions, No. 9, 11, p. 383.

[Ku 81 a] *Kukulin, V.I., V.M. Krasnopol'sky* and *M.A. Myselhi*: Yad. Fiz., **4**, (1981).

[Ku 81 b] *Kukulin, V.I.* and *V.N. Pomerantsev*: Theor. Math. Phys. **47**, 244 (1981).

[Le 74] *Levinger, J.S.*: "The Two- and Three Body Problem", in Springer Tracts in Modern Physics, **71**, 1974, Springer Verl. Heidelberg, N.Y.

[Li 61] *Lippmann, B.A.* and *H.M. Schey*: Phys. Rev. **121**, 1112 (1961).

[Li 66] *Lipperheide, R.*: Nucl. Phys. **89**, 97 (1966).

[Li 70] *Light, A.L.* and *A. Pagnamenta*: Phys. Rev. **D 2**, 1150 (1970).

[Li 75] *Lichtenberg, D.B.* and *J.G. Wills*: Phys. Rev. Lett., **35**, 1055 (1975).

[Li 77] *Liberman, D.A.*: Phys. Rev. **D 16**, 1542 (1977).

[Lo 72] *Lobashev, V.M.* et al: Nucl. Phys. **A 197**, 241 (1972).

[Lo 75] *Lovas, I.*: Ann. Phys. **89**, 96 (1975).

[Ma 67] *Mailing, L., V.I. Kukulin* and *Yu.F. Smirnov*: Nucl. Phys. **A 103**, 681 (1967).

[Ma 69] *Mahaux, C.* and *H.A. Weidenmüller*: Shell-model approach to nuclear reactions (North-Holland, Amsterdam, 1969).

[Ma 77] *Matveev, V.A.* and *P. Sorba*: Nuovo Cim. Lett., **20**, 145 (1977);
Matveev, V.A. and *P. Sorba*: Nuovo Cimento **45 A**, 357 (1978).

[Ma 78] *Marciano, W.* and *H. Pagels*: Phys. Repts. 36, 137 (1978)

[Me 76] *Merkuriev, S.P., A. Laverne* and *G. Gignoux*: Ann. Phys. (N.Y.) **99**, 30 (1976); Proc. 1977

[Me 79] *Merkuriev, S.P.* and *S.A. Posdneev*: Yad. Fiz. **30**, 941 (1979).

[Mi 76] *Miller, D.* et al.: Phys. Rev. Lett. **36**, 763 (1976);
Miller, D. et al.: Phys. Rev. **D 16**, 2016 (1977).

[Mi 78] *Mitra, A.N.*: Phys. Rev. **D 17**, 729 (1978).

[Mo 69] *Moshinsky, M.*: Harmonic Oscillator in Modern Physics: from Atoms to Quarks, Gordon and Breach, N.Y. 1969.

[Mo 70] *Moshinsky, M.* and *C. Quesne*: J. Math. Phys., **11**, 1631 (1970).

[Mu 78] *Mulders, P.J.G., A.Th.M. Aert* and *J.J. de Swart*: THEF Preprint NYM-78. The Netherlands, Nijmegen (1978).

[Ne 66] *Newton, R.*: "Scattering theory of waves and particles", McGraw Hill Book Co. N.Y. 1966.

[Ne 69] *Neudatchin, V.G.* and *Yu.F. Smirnov*: "Nucleon Clusters in Light Nuclei", Nauka, Moscow, 1969 (in Russian).

[Ne 71] *Neudatchin, V.G., V.I. Kukulin, V.L. Korotkikh* and *V.P. Korennoy*: Phys. Lett. **34 B**, 581 (1971).

[Ne 72] *Neudatchin, V.G., V.I. Kukulin, A.N. Bojarkina* and *V.P. Korennoy*: Nuovo Cim. Lett., **5**, 834 (1972).

[Ne 75] *Neudatchin, V.G., I.T. Obukhovsky, V.I. Kukulin* and *N.F. Golovanova*: Phys. Rev. **C 11**, 128 (1975).

[Ne 77] *Neudatchin, V.G., Yu.F. Smirnov* and *R. Tamagaki*: Progr. Theor. Phys., **58**, 1072 (1977); Smirnov, Yu.F., Obukhovsky, I.T., V.G. Neudatchin and R. Tamagaki: Yad. Fiz., **27**, 860 (1978).

[Ne 78] *Neudatchin, V.G.*: in "Clustering Aspects of Nuclear Structure and Nuclear Reactions" (Proc. III Intern. Conf., Winnipeg, 1978), p. 469, W.T.H. van Oers, I.P. Svenne, J.S.C. McCee, W.R. Falk Editors, AIP Conference Proceedings No. 47, AIP, N.Y., 1978.

[Ne 79] *Neudatchin, V.G., Yu.F. Smirnov* and *N.F. Golovanova*: Clustering Phenomena and High-Energy Reactions, in "Advances in Nuclear Physics", Vol. 11 ed. by J.W. Negele and E. Vogt, Plenum Press, N.Y. (1979), p.p. 1–133.

[Ni 79] *Nishika, H., S. Saito, H. Kanada* and *T. Kaneko*: Preprint Nagoya Univ. DPNU 24–79, 1979.

[No 65] *Noyes, H.P.*: Phys. Rev. Lett. **15**, 538 (1965).

[Ob 79a] *Obukhovsky, I.T., V.G. Neudatchin, Yu.F. Smirnov* and *Yu.M. Tchuvil'sky*: Phys. Lett. **88 B**, 231 (1979).

[Ob 79b] *Obukhovsky, I.T.* and *Yu.F. Smirnov*: 30th All-Union Conference of Nuclear Spectroscopy, Leningrad, March 1980, Abstracts of Contributed Papers; I.T. Obukhovsky and V.M. Dubovik, JINR Preprint P 2-80-501, Dubna (1980)

[Ok 66] *Okai, S.* and *S. C. Park*: Phys. Rev. **145**, 787 (1966).

[Ok 80] *Oka, M.* and *K. Yazaki*: Progr. Theor. Phys. **66**, 556 (1981).

[Ob 82] *Obukhovsky, I.T., Yu.F.Smirnov, Yu.M.Tchuvil'sky*: J.Phys.A: Math. and General **15**, 7 (1982).

[Pi 71] *Picker, H.S., E.F. Redish* and *G.J. Stephenson*: Phys. Rev. **C 4**, 287 (1971).

[Po 75] *Pong, W.S.* and *N. Austern*: Ann. Phys. **93**, 369 (1975).

[Po 76] *Pomerantsev, V.N.*: Theor. Math. Phys. **22**, 94 (1976).

[Po 77] *Pomerantsev, V.N.*: Ghost states in four-body scattering Prepring MSU 1977.

[Po 78] *Pomerantsev, V.N.*: Dissertation, Moscow State Univ., 1978.

[Pr 77] See a special volume: Progr. Theor. Phys., Suppl. 62 (1977).

[Ra 49] *Racah, G.*: Phys. Rev. **76**, 1352 (1949).

[Ra 51] *Racah, G.*: Group Theory and Spectroscopy, Princeton, N.J. (1951).

[Re 68] *Reid, R. V.*: Ann. Phys. (N.Y.) **50**, 411 (1968).

[Re 70] *Redish, E.F., G.J. Stephenson* and *G.M. Lerner*: Phys. Rev. **C 2**, 1665 (1970).

[Ri 78] *Ribeiro, J.E.*: CFMC Preprint E-6178, Lisboa (1978); J.D. de Deus and J.E. Ribero, CFMC Preprint, Lisboa (1978). See also, J.E. Ribeiro , D. Phil. Thesis, Oxford (1978).

[Ro 65] *Rotter, I.*: Ann. der Phys. **16**, 242 (1965).

[Ro 78] *Romo, W.J.*: Nucl. Phys. **A 302**, 61 (1978).

[Ru 75] *DeRujula, A., H. Georgi* and *S. Glashow*: Phys. Rev. **D 12**, 147 (1975).

[Sa 54] *Sack, S., L.S. Beidenharn* and *G. Breit*: Phys. Rev. **93**, 321 (1954); see also: W. Pearce and P. Swan: Nucl. Phys., **78**, 433 (1966).

[Sa 68] *Saito, S.*: Progr. Theor. Phys. **40**, 893 (1968).

[Sa 69a] *Saito, S.*: ibid **41**, 705 (1969).

[Sa 69b] *Satchler, G.R.*, et al.: Nucl. Phys. **A 112**, 1 (1969).

[Sa 77] *Saito, S.*: Progr. Theor. Phys. Suppl., **62**, 11 (1977).

[Sc 74a] *Schmid, E.*: "The quantum mechanical three-body problem", Vieweg, Braunschweig, 1974.

[Sc 74b] *Scheerbaum, R.R.* and *C.M. Shakin*: Phys. Rev. **C 9**, 116 (1974).

[Sc 74a] *Schmid, E.*: "The quantum mechanical three-body problem", Vieweg, Braunschweig, 1974.

[Sc 77] *Schmid, E.*: Preprint "Two- and Three-Body Interactions of Composite Particles", Tubingen Univ., 1977.

[Sc 79] *Schmid, E.*: Talk at the Int. Conf. on Few-Body Problem, Dubna, 1979.

[Sh 63] *De Shalit, A* . and *I. Talmi*: Nuclear Shell Theory, N.Y. (1963).

[Sh 69] *Shanley, P.E.*: Phys. Rev. **187**, 1328 (1969).

[Sh 73] *Sheerbaum, R.R., C.M. Shakin* and *R.M. Thaler*: Ann. Phys., **76**, 333 (1973).

[Sh 77] *Shakin, C.M.* and *R.M. Thaler*: Phys. Rev. C 7, 494 (1973).

[Sl 71] *Sloan, I.H.*: Nucl. Phys. **A 168**, 211 (1971).

[Sl 78] *Slavnov, A.A.* and *L.D. Faddeev*: An Introduction to the Quantum Theory of Gauge Fields. Nauka, Moscow (1978) (in Russian).

[Sm 61] *Smirnov, Yu.F.* and *D. Chlevovska*: Nucl. Phys. **26**, 306 (1961); **27**, 177 (1961); **39**, 346 (1962).

[Sm 63] *Smirnov, Yu.F.* and *K. V. Shitikova*: Izv. Akad. Nauk SSSR, Ser. Fiz., **27**, 1442 (1963); I.V. Kurdumov, Yu.F. Smirnov, K.V. Shitikova and S.El. Samarai: Nucl. Phys. **A 145**, 593 (1970).

[Sm 74] *Smirnov, Yu.F., I.T. Obukhovsky, Yu.M. Tchuvilsky* and *V.G. Neudatchin*: Nucl. Phys. **A 235**, 289 (1974).

[Sm 78] *Smirnov, Yu.F.* and *Yu.M. Tchuvil'sky*: J. Phys. G: Nucl. Phys., **4**, L1 (1978).

[So 79] *So, S.I.* and *D. Strottman*: J. Math. Phys. **20** (1), 153 (1979); D. Strottman, J. Math. Phys. **20** (8), 1643 (1979).

[St 74a] *Stamp, A.P.*: Nucl. Phys. **A 220**, 137 (1974).

[St 74b] *Strauer, M.R.* and *P.U. Sauer*: Nucl. Phys. **A 131**, 1 (1974).

[Sw 55] *Swan, P.*: Proc. Roy. Soc. **A 228**, 10 (1955).

[Sw 71] *De Swart, J.J.* et al.: Springer Tracts in Modern Physics, ed. by Höhler, Springer, N.Y. (1971), Vol. **60**, p. 138.

[Su 77] *Suzuki, Y.*: Proc. Intern. Conf. on Nucl. Structure, Tokyo, September, 1977.

[Ta 72] *De Takaszy, N.*: Nucl. Phys. **A 178**, 469 (1972).

[Ta 78] *DeTar, C.*: Phys. Rev. **D 17**, 302 (1978); **D 17**, 323 (1978); **D 19**, 1028 (1979).

[To 78] *Tohsaki-Suzuki, A.*: Progr. Theor. Phys., **60**, 1013 (1978).

[Va 71] *Vanagas, V.V.*: "Algebraic Methods in the Theory of Nucleus", Vilnyus, Mintis, 1971 (in Russian).

[Wa 62] *Wackman, P.H.* and *N. Austern*: Nucl. Phys., **30**, 529 (1962).

[We 46] *Weyl, H.*: The Classical Groups, Princeton University Press, Princeton, N.J.; (1946).

[We 67] *Weinberg, S.*: Phys. Rev. Lett. **19**, 1264 (1967); Salam A. In: Elementary Particle Theory. Ed. N. Svartholm, Stockholm, 1968, p. 367.

[We 77] *Weiss, M.S.*: Fizika **9**, Suppl 4, 315 (1977) (Proceed. Int. Symp. Nucl. Coolisions and Their Micr. Description).

[Wi 37] *Wigner, E.P.*: Phys. Rev. **51**, 106 (1937).

[Wi 77a] *Wildermuth, K.* and *Y.C. Tang*: An Unified Theory of the Nucleus, Vieweg, Braunschweig, 1977.

[Wi 77b] *Wills, J.G., D.B. Lichtenberg* and *J.T. Keihl*: Phys. Rev. **D 15**, 3358 (1977).

[Wo 77] *Wong, C.W.*: Phys. Rev. **D 16**, 1590 (1977).

[Wo 80] *Woronchev, V.P., V.I. Kukulin* and *V.M. Krasnopol'sky*: J. Phys. G: Nucl. Phys. 1981, in press.

[Wr 71] *Wright, J.*: Phys. Rev., **D 4**, 1844 (1971).

[Ya 37] *Yamanouchi, T.A.*: Proc. Phys.-Math. Soc. Japan, **19**, 436 (1937).

[Ya 67] *Yakubovski, O.A.*: Yad. Fiz. **5**, 1312 (1967).

[Ya 75] *Yang, K.H., W.J. Gerace* and *J.F. Walker*: Nucl. Phys. **A 242**, 365 (1975).

[Zi 64] *Ziman, J.M.*: "Principles of the theory of solids" Cembridge, Univ. Press 1964.

Daphne F. Jackson

Department of Physics, University of Surrey, Guildford, UK

Direct Cluster Reactions – Progress Towards a Unified Theory

Contents

1 Introduction

1.1 Definition of Direct Cluster Reactions

This article is intended to give a very simple and straightforward discussion of what we mean when we talk about "clusters" in the context of direct reactions and scattering. Although three very informative international conferences have been held up to now there remains a good deal of uncertainty about the significance of clustering in nuclei and some conflict and confusion over the techniques necessary to describe the reactions which are designed to probe this effect. It is likely that this article will pose more questions than it succeeds in answering because the principle objective will be to clarify the problems and understand more clearly why difficulties and disagreements have arisen. No attempt has been made to give a complete review of theoretical work or of experimental data; instead references have been chosen to illustrate the key points.

The term "cluster" is usually used to describe a small group of nucleons within a nucleus which may be described, partially at least, in terms of the properties of the free composite particle. This approach is usually restricted to composite particles with mass number 2, 3 or 4, although heavier clusters are sometimes considered. The term "clustering in nuclei" is used in the literature with a wider variety of meanings. Some authors take the view that the appearance of groups of nucleons within the nucleus with the quantum numbers and other properties of the free composite particle is a straightforward indication of departures from the shell model, while others take the view that this phenomenon is not necessarily in conflict with the shell model and that extended shell model calculations can predict the relevant probabilities, even when these probabilities seem large. A third group of authors take the intermediate view that the presence of clustering is indicated only when the probabilities for observing clusters or cluster properties are enhanced above the predictions of the shell model. Even if the second view is the correct one it may still be convenient to work within the framework of the cluster model if that model provides a more transparent formalism. Indeed, it will be shown later that the cluster model does provide a particularly convenient formalism for the description of certain kinds of reactions, such as multi-particle knock-out and transfer reactions, and also certain break-up reactions, but we do not assume *a priori* that this implies any fundamental deficiency in the shell model. Instead we look for tests of the models which can be made with the available experimental information.

Some of the different attitudes to clustering arise from the use of limited or partial representations of both the shell and cluster models. At the most complete and fundamental level there can be no contradiction between these models since both can provide complete descriptions of the A-nucleon Hilbert space. In practice, however, many restrictions and limitations are imposed in order to obtain manageable descriptions of selected features of nuclear structure or nuclear reactions. For detailed discussions of the content of the cluster model and the shell model we refer the reader to the first volume in this series [Wi 77] and to the reviews by Towner [To 77 a] and Macfarlane [Ma 78].

By "direct cluster reaction" we mean a reaction in which the many-body interaction of the projectile with the target nucleus can be represented by the interaction of the projectile with a subset of nucleons (the cluster) treated as a single entity. This implies that the interactions of the projectile with the nucleons in the cluster can be treated, not one-by-one, but coherently. We would therefore expect that the cluster is excited or removed from the nucleus, or both, in a one-step process. Thus we would include the (p, α) reaction if it arises from knock-out of the α-particle or one-step pick-up of three nucleons, but not if it arises from pick-up of three nucleons in three successive collisions. We would not necessarily exclude from the class of direct cluster reactions those two-step processes in which the cluster is excited, de-excited or broken-up in the exit channel. In addition, the reaction $A(\alpha, 2\alpha)B^*$ could proceed by direct knock-out of an α-particle leaving the residual nucleus in an excited state or it could proceed indirectly by inelastic scattering followed by decay of the excited nucleus A^* to $\alpha + B^*$. The latter process is an example of a sequential reaction which can usually be distinguished experimentally by a careful study of the kinematics. At low projectile energies excitation of the target nucleus to a compound state may be followed by particle decay of the compound nucleus. We choose to exclude these compound and sequential reactions from our classification, which implies that we are concerned with reactions at projectile energies above a certain lower limit; we must not disregard them completely, however, because an overwhelming preponderance of sequential over direct reactions for a range of energies may indicate that the structure of the target nucleus is such that direct cluster reactions are not favoured.

In some cases, such as the example given above, direct cluster reactions can be identified experimentally, usually by means of examination of the energy or angular correlations. In other cases, for example particle emission following pion absorption, the experimental measurements may not allow such separation to be made. In such a case, a careful comparison of various theoretical predictions with the data is required before any conclusions can be drawn about the reaction mechanism. It is too naive to take the view that the copious emission of, for example, α-particles or deuterons in a nuclear reaction, and particularly in inclusive reactions, is a signature of clustering in nuclei in the sense of a breakdown of the shell model.

1.2 Examples of Direct Cluster Reactions

The list of reactions which satisfy the definition given in the preceding section and have been studied experimentally and theoretically is surprisingly long. This is not so much an indication of interest in clustering phenomena as such, but more an indication of the importance of a more general problem in nuclear physics – namely, how and why do nuclei fragment?

(a) Quasi-elastic Knock-out Reactions

In these reactions a particle identical to the projectile is detected *in coincidence* with the free particle equivalent to the cluster. Most studies have been carried out with incident protons, e.g. (p, pd), (p, pt), (p, pα), but studies with incident deuterons and α-particles in such reactions as (α, 2α), (d, dα), have given important results. Electron-induced knock-out reactions have been particularly valuable in studies of proton single-particle states in nuclei while pion-induced knock-out of nucleons yields important information on the pion-nucleon interaction. Knock-out of heavier particles by these projectiles has not been the subject of extensive study to date.

Because these reactions lead to the appearance of the cluster as a free particle in the final state, studies of the cross-sections lead to a fairly direct comparison of the wavefunction of the free particle and the wavefunction of the cluster in the target nucleus. Hence, in the absence of special difficulties, we would expect these reactions to be among the most important, possibly *the* most important, in providing information on clustering in nuclei.

(b) Quasi-elastic Transfer Reactions

These reactions constitute a special case of quasi-elastic reactions in which the interaction of the projectile with the cluster leads to correlated emission of the products of a transfer reaction, for example the reaction, (p, d ^{3}He) may be interpreted as arising from the reaction $p + \alpha \rightarrow d + {}^3He$ on an α-cluster.

(c) Multi-nucleon Transfer Reactions

These widely-studied reactions have produced much of the available information on nuclear structure at a level more complex than single-particle properties. They have been studied with incident protons, e.g. (p, α) (p, ^{6}Li), with light ions, e.g. (d, α), (t, p), (^{6}Li, d), and with heavy ions, e.g. (^{16}O, ^{12}C), (^{16}O, ^{20}Ne).

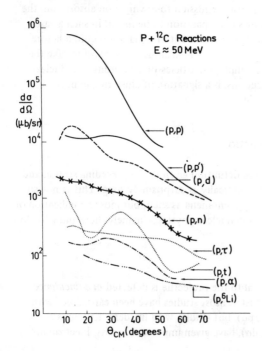

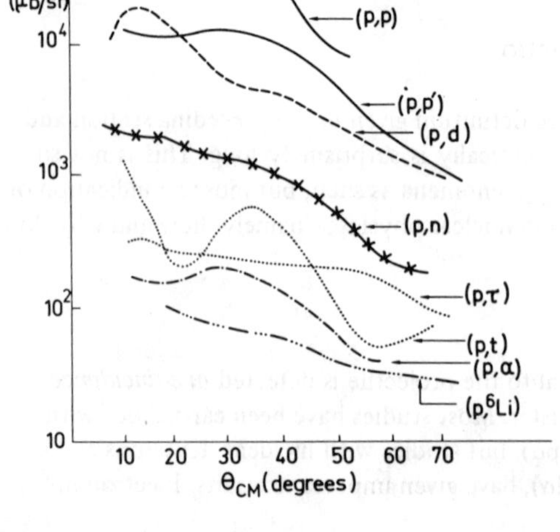

Fig. 1

Cross-sections induced by 50 MeV protons incident on ^{12}C [Re74, 77].

Some comparisons have been made of the relative importance of various transfer processes. Figure 1 shows the data collected by Redish [Re 74, 77] on proton-induced reactions on ^{12}C at an incident energy of around 50 MeV. Although elastic and inelastic scattering, charge-exchange, and one-nucleon transfer are the dominant reactions, the two-, three- and five-nucleon transfer reactions all have comparable magnitudes. Another study has been carried out [Wa 71 a] on reactions induced in ^{40}Ca by ^{16}O ions incident at 70 MeV which is well above the Coulomb barrier. Cross-sections were obtained at a fixed angle and summed up to 12–15 MeV in excitation. In general the transfer of each additional nucleon results in a decrease in the cross-section by an order of magnitude. The exception to this rule is transfer of two protons and two nucleons (i.e. a possible α-cluster) which yields cross-sections which are two orders of magnitude higher than this simple rule suggests. Double α-particle transfer also yields high cross-sections.

(d) Photonuclear Reactions

The application of the quasi deuteron model for the (γ, np) reactions is one of the oldest examples of a cluster model approach to nuclear reactions [Le 51, Le 60, Go 58]. This model has also been used [Ma 69, Sc 74] to describe (γ, p) and (γ, n) reactions by integrating over the variables relating to the undetected nucelon. Alternatively, it has been assumed that the photon is absorbed by an α-particle cluster rather than a deuteron cluster in the nucleus and this model appears to give a useful description of some (γ, n) data [Ra 68, Mi 71].

In principle, photonuclear reactions give similar information to knock-out reactions but with more emphasis on high momentum components. We do not yet know how to handle high momentum components in nuclear wavefunctions very consistently and, until very recently, the quality of the data has not been adequate to reveal detailed nuclear structure information.

(e) Pion Capture

Fairly extensive studies have been made of singles spectra of neutrons, charged particles, or prompt γ-rays emitted following interaction of fast π^+ and π^- or stopped π^- with nuclei. Some coincidence measurements have been made. Relatively large proportions of fast particles heavier than the nucleon are seen. The detection of nuclear γ-rays gives a means of identifying the residual nuclei. Removal of (pn), (2p2n) and multiples of (2p2n) seem to be important but these data do not reveal whether the excitation involves one-step interaction with an α-cluster or whether an interaction with two nucleons followed by a sequence of nucleon-nucleon collisions in the pre-equilibrium phase leads to removal of additional nucleons by knock-out or pick-up.

Moderate agreement with selected data has been achieved starting from pion capture on two nucleons [Wa 78] or on an α-cluster [Ko 66, Br 79]. The latter method may embrace the former to some extent, but was chosen partly in view of the importance of the production of deuterons and tritons in pion capture on ^{16}O and ^{12}C, and, even in this method, final state interactions must be involved to account for the strong production of α-particles. A review of experimental information on stopped pion capture in these light nuclei and a comparison of theoretical models, including pre-equilibrium models, has recently been completed [Ja 79].

The rapid development of pion-nuclear physics could make pion capture an important source of information on clustering in nuclei, complementary to that obtained from knock-out and transfer reactions.

(f) Inclusive Reactions

These reactions can be represented by the expression

projectile + target → x + anything

where x is the detected particle. When x is of the same species as the projectile the process is sometimes called summed inelastic scattering. The fast pion capture process discussed in the preceding section is an example of an inclusive reaction when only the singles spectrum is measured.

Although coincidence measurements have received most attention in the study of clustering in nuclei, inclusive studies of α-particle emission induced by electrons [Fl 79a] and heavy ions [Br 61 a, Gr 75, Bu 76] have produced interesting results.

(g) Heavy Ion Peripheral Collisions

The wide range of heavy ion reactions contains many features outside the context of direct cluster reactions, but peripheral heavy ion collisions at energies well above the Coulomb barrier do show some simple features which may be described in terms of projectile break-up accompanied by transfer of a small number of nucleons [Ge 78, Sc 78]. Preliminary explorations of the relationship between these processes and knock-out or capture reactions have been made [Ja 77a] and rely on some knowledge of the probabilities of emission of clusters by the projectile or target or possibly by a heavy ion ejectile.

(h) Alpha-decay

The evolution of an excited or unstable nucleus into a system consisting of a tightly bound α-particle in its ground state and a residual nucleus presents the oldest and possibly still the most searching test of our understanding of nuclear fragmentation. The prediction of absolute α-decay rates and branching ratios provides a particularly severe test of the ability of the shell model to describe the probability of preformation of an α-particle. Until recently, the calculation of absolute decay rates had been confused by uncertainties over the correct treatment of the problem of barrier penetration and the α-nucleus interaction. Very substantial advances have now been made.

1.3 Special Features of Direct Cluster Reactions

In all of the examples discussed in the preceding section at least one, and more often two or more, of the incident and outgoing particles are light or heavy ions and are therefore likely to be strongly absorbed. In the case of light ions this means that interpretation of the data may be influenced by the ambiguities in the optical potential used to describe the incoming or outgoing distorted waves and in the case of heavy ions a more serious uncertainty may result from the present incomplete knowledge of the optical potential, particularly with regard to its energy dependence. In all cases we may expect that the strong absorption will give rise to localization of the reaction in the surface or in the extreme periphery of the nucleus, and to a substantial reduction in the magnitude of the cross-section compared to the plane wave value. An indication of the reduction in the cross-section for the (p, pα) reaction due to distortion is given in Figure 2.

We may also expect that the exclusion principle will inhibit clustering in the interior of nucleus so that the probability of finding nucleons correlated as in a free composite particle may be largest in the nuclear surface. This effect will reinforce the surface localization as can be seen from Figure 3. There is clearly a danger that errors in the

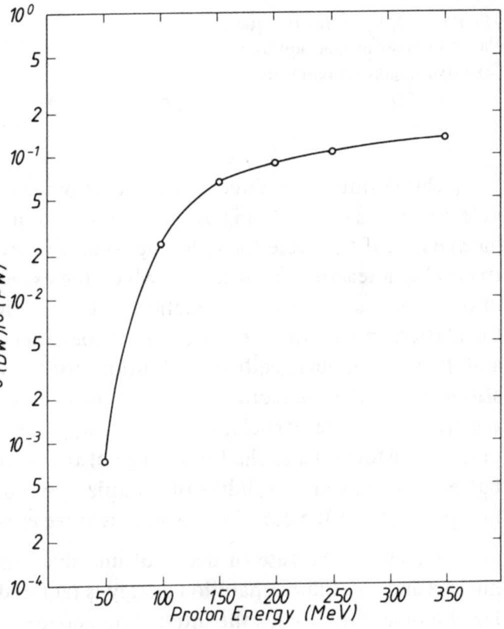

Fig. 2
The ratio of the cross-section for the reaction $^{24}Mg(p, p\alpha)$ ^{20}Ne(gs) calculated in DWIA to that calculated in PWIA as a function of proton bombarding energy for $\theta_p = 60°$ and $E_R = 0$ [CH77].

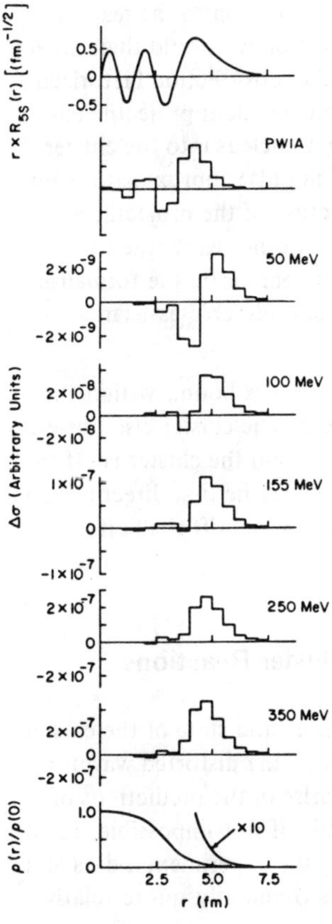

Fig. 3
The radial distribution of contributions to the cross-section for the reaction $^{24}Mg(p, p\alpha)$ ^{20}Ne(gs) for the PWIA cross-section, which is the same at all energies, and for the DWIA cross-sections at several energies [CH77].

choice of optical potentials and in the treatment of antisymmetrization could have dramatic effects on the shape and magnitude of predicted cross-sections and hence on deductions arising from comparision of theoretical cross-sections with experimental data and, further, that it may be difficult to disentangle errors arising in these two aspects of the reaction theory.

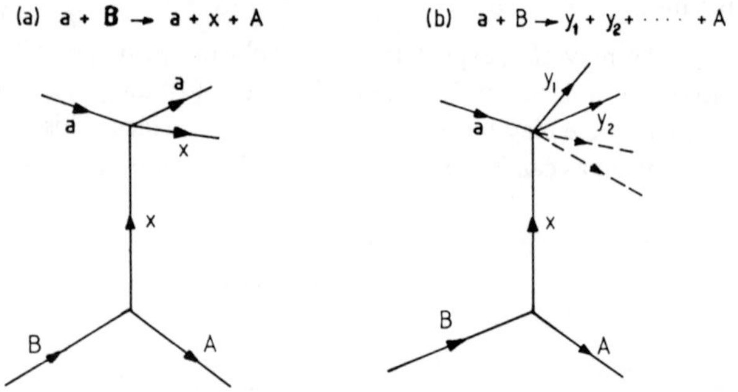

Fig. 4
The pole graph for (a) the quasi-elastic knock-out reaction and (b) a disintegration reaction.

Our definition of direct cluster reactions suggests that we may expect to see the pole graph, illustrated in Figure 4, providing an important contribution to the reaction mechanism. If this were the only important feature of the reaction we would then expect to develop a reaction formalism in which the expression for the cross-section factorized into a term describing the interaction vertex, involving only the incident projectile and the cluster, and a term describing the dissociation of the target nucleus into the cluster and the residual nucleus. If the conditions for the validity of impulse approximation are also satisfied, the interaction could be replaced by the interaction of the projectile with the free composite particle, which in principle is known from other experiments. In practice, distortion and the finite range of the interaction will complicate the formalism, but even approximate validity of factorization could be of considerable assistance in interpretation of the coincidence and inclusive experiments.

Except in the case of decay of unstable nuclei, the cluster is bound within the nucleus and a definite separation energy is required to release it. The cluster also possesses fermi momentum. Hence the interaction between the projectile and the cluster is off the energy-shell for the free interaction between these particles. It may be that direct cluster reactions, as we have defined them, are only clearly identifiable when off-shell effects are small.

1.4 Problems Arising in the Description of Direct Cluster Reactions

Because of uncertainties arising in the description of the bound state of the cluster in the nucleus and in the choice of optical potentials to generate the distorted wavefunctions of the incoming and outgoing particles large variations arise in the predictions of absolute magnitudes for cross-sections. This makes it difficult, if not impossible, to extract absolute values of spectroscopic factors by comparing the experimental data and the theoretical predictions. For this reason, experimentalists frequently quote relative spectroscopic factors for transitions to different states of the same residual nucleus. Since the surface localization of these reactions is widely accepted, the use of reduced widths evaluated at a fixed radius has been advocated [Ja 79a, 79b], and such reduced widths do indeed show much less variation with choice of potential parameters than the extracted spectroscopic factors. This procedure, however, has the unwelcome feature of introducing an arbitrary radius which cannot be defined within the context of any theory or model. For this reason, substantial efforts have been made to eliminate the arbitrary radii required in the R-matrix theory of α-decay, and a consistent description of α-decay and α-transfer has been sought.

The spectroscopic factor is, in any case, more closely related to nuclear structure calculations than is the reduced width. Spectroscopic factors are usually obtained theoretically in calculations based on the shell model, SU(3), the pairing model, or some similar model. The cluster model is very rarely used for this purpose. One of the few cases for which a comparison of shell and cluster model predictions can be made involves reactions on ^{16}O leading to various states in ^{12}C. Figure 5 shows this comparison for capture of stopped π^- on ^{16}O leading to the g.s. and first 2^+ state of ^{12}C accompanied by emission of a triton [Ja 79].

Some spectroscopic information can be obtained within the framework of the shell model without very explicit consideration of radial functions. Where radial functions are introduced they are usually taken to be of oscillator form. Techniques introduced by Talmi [Ta 52] and by Brody and Moshinsky and others [Br 61, Sm 61, 62] make it possible to rearrange oscillator wavefunctions for individual nucleons into an internal cluster wavefunction and a relative function but this relative function does not possess the correct radial behaviour in the important surface region to which direct cluster reac-

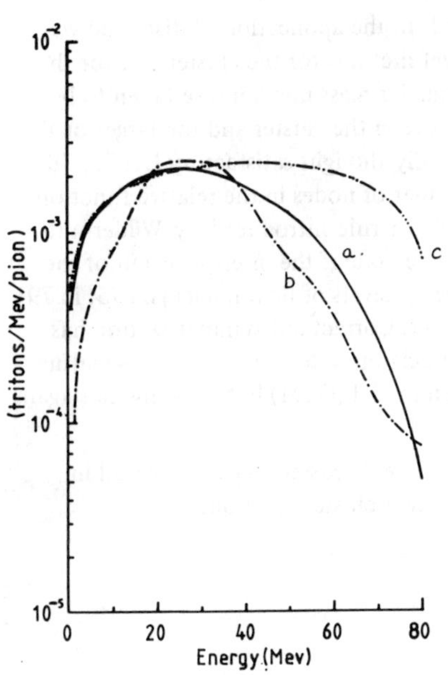

Fig. 5
The triton spectrum arising from stopped pion capture in ^{16}O calculated (a) using shell model spectroscopic factors, (b) using cluster model spectroscopic factors, and (c) using phase space only [Ja 79].

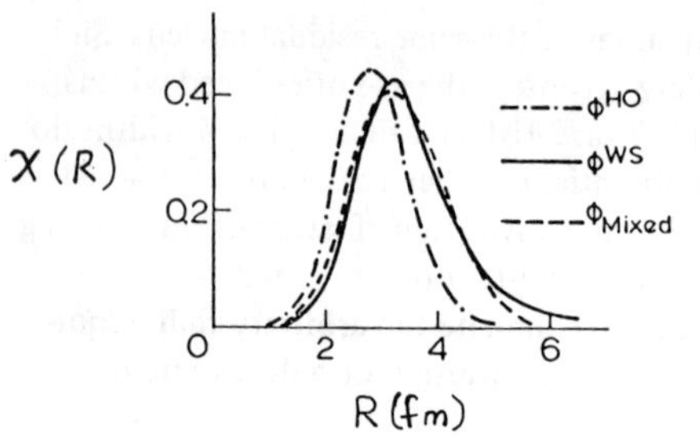

Fig. 6
The relative wavefunction for the decay ^{20}Ne $(8^+) \to \alpha + {}^{16}O$(gs) using an oscillator function, a function generated in a Saxon-Woods potential, and using oscillator functions with configuration mixing [Ar 73].

tions are sensitive. In principle, sufficient configuration mixing should lead to the correct asymptotic behaviour but this is rarely achieved in practice, except for very light nuclei. For example, admixtures up to $8\hbar\omega$ must be included in the oscillator model of the excited states of ^{20}Ne in order to obtain a relative function with the correct asymptotic behaviour [Ar 73], as shown in Figure 6.

The formalism of the cluster model readily yields the required relative function, and is therefore generally used in the development of the reaction theory for direct cluster reactions. It is feasible to introduce spectroscopic factors into the cluster formalism and to define them properly within the framework of the cluster model but, as we have already noted, this procedure is rarely adopted. Thus, in most cases, there is no adequate connection between the reaction theory which is used to analyse experimental data and hence to derive the experimental spectroscopic factors and the nuclear structure theory which is used to predict theoretical spectroscopic factors. *The cluster model has not been developed to give spectroscopic factors while the shell model has not been developed to give adequate radial functions.*

Shell model calculations are antisymmetrized. In the application of distorted wave theory to direct cluster reactions, the internal wavefunctions for the cluster and for the target or the residual nucleus (whichever has the smaller mass number) are taken to be antisymmetrized, but the exchange of nucleons between the cluster and the target or the residual nucleus is rarely treated correctly. It is usually thought satisfactory [Ar 73, 78, Ch 75, 77, Mi 77 a] to determine the minimum number of nodes in the relative function consistent with the Pauli principle using the oscillator rule introduced by Wildermuth and collaborators [Wi 66]. This procedure may be adequate if the interior region of the nucleus is most important, as in calculations of energy levels of light nuclei [Bu 75, Pi 79], but it is unlikely to be sufficient for surface reactions. Correct antisymmetrization has been achieved in scattering and reactions in few-nucleon systems using the resonating group model (RGM) or the orthogonality condition model (OCM) but these models again make use of simple analytic radial functions.

Each of the problems raised in this section will be discussed in greater detail in Chapters 2 and 3 where we develop the theory of direct cluster reactions.

2 Reaction Theory

2.1 Cluster Formalism

We consider first a bound system consisting of a core A containing A nucleons and a cluster x containing x nucleons. A convenient choice of relative coodinates for this system, expressed in terms of the shell model coordinates r_j ($j = 1, 2, \ldots A + x$) is

$$\underset{\sim}{\xi}_k = \underset{\sim}{r}_{A+k} - \underset{\sim}{R}_x \tag{2.1a}$$

$$\underset{\sim}{\eta}_i = \underset{\sim}{r}_i - \underset{\sim}{R}_A \tag{2.1b}$$

$$\underset{\sim}{R} = \underset{\sim}{R}_x - \underset{\sim}{R}_A \tag{2.1c}$$

where

$$\underset{\sim}{R}_x = \frac{1}{M_x} \sum_{k=1}^{x} M_k \underset{\sim}{r}_{A+k} \cong \frac{1}{x} \sum_{k=1}^{x} \underset{\sim}{r}_{A+k}, \tag{2.2a}$$

$$\underset{\sim}{R}_A = \frac{1}{M_A} \sum_{i=1}^{A} M_i \underset{\sim}{r}_i \cong \frac{1}{A} \sum_{i=1}^{A} \underset{\sim}{r}_i. \tag{2.2b}$$

Thus the $\underset{\sim}{\xi}_k$, $\underset{\sim}{\eta}_i$ are not all independent since

$$\sum_{k=1}^{x} \underset{\sim}{\xi}_k = 0, \qquad \sum_{i=1}^{A} \underset{\sim}{\eta}_i = 0. \tag{2.3}$$

This system of coordinates produces a convenient simplification in notation for the purposes of this article. For detailed calculations it is frequently more convenient to use the Jacobi coordinate system [Wi 77].

In many applications of the cluster model to reaction theory it is assumed that the Hamiltonian for the bound system can be approximated by

$$H_{A+x} = H_x(\xi) + H_A(\eta) + T_{xA} + V_{xA} + V_{xA}^C \tag{2.4}$$

where ξ, η represent all the internal coordinates $\xi_1 \ldots \xi_x$, $\eta_1 \ldots \eta_A$, respectively, T_{xA} is the kinetic energy operator for the motion of x relative to A, V_{xA} is the sum of all two-body nuclear interactions between the nucleons in the cluster and the nucleons in the core, i.e.

$$V_{xA} = \sum_{k=1}^{x} \sum_{i=1}^{A} V_{nn}(\underset{\sim}{r}_{A+k} - \underset{\sim}{r}_i) = \sum_k \sum_i V_{nn}(\underset{\sim}{\xi}_k - \underset{\sim}{\eta}_i + \underset{\sim}{R}), \tag{2.5}$$

and V^C_{xA} is the Coulomb interaction between x and A. The nuclear wavefunction can now be written as an expansion in a complete set of states of the core, i.e.

$$\Psi_{A+x} = \mathscr{A} \sum_{n_A n_x n} C(n\, n_A n_x)\, \Phi_A^{n_A}(\underset{\sim}{\eta})\, \phi_x^{n_x}(\underset{\sim}{\xi})\, \bar{\chi}^n_{xA}(\underset{\sim}{R}) \tag{2.6}$$

where n n_A n_x represent all the relevant quantum numbers for the relative motion, the core, and the cluster, respectively, $\mathscr{A}$ is the antisymmetrization operator which exchanges nucleons between the cluster and the core, and C is a constant including the fractional parentage coefficient and Clebsch-Gordan coefficients which ensure that Ψ has definite angular momentum, isospin, etc. It is assumed that ϕ_x, Φ_A are antisymmetrized with respect to exchange of the internal coordiates within the cluster and the core, respectively.

Following Park *et al* [Pa 72] we write Ψ in a simplified notation as

$$\Psi_{A+x} = \sum_m [\Phi_d(m) + \sum_{\nu=1}^{x} \sum_S (-1)^\nu\, \Phi^S_{ex,\nu}(m)] \tag{2.7}$$

where m represents the set of quantum number symbols n n_A n_x and $\Phi_d(m)$ is a "direct" term of the form

$$\Phi_d(m) = C(m)\, \phi_x\, \Phi_A\, \bar{\chi}_{xA}, \tag{2.8}$$

while $\Phi_{ex,\nu}$ is an exchange term which arises from exchange of ν particles and the sum over S denotes all the different terms arising from a given ν-particle exchange. A set of coupled integro-differential equations can be obtained for the $\bar{\chi}^n$ starting from the Hill-Wheeler equation. This gives

$$\begin{aligned}
[T_{xA} + \bar{U}^{mm}_{xA} - \epsilon^m_{xA}]\, \bar{\chi}^n_{xA} = &- \sum_{m' \neq m} \bar{U}^{mm'}_{xA}\, \bar{\chi}^{n'}_{xA} \\
&- \sum_m \sum_{\nu S} (-1)^\nu \int \Phi_A^{n_A *}\, \phi_x^{n_x *}\, [V_{xA} + V^c_{xA}]\, \Phi^S_{ex,\nu}(m')\, d\underset{\sim}{\eta}\, d\underset{\sim}{\xi} \\
&- \sum_{m'} \sum_{\nu S} (-1)^\nu \int \Phi_A^{n_A *}\, \phi_x^{n_A *}\, [T_{xA} - \epsilon^{m'}_{xA}]\, \Phi^S_{ex,\nu}(m')\, d\underset{\sim}{\eta}\, d\underset{\sim}{\xi}
\end{aligned} \tag{2.9}$$

where

$$\bar{U}^{mm'}_{xA} = \int \Phi_A^{n_A *}\, \phi_x^{n_x *}\, [V_{xA} + V^C_{xA}]\, \phi_x^{n'_x}\, \Phi_A^{n'_A}\, d\underset{\sim}{\eta}\, d\underset{\sim}{\xi} \tag{2.10}$$

$$\epsilon^m_{xA} = E_{A+x} - E_A^{n_A} - E_x^{n_x} \tag{2.11}$$

We now take m = 0 to represent the term in the cluster expansion in which the internal states of the core, the cluster, and the quantum state of the relative motion each take the lowest allowed value, and set

$$\epsilon_{xA} \equiv \epsilon^{0}_{xA}, \quad \bar{\chi}_{xA} \equiv \bar{\chi}^{0}_{xA} \tag{2.12}$$

$$-\int K^{om'}_{xA}(\underset{\sim}{R}, \underset{\sim}{R}')\, \bar{\chi}^{n'}_{xA}(\underset{\sim}{R}')\, d\underset{\sim}{R}' = \sum_{\nu S}{}'(-1)^{\nu} \int \Phi^{0*}_{A}\, \phi^{0*}_{A}\, \Phi^{S}_{ex,\nu}(m')\, d\underset{\sim}{\eta}\, d\underset{\sim}{\xi} \tag{2.13}$$

$$-\int G^{om'}_{xA}(\underset{\sim}{R}, \underset{\sim}{R}')\, \bar{\chi}^{n'}_{xA}(\underset{\sim}{R}')\, d\underset{\sim}{R}' = \sum_{\nu S}(-1)^{\nu} \int \Phi^{0*}_{A}\, \phi^{0*}_{x}\, [V_{xA} + V^{q}_{xA}]\, \Phi^{S}_{ex,\nu}(m')\, d\underset{\sim}{\eta}\, d\underset{\sim}{\xi} \tag{2.14}$$

so that the equation for $\bar{\chi}_{xA}$ becomes

$$[T_{xA} + \bar{U}_{xA} - \epsilon_{xA}]\, \bar{\chi}_{xA}(\underset{\sim}{R}) = -\sum_{m' \neq 0} \bar{U}^{om'}_{xA}\, \bar{\chi}^{n'}_{xA}(\underset{\sim}{R}) + \sum_{m'} \int G^{om'}_{xA}(\underset{\sim}{R}, \underset{\sim}{R}')\, \bar{\chi}^{n'}_{xA}(\underset{\sim}{R}')\, d\underset{\sim}{R}'$$
$$+ \sum_{m'} \int K^{om'}_{xA}(\underset{\sim}{R}, \underset{\sim}{R}')\, [T_{xA}(\underset{\sim}{R}') - \epsilon^{m'}_{xA}]\, \bar{\chi}^{n'}_{xA}(\underset{\sim}{R}')\, d\underset{\sim}{R}'. \tag{2.15}$$

Several approximate forms of equation (2.15) may be derived.

(i) *Neglect exchange and coupling to higher states*

This leads to the simplest possible equation of the form

$$[T_{xA} + \bar{U}_{xA}]\, \bar{\chi}_{xA}(\underset{\sim}{R}) = \epsilon_{xA}\, \bar{\chi}_{xA}(\underset{\sim}{R}) \tag{2.16}$$

where $\bar{U}_{xA}$ is defined through equation (2.10) as the expectation value of $V_{xA} + V^{C}_{xA}$ evaluated with the lowest allowed "direct" term in the cluster wavefunction (2.6). The construction of this potential is discussed in section 2.3.

(ii) *Neglect exchange but include coupling to higher states*

This leads to a set of coupled equations of the form

$$[T_{xA} + \bar{U}_{xA} - \epsilon_{xA}]\, \bar{\chi}_{xA}(\underset{\sim}{R}) = -\sum_{m \neq 0}{}' \bar{U}^{om}_{xA}\, \bar{\chi}^{n}_{xA}(\underset{\sim}{R}). \tag{2.17}$$

An equation of this kind is frequently solved in an approximate way by the introduction of an effective potential, which may be state-dependent, to give

$$[T_{xA} + \bar{U}^{eff}_{xA} - \epsilon_{xA}]\, \bar{\chi}_{xA}(\underset{\sim}{R}) = 0. \tag{2.18}$$

(iii) *Neglect coupling but include exchange*

This yields the equation [Ar74]

$$[T_{xA} + \bar{U}_{xA}]\, \bar{\chi}_{xA} = \epsilon_{xA}(1 + K)\, \bar{\chi}_{xA} + K\, T_{xA}\, \bar{\chi}_{xA} + G\, \bar{\chi}_{xA} \tag{2.19}$$

where K is the exchange operator introduced by Feshbach [Fe73a] with matrix elements $K^{oo}(\underset{\sim}{R}, \underset{\sim}{R}')$ and G is the corresponding potential operator with matrix elements $G^{oo}(\underset{\sim}{R}, \underset{\sim}{R}')$.

The solution of equation (2.19), or simplified versions of it, has received considerable attention in recent years. Some of the methods used are discussed in section 2.4. The role of the coupling terms has received very little attention for the bound state problem in the cluster model.

The cluster wavefunction (2.6) is expressed entirely in relative coordinates and is therefore translationally invariant. Reference to an arbitrary origin, such as the origin of shell model coordinates, introduces a centre-of-mass function $Z(\underset{\sim}{R}_{A+x})$ where

$$\underset{\sim}{R}_{A+x} = \frac{1}{A+x} \sum_{j=1}^{A+x} \underset{\sim}{r}_j .$$

It has been shown by Wildermuth and collaborators [Wi58, Wi66, 77] that, when this centre-of-mass function is included, each term in the expansion (2.6) can be related to a shell model wavefunction, provided that oscillator functions are used and that a common length parameter is used throughout. With the same restrictions, different cluster representations of the same nuclear state are equivalent [Wi66, Wi77, Ta78], i.e. the expansion of the wavefunction for A+x nucleons can be made in terms of the clusters A+x, or D+y, etc. In practice, cluster model wavefunctions with a single oscillator parameter rarely give an adequate description of the ground state properties of nuclei, particularly when the expansion (2.6) is truncated after a single term. The choice of different oscillator parameters for the internal wavefunctions of x and A can be shown to be equivalent to configuration mixing in the principal quantum number [Ta62, Wi62], i.e. configuration mixing of major shells, but this produces a relatively small improvement in the description of the nuclear properties. It is often more effective to choose both the functional form and the parameters of each function ϕ_x, Φ_A and $\bar{\chi}_{xA}$ to give the best possible description of the relevant physical properties. This representation is known as the generalized cluster model.

It can be seen from equations (2.6), (2.7) and (2.13) that the normalization of Ψ is given by

$$<\Psi_i|\Psi_j> = <\bar{\chi}_i|1 - K|\bar{\chi}_j> = \delta_{ij}. \tag{2.20}$$

Also, assuming that the operators U, G, K in equation (2.19) are all short-range, the asymptotic behaviour of $\bar{\chi}$ must be [Be65a, Pi65]

$$\bar{\chi}(\underset{\sim}{R}) \underset{R\to\infty}{\longrightarrow} \frac{e^{-\alpha R}}{R}, \quad \alpha^2 = \frac{2\mu}{\hbar^2} S_{xA} \tag{2.21}$$

where $S_{xA} = -\epsilon_{xA}$ is the separation energy required for fragmentation of the system A+x leaving nuclei x and A in definite final states. The correct description of this asymptotic behaviour is of the greatest importance in analyses of direct reactions and particularly of those which may be localised in the surface of the target nucleus. For this reason, we are unable to make very much use of oscillator functions or of approximate methods which depend on the use of oscillator functions and oscillator potentials.

2.2 Distorted Wave Formalism

In this section some standard expressions for transition matrix elements are noted, in a form convenient for the description of direct cluster reactions. Full derivations of the formulae are given in standard works on direct reaction theory [McC68, Au70, Ja70a].

We first consider the process

$$a + A \rightarrow b + B. \tag{2.22}$$

The total Hamiltonian for the initial system (excluding centre-of-mass motion) is given by

$$H_{a+A} = H_a + H_A + T_{aA} + V_{aA} + V_{aA}^C \tag{2.23a}$$

where H_a, H_A are the internal Hamiltonians for the projectile and the target A, respectively, T_{aA} describes the relative motion, V_{aA} is the total nuclear interaction and V_{aA}^C is the Coulomb interaction. Equation (2.23a) can also be written in the form

$$H_{a+A} = K_{a+A} + W_{aA} \tag{2.23b}$$

where K_{a+A} is a model Hamiltonian for the initial system in which $V_{aA} + V_{aA}^C$ is replaced by the optical potential $U_{aA} + U_{aA}^C$ for scattering of a from A.

(From now on we use U and χ to represent the potential and wavefunction of a scattering state of the relative motion, and $\overline{U}$ and $\overline{\chi}$ to represent the same quantities for a bound state of the relative motion.) In a similar way, the total Hamiltonian for the final state can be written as

$$H_{b+B} = K_{b+B} + W_{bB}. \tag{2.24}$$

The transition matrix element can now be written in the post or prior forms

$$\textbf{prior} \quad T_{fi} = \langle \Psi_{bB}^- | W_{aA} | \Psi_A \phi_a \chi_a^+ \rangle \tag{2.25a}$$

$$\textbf{post} \quad T_{fi} = \langle \chi_b^- \phi_b \Psi_B | W_{bB} | \Psi_{aA}^+ \rangle \tag{2.25b}$$

where the distorted wavefunctions χ obey the equations

$$(T_{aA} + U_{aA} + U_{aA}^C)\, \chi_{aA} = (E - E_A^i - E_a^i)\, \chi_{aA} \tag{2.26a}$$

$$(T_{bB} + U_{bB} + U_{bB}^C)\, \chi_{bB} = (E - E_B^f - E_b^f)\, \chi_{bB}, \tag{2.26b}$$

and ϕ_a, ϕ_b are the internal wavefunctions of the projectile a and reaction product b, respectively. The functions Ψ_{aA}^+, Ψ_{bB}^- are exact solutions of the Hamiltonians H_{a+A}, H_{b+B} with outgoing and incoming boundary conditions, respectively, and in consequence the matrix elements (2.25a), (2.25b) are exact and equivalent. When approximations are made for $\Psi^\pm$, however, the post and prior forms may give different results.

When an iterative solution for Ψ^+ is obtained, the post form of the matrix element becomes

$$\begin{aligned} T_{fi} = &\langle \chi_b^- \phi_b \Psi_B | W_{bB} | \Psi_A \phi_a \chi_a^+ \rangle \\ &+ \langle \chi_b^- \phi_b \Psi_B | W_{bB}\, G_U^+\, W_{aA} | \Psi_A \phi_a \chi_a^+ \rangle + \ldots \end{aligned} \tag{2.27}$$

where G_U is the Green's function

$$G_U^+ = (E - H_a - H_A - T_{aA} - U_{aA} - U_{aA}^C + i\epsilon)^{-1}. \tag{2.28}$$

The first term in equation (2.27) is the distorted wave Born approximation (DWBA) which should be a good approximation if W is a small perturbation and second order processes are not significant. The higher order terms in equation (2.27) represent multiple scattering processes in which the initial or final systems, or both, pass through intermediate states.

This formalism may be extended to allow for a cluster model description of the particles involved in the reaction. We assume that a cluster x is stripped from projectile a so that

$$a = b + x \qquad B = A + x$$

The exact Hamiltonians for these system are

$$H_a = H_b + H_x + T_{xb} + V_{xb} + V_{xb}^C \tag{2.29}$$

$$H_B = H_A + H_x + T_{xA} + V_{xA} + V_{xA}^C \tag{2.30}$$

but in practice model Hamiltonians will be used with the probable form

$$K_a = H_b + H_x + T_{xb} + \overline{U}_{xb}^{eff} \tag{2.31}$$

$$K_B = H_x + H_A + T_{xA} + \overline{U}_{xA}^{eff} \tag{2.32}$$

so that the perturbation becomes

$$\begin{aligned} W_{aA} &= H_{a+A} - K_{a+A} \\ &= V_{xb} + V_{xb}^C - \overline{U}_{xb}^{eff} + V_{aA} + V_{aA}^C - U_{aA} - U_{aA}^C \\ &= (V_{xb} + V_{xb}^C - \overline{U}_{xb}^{eff}) + V_{xA} + (V_{bA} - U_{aA}) + (V_{aA}^C - U_{aA}^C) && (2.33\,a) \\ &= \Delta V_{aA}^1 + \Delta V_{aA}^2 + \Delta V_{aA}^3 + \Delta V_{aA}^4 && (2.33\,b) \end{aligned}$$

where we have set $V_{aA} = V_{xA} + V_{bA}$. Similarly,

$$\begin{aligned} W_{bB} &= H_{b+B} - K_{b+B} \\ &= (V_{xA} + V_{xA}^C - \overline{U}_{xA}^{eff}) + V_{xb} + (V_{bA} - U_{bB}) + (V_{bB}^C - U_{bB}^C) && (2.34\,a) \\ &= \Delta V_{bB}^1 + \Delta V_{bB}^2 + \Delta V_{bB}^3 + \Delta V_{bB}^4 . && (2.34\,b) \end{aligned}$$

The terms ΔV^1 are usually neglected although it has been shown [Ja77b] that inclusion of ΔV_{bB}^1 increases the cross-section for the ^{208}Pb (^{16}O, ^{12}C) ^{212}Po(gs) reaction by a factor of 1.7. The terms ΔV_{aA}^2 or ΔV_{bB}^2 are usually dominant in DWBA. The terms ΔV^3 and ΔV^4 are frequently neglected, but the importance of the Coulomb terms has been demonstrated in two-nucleon transfer reactions. For example, in the ^{208}Pb(^{16}O, ^{14}C) ^{210}Po(gs) reaction inclusion of ΔV_{aA}^4 reduces the cross-section by a factor of 0.51 and ΔV_{bB}^4 reduces the cross-section by 0.69, in both cases without significant change in shape [DeV73, 75]. The terms ΔV_{bB}^1, ΔV_{bB}^3 and ΔV_{aA}^3 lead to core excitation while the terms ΔV^4_{aA}, ΔV^4_{bB} lead to Coulomb excitation of the core and the residual nucleus, respectively, and the term ΔV^1_{aA} leads to excitation of the outgoing particle.

The post form of the matrix element now becomes

$$T_{fi}(\text{DWBA}) = \langle \chi_b^-(\mathbf{k}_b, \mathbf{r}_f)\, \phi_b(\lambda)\, \Psi_B(\xi, \eta, \mathbf{R}) | \sum_{i=1}^{4} \Delta V^i_{bB}(\xi, \eta, \lambda, \mathbf{r}, \mathbf{R}) \times$$
$$\times | \chi_a^+(\mathbf{k}_a, \mathbf{r}_i)\, \phi_a(\xi, \lambda, \mathbf{r})\, \Phi_A(\eta) \rangle \tag{2.35}$$

where (see Figure 7)

$$\mathbf{r}_f = \frac{M_A}{M_B}\mathbf{R} - \mathbf{r} = \gamma \mathbf{R} - \mathbf{r} \tag{2.36a}$$

$$\mathbf{r}_i = \mathbf{R} - \frac{M_b}{M_a}\mathbf{r} = \mathbf{r} - \delta \mathbf{r} \tag{2.36b}$$

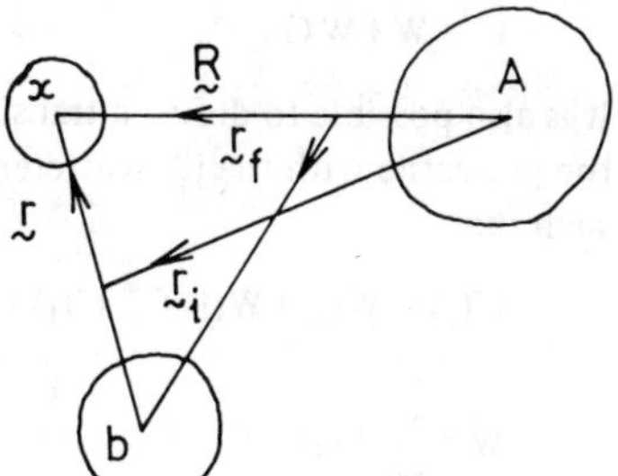

Fig. 7
The coordinate system for transfer reactions.

We thus obtain a multi-dimensional integral which can be simplified only with certain special assumptions [Sa73]. When projectile, cluster and core excitation are neglected, the matrix element reduces to

$$T_{fi}(\text{DWBA}) = \int \chi_b^{-*}(\mathbf{k}_b, \gamma\mathbf{R} - \mathbf{r})\, O(\mathbf{r}, \mathbf{R})\, \chi_a^+(\mathbf{k}_a, \mathbf{R} - \delta\mathbf{r})\, d\mathbf{r}\, d\mathbf{R} \tag{2.37}$$

where O is the overlap integral

$$O(\mathbf{r}, \mathbf{R}) = \int X^*(\xi, \mathbf{R})\, Y(\xi, \mathbf{r})\, d\boldsymbol{\xi} \tag{2.38a}$$

$$X^*(\xi, \mathbf{R}) = \int \Psi_B^*(\xi, \eta, \mathbf{R})\, \Phi_A(\eta)\, d\boldsymbol{\eta} \tag{2.38b}$$

$$Y(\xi, \mathbf{r}) = \int \phi_b^*(\lambda)\, \phi_a(\lambda, \xi, \mathbf{r})\, d\boldsymbol{\lambda}\,. \tag{2.38c}$$

The evaluation of overlap integrals is greatly facilitated by use of the cluster formalism as will be shown in section 3.1.

Using the same notation, knock-out and disintegration processes can be represented by the expressions

$$b + B \rightarrow b + x + A \qquad \text{knock-out} \qquad (2.39\,a)$$

$$b + B \rightarrow c + d + \ldots + A \qquad \text{disintegration, etc.} \qquad (2.39\,b)$$

In these reactions there are at least two fast particles in the final state in addition to the residual, recoiling nucleus. This has two very important effects on the reaction formalism.

We first obtain some formulae for two-body scattering and then discuss their applicability to three-body systems. In the case of inelastic scattering for which we may drop the labels on W in equations (2.55) and (2.27) it is easy to show that the repeated operation of the potential W may be represented by a transition operator t^+, where

$$t^+ = W + W G_U^+ t^+.$$

It is also possible to define a transition operator $t^+(j)$ which describes the interaction of the projectile with the jth scatterer in the nucleus (which may be a nucleon or a cluster), such that

$$t^+(j) = W(j) + W(j)\, G_U^+\, t^+(j)$$

$$W = \sum_j W(j)$$

After some manipulation it is possible to express the transition matrix element in the form

$$T_{fi} = \langle \chi_f^- \phi_f \Psi_B | \sum_j t(j) | \Psi_i \phi_i \chi_i^+ \rangle$$

$$+ \langle \chi_f^- \phi_f \Psi_f | \sum_j t(j)\, G_U^+ \sum_{k \neq j} t(k) | \Psi_i\, \phi_i\, \chi_i^+ \rangle + \ldots \qquad (2.4))$$

The approximation obtained by terminating this expansion after the first term is called the multiple scattering approximation because the operator t(j) sums to all orders the interaction of the projectile with the jth scatterer in the presence of the potential U. The higher terms correspond to successive scatterings from different nucleons or clusters, as illustrated in Figure 8.

An expression for the potential U which describes elastic scattering from a system of bound projectiles can be derived [Wa 57, 58] by introducing the operator

$$\tilde{t}(j) = V(j) + V(j)\, Q\, G_0 \tilde{t}(j)$$

$$V = \sum V(j)$$

where Q is an operator which projects off the ground state of the target nucleus. The multiple scattering series for U is then given by

$$U = (\phi_i \Psi_i | \sum_j \tilde{t}(j) | \phi_i\, \Psi_i) + (\phi_i \Psi_i | \sum_{j \neq k} \tilde{t}(j)\, Q\, G_0\, \tilde{t}(k) | \phi_i \Psi_i) \qquad (2.41)$$

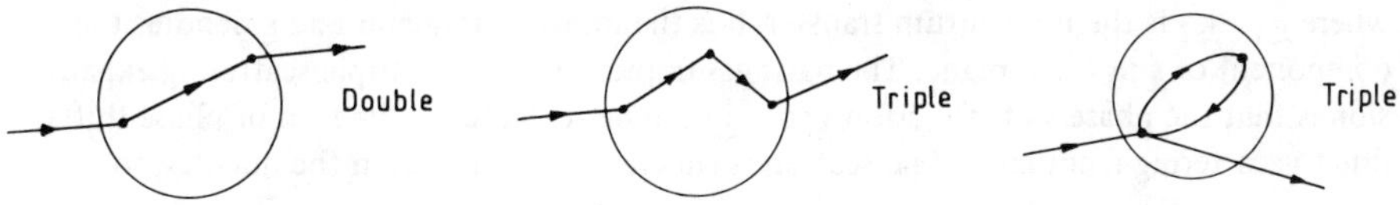

Fig. 8 Successive two-body scattering processes described by higher order terms in the multiple scattering approximation.

where G_0 is the Green's function for the free particle and $\phi_i \Psi_i$ represent the ground states of the composite projectile and the target nucleus. Apart from the presence of the projection operator Q the expansion for U has the same form as equation (2.40); the presence of Q implies that the intermediate states in the multiple scattering series involve excited states of the target nucleus. The elastic scattering amplitude is given by

$$f(\underset{\sim}{k}_i, \underset{\sim}{k}_f) = -(\mu/2\pi\hbar^2) < \phi_i \Psi_i \underset{\sim}{k}_f | U + U P G_0 U + \ldots | \phi_i \Psi_i \underset{\sim}{k}_i >$$

where P = 1 − Q. Hence exact solution of the Schrödinger equation with potential U takes into account all multiple scattering processes in which the intermediate state is the ground state of the target.

The transition operators t(j) and $\tilde{t}$(j) are complicated many-body operators. It is frequently very convenient to use the impulse approximation and replace t(j) or $\tilde{t}$(j) by the two-body operator τ(j) for free scattering of the projectile from a nucleon or composite particle. This procedure may be valid if the amplitude of the incident wave reaching each scatterer is unaffected by the presence of the rest of the nucleus and if the binding energies of the scatterers are very much less than the incident energy so that the scatterers appear effectively free. However, the energy and momentum conditions for the free projectile-scatterer interaction are different from the interaction in the nucleus, owing to the binding enregy and fermi momentum of the scatterer in the nucleus, so that the transition matrix element for the latter is off the energy shell for free scattering. This approximation yields the distorted wave impulse approximation (DWIA)

$$T_{fi} = < \chi_f^- \phi_f \Psi_f | \sum_j \tau(j) | \Psi_i \phi_i \chi_i^+ > \qquad (2.42)$$

Multiple scattering at medium and high energies has been studied extensively using the semi-classical approximation along the lines developed by Glauber [Gl59]. The transition matrix element is given by

$$T_{fi} = -i\hbar v \int e^{i(\underset{\sim}{k}_i - \underset{\sim}{k}_f) \cdot \underset{\sim}{b}} G_{fi}'(b)\, d^2 b$$

$$G_{fi}(b) = \int \Psi_f^*(\underset{\sim}{r})\, G(\underset{\sim}{b} - \underset{\sim}{s})\, \Psi_i(\underset{\sim}{r})\, d\underset{\sim}{r} \qquad (2.43\,a)$$

$$G(\underset{\sim}{b} - \underset{\sim}{s}) = 1 - e^{i\chi(\underset{\sim}{b} - \underset{\sim}{s})} \qquad (2.43\,b)$$

where $\underset{\sim}{k}_i - \underset{\sim}{k}_f$ is the momentum transfer, b is the impact parameter and $\underset{\sim}{s}$ denotes the component of $\underset{\sim}{r}$ in the b-plane. The basic assumption for the multiple scattering expansion is that the phase shift function $\chi(\underset{\sim}{b} - \underset{\sim}{s})$ can be replaced by the sum of phase shifts due to scattering from individual scatterers (nucleons or clusters) in the nucleus, so that

$$\chi(\underset{\sim}{b} - \underset{\sim}{s}) \simeq \sum_j{}' \chi_j(\underset{\sim}{b} - \underset{\sim}{s}_j). \tag{2.43c}$$

Hence

$$\begin{aligned} G(\underset{\sim}{b} - \underset{\sim}{s}) &\simeq 1 - \exp\left[i \sum{}' \chi_j(\underset{\sim}{b} - \underset{\sim}{s}_j)\right] \\ &\simeq 1 - \Pi\{1 - \Gamma_j(\underset{\sim}{b} - \underset{\sim}{s}_j)\} \\ &\simeq \sum_j{}' \Gamma_j(\underset{\sim}{b} - \underset{\sim}{s}_j) - \sum_{j \neq k}{}' \Gamma_j(\underset{\sim}{b} - \underset{\sim}{s}_j)\, \Gamma_k(\underset{\sim}{b} - \underset{\sim}{s}_k) + \sum_{j \neq k \neq l}{}' \Gamma_j \Gamma_k \Gamma_l + \ldots \end{aligned} \tag{2.44}$$

Equation (2.44) represents the multiple scattering expansion for the operator G(b − s) in terms of the profile functions $\Gamma_j(\underset{\sim}{b} - \underset{\sim}{s}_j) = 1 - \exp[i\, \chi_j(\underset{\sim}{b} - \underset{\sim}{s}_j)]$. It permits successive small angle scattering from different scatterers but does not allow the large angle scatterings necessary for scattering to occur more than once from the same scatterer, in contrast to the expansion (2.40). In both cases, the double-scattering term will involve a two-particle density function, and so on.

It is sometimes implied that the formulae derived for two-body scattering are automatically applicable to three-body systems, at least to first order. This is not strictly correct as it has been shown [Aa 61, Gr 66] that the Born series and the distorted wave Born series obtained from the Lippmann-Schwinger equation are divergent for the three-body system. This is due to the presence of "disconnected diagrams" in which one of the particles does not interact with the others [Lo 64, We 64, Do 66]. Such processes are not true three-body interactions and their presence means that the Lippmann-Schwinger equation does not possess a unique solution satisfying the Schrödinger equation. A satisfactory treatment of this problem has been given by Fadeev and others [Fa 61, Lo 64, Am 63, We 64, Do 66]. In Fadeev theory the transition operator T is written as a sum of three operators T^i such that

$$T(E) = \sum_{i=1}^{3}{}' T^i(E)$$

$$T^i(E) = t_i(E) + t_i(E)\, G_0(E) \sum_{j \neq i} T^j(E)$$

$$t_i = v_{jk} + v_{jk}\, G_0\, t_i$$

where G_0 is the free Green's function for the three particles and $t_i(E)$ is the operator for three-body scattering in which particle i is a non-interacting spectator of the interaction of the interaction between the other two particles. The Fadeev equations are a set of coupled equations but the iterated solution includes only those processes in which successive interactions take place between different particles. A knowledge of the two-body matrix elements off the energy shell is required. Several authors [Re70, Do66, Me69, Ka70] have considered the application of Fadeev theory to knock-out reactions and have shown that the lowest order term for the reaction denoted by equation (2.37a) is given by the operator

$$T = \Omega^-_{\tilde{b}x}\, t_A(bx)\, \Omega^+_{\tilde{b}} \tag{2.45}$$

where $\Omega^+_{\tilde{b}}$ operates on the initial state to give a distorted wave for the incident particle b and $\Omega^-_{\tilde{b}x}$ operates on the final state to give the distortion of the two outgoing particles by the residual nucleus. McCarthy [McC68] has called the corresponding transition matrix element the distorted wave t-matrix approximation (DWTA) and has developed this method with co-workers [Li64, Li66, De68, McC69]. Using a plane wave approximation, Redish *et al* [Re70] have shown that the required two-body matrix element is the half-off-shell matrix element

$$t_A(bx) = \langle \underset{\sim}{P}_{on} | t_{bx}(\hbar^2 p^2_{on}/2\mu_{bx}) | \underset{\sim}{P}_{off} \rangle \tag{2.46}$$

where $\underset{\sim}{P}_{on}, \underset{\sim}{P}_{off}$ are the momenta of the scattered particle in the two-body bx centre-of-mass system before and after collision, respectively, and μ_{bx} is the corresponding reduced mass. In a distorted wave theory the required matrix element is fully-off-shell but is usually approximated by a half-off-shell or even an on-shell matrix element (see below).

The construction of a distorted wavefunction for the final state still presents problems. The interaction of the two outgoing particles with the residual nucleus (the core) are functions of relative coordinates and it is therefore necessary to transform to a set of suitable relative coordinates and to determine the corresponding set of conjugate momenta [Ja65a, Ja70]. If the set of relative coordinates is denoted by the column vector $[\underset{\sim}{x}]$ and is obtained from the set of basic coordinates $[\underset{\sim}{r}]$ by the transformation

$$[\underset{\sim}{x}] = [a]\,[\underset{\sim}{r}] \tag{2.47}$$

where the transformation matrix [a] depends on the masses of the particles involved, the momenta $[\underset{\sim}{k}]$ conjugate to the coordinates $[\underset{\sim}{x}]$ are given by

$$[\underset{\sim}{q}]^T = [\underset{\sim}{k}]^T\,[a] \tag{2.48}$$

where $[\underset{\sim}{k}]^T$ is a row vector and $[\underset{\sim}{q}]$ represents the set of momenta in the lab frame. If the residual nucleus can be taken to be infinitely heavy no difficulties arise but in general a coupling term occurs in the Hamiltonian for the unperturbed final state when expressed in relative coordinates, so that the total wavefunction cannot be factorized into a product of two distorted wavefunctions. Here we follow the standard procedure of using the symmetric relative coordinates

$$\underset{\sim}{r}_{xA} = \underset{\sim}{R}_x - \underset{\sim}{R}_A,\; \underset{\sim}{r}_{bA} = \underset{\sim}{R}_b - \underset{\sim}{R}_A, \tag{2.49}$$

so that $\underset{\sim}{r}_{xA} \equiv \underset{\sim}{R}$, as defined in equation (2.1c). The coupling now appears in the operator

$$T_{coup} = \frac{\hbar^2}{M_A}\, \underset{\sim}{k}_{bA} \cdot \underset{\sim}{k}_{xA}\,. \tag{2.50}$$

This operator is either neglected completely or replaced by its expectation value in plane wave states which gives a correction to the energy [Li64, Ch77]. The final state wave-function, including centre-of-mass motion, now factorizes into

$$\Psi_f^- \simeq \chi_{bA}^-(\underset{\sim}{k}_{bA}, \underset{\sim}{r}_{bA})\, \chi_{xA}^-(\underset{\sim}{k}_{xA}, r_{xA})\, e^{i\underset{\sim}{K}^f_{B+b}}\, \phi_x^f(\xi)\, \Phi_A^f(\eta) \tag{2.51}$$

where the distorted wavefunctions obey the equations

$$(T_{bA} + U_{bA})\, \chi_{bA} = \epsilon_{bA}\, \chi_{bA} \tag{2.52a}$$

$$(T_{xA} + U_{xA})\, \chi_{xA} = \epsilon_{xA}\, \chi_{xA} \tag{2.52b}$$

where

$$\epsilon_{bA} = \hbar^2 k_{bA}^2/2\mu_{bA}, \; \epsilon_{xA} = \hbar^2 k_{xA}^2/2\mu_{xA} \tag{2.53}$$

$$\underset{\sim}{k}_{bA} = \underset{\sim}{q}_b - \frac{M_b}{M_b + M_x + M_A}\underset{\sim}{p}_b, \; \underset{\sim}{k}_{xA} = \underset{\sim}{q}_x - \frac{M_x}{M_b + M_x + M_A}\underset{\sim}{p}_b, \tag{2.54}$$

$$\underset{\sim}{K}_{b+B} = \underset{\sim}{q}_b + \underset{\sim}{q}_x + \underset{\sim}{q}_A \tag{2.55}$$

and $\underset{\sim}{p}_b, \underset{\sim}{q}_b, \underset{\sim}{q}_x, \underset{\sim}{q}_A$ represent the momenta of the incident particle, the outgoing particles and the residual nucleus A, respectively, in the lab frame. In the initial state no such problem of coupling arises and the wavefunction can be expressed in relative coordinates in the form

$$\Psi_i^+ = \chi_{bB}^+(\underset{\sim}{k}_{bB}, \underset{\sim}{r}_{bB})\, e^{i\underset{\sim}{K}^i_{b+B}\cdot \underset{\sim}{R}_{b+B}}\, \Psi_B^i(\xi, \eta, \underset{\sim}{r}_{xA}) \tag{2.56}$$

where

$$\underset{\sim}{r}_{bB} = \underset{\sim}{R}_b - \underset{\sim}{R}_B = R_b - \frac{1}{M_B}(M_x \underset{\sim}{R}_x + M_A \underset{\sim}{R}_A) \tag{2.57}$$

$$\underset{\sim}{K}^i_{b+B} = \underset{\sim}{p}_b \tag{2.58}$$

$$\underset{\sim}{k}_{bB} = \frac{M_B}{M_b + M_B}\underset{\sim}{p}_b \simeq \frac{M_x + M_A}{M_b + M_x + M_A}\underset{\sim}{p}_b. \tag{2.59}$$

The DWIA matrix element now becomes

$$T_{fi}(\mathrm{DWIA}) = \delta(\underset{\sim}{p}_b - \underset{\sim}{q}_b - \underset{\sim}{q}_a - \underset{\sim}{q}_A) < \phi_x^f(\xi)\, \Phi_A^f(\eta)\, \chi_{bA}^-(\underset{\sim}{k}_{bA}, \underset{\sim}{r}_{bA})\, \chi_{xA}^-(\underset{\sim}{k}_{xA}, \underset{\sim}{r}_{xA})$$

$$\times\; |\tau_{bx}(\underset{\sim}{R}_b - \underset{\sim}{R}_x)|\, \Psi_B^i(\xi, \eta, \underset{\sim}{r}_{xA})\, \chi^+(\underset{\sim}{k}_{bB}; \underset{\sim}{r}_{bB}) >. \tag{2.60}$$

The coordinates can be simplified by setting

$$\underset{\sim}{r}_{xA} = \underset{\sim}{R}, \; \underset{\sim}{R}_b - R_x = \underset{\sim}{s} \tag{2.61a}$$

$$r_{bA} = \underset{\sim}{s} + \underset{\sim}{R} \tag{2.61b}$$

$$r_{bB} = \underset{\sim}{R}_b - \underset{\sim}{R}_x - \frac{1}{M_B}[M_x \underset{\sim}{R}_x + M_A \underset{\sim}{R}_A - M_B \underset{\sim}{R}_X] \tag{2.61c}$$

$$\simeq \underset{\sim}{s} + \frac{M_A}{M_B}\underset{\sim}{R} \simeq \underset{\sim}{s} + \gamma \underset{\sim}{R}. \tag{2.61d}$$

If it can be assumed that the interaction $\tau_{bx}(\underset{\sim}{s})$ is sufficiently short ranged that the distorted wavefunctions do not change significantly over this range, we may use the WKB formalism to write

$$\chi_i^+(\underset{\sim}{k}_{bB}, \underset{\sim}{r}_{bB}) = D^+(\underset{\sim}{k}_{bB}, \underset{\sim}{s} + \gamma\underset{\sim}{R})\, e^{i\underset{\sim}{k}_{bB}\cdot(\underset{\sim}{s}+\gamma\underset{\sim}{R})}$$

$$\simeq e^{i\underset{\sim}{k}_{bB}\cdot\underset{\sim}{s}}\, \chi^+(\underset{\sim}{k}_{bB}, \gamma\underset{\sim}{R}) \tag{2.62}$$

$$\chi^{-*}(\underset{\sim}{k}_{bA}, \underset{\sim}{r}_{bA}) = D^{-*}(\underset{\sim}{k}_{bA}, \underset{\sim}{s} + \underset{\sim}{R})\, e^{-i\underset{\sim}{k}_{bA}\cdot(\underset{\sim}{s}+\underset{\sim}{R})}$$

$$\simeq e^{-i\underset{\sim}{k}_{bA}\cdot\underset{\sim}{s}}\, \chi^{-*}(\underset{\sim}{k}_{bA}, \underset{\sim}{R}). \tag{2.63}$$

and hence the matrix element (2.60) becomes

$$T_{fi}(\mathrm{DWIA}) \simeq \delta(\underset{\sim}{p}_b - \underset{\sim}{q}_b - \underset{\sim}{q}_x - \underset{\sim}{q}_A)\, T_{bx} \times$$

$$\times \int \phi_x^{f*}(\xi)\, \Phi_A^{f*}(\eta)\, \chi_{bA}^{-*}(k_{bA}, \underset{\sim}{R})\, \chi_{xA}^{-*}(\underset{\sim}{k}_{xA}, \underset{\sim}{R})\, \Psi_B^i(\xi, \eta, \underset{\sim}{R}) \times$$

$$\times\, \chi_{bB}^+(\underset{\sim}{k}_{bB}, \gamma\underset{\sim}{R})\, d\eta\, d\xi\, d\underset{\sim}{R} \tag{2.64}$$

where

$$T_{bx} = \int e^{-i\underset{\sim}{k}_{bA}\cdot\underset{\sim}{s}}\, \tau_{bx}(s)\, e^{i\underset{\sim}{k}_{bB}\cdot\underset{\sim}{s}}\, d\underset{\sim}{s}. \tag{2.65}$$

Equation (2.64) represents the factorized form of the matrix element. It is an approximation because of the assumption used to obtain equation (2.62) and (2.63). Using the conservation of momentum contained in $\delta(\underset{\sim}{p}_b - \underset{\sim}{q}_b - \underset{\sim}{q}_x - \underset{\sim}{q}_A)$ and the definitions (2.54) and (2.59) of $\underset{\sim}{k}_{bB}$ and $\underset{\sim}{k}_{bA}$, the two-body matrix element T_{bx} can be written in the same form as the half-off-shell matrix element $t_A(bx)$ defined in equation (2.46) with

$$\underset{\sim}{P}_{on} = \frac{1}{M_b + M_x}(M_x \underset{\sim}{q}_b - M_b \underset{\sim}{q}_x) \tag{2.66a}$$

$$\underset{\sim}{P}_{off} = \frac{1}{M_b + M_x}(M_x \underset{\sim}{p}_b + M_b \underset{\sim}{q}_A) = \frac{1}{M_b + M_x}(M_x p_b - M_b \underset{\sim}{p}_x) \tag{2.66b}$$

where

$$\underset{\sim}{p}_x = -\underset{\sim}{q}_A \tag{2.66c}$$

is the momentum of the cluster x in the target nucleus and $\underset{\sim}{q}_A$ is the recoil momentum of the residual nucleus.

When particles b and x have zero intrinsic spin the matrix element (2.64) leads to a factorized cross-section of the form

$$\frac{d^3\sigma}{d\Omega_b\, d\Omega_x\, dE} \propto |T_{bx}|^2\, |g|^2 \tag{2.67}$$

where g is distorted momentum distribution defined by the integral in equation (2.64). For particles of non-zero spin, the inclusion of spin-orbit distortion prevents factorization of the cross-section [Ja 76]. Recent studies of the (p, 2p) reaction at intermediate energies [Ch 79, Ja 79 c] have shown that the main effect of spin-orbit distortion arises in the final state and that the correction to the cross-section is ∼ 10–20 %. Spin-orbit effects could therefore be important in such reactions as (p, pd), (p, d^3He).

The distorted momentum distribution is defined as

$$g = \int \chi_{bA}^{-*}(\underset{\sim}{k}_{bA}, \underset{\sim}{R})\, \chi_{xA}^{-*}(\underset{\sim}{k}_{xA}, \underset{\sim}{R})\, \mathcal{O}^{fi}(\underset{\sim}{R})\, \chi_{bB}^{+}(\underset{\sim}{k}_{bB}, \gamma \underset{\sim}{R})\, d\underset{\sim}{R} \tag{2.68}$$

where $\mathcal{O}(\underset{\sim}{R})$ is the overlap integral.

$$\mathcal{O}^{fi}(\underset{\sim}{R}) = \int \phi_x^{f*}(\xi)\, \chi^{fi}(\xi, \underset{\sim}{R})\, d\underset{\sim}{\xi} \tag{2.69}$$

$$\chi^{fi}(\xi, \underset{\sim}{R}) = \int \Phi_A^{f*}(\eta)\, \Psi_B^i(\xi, \eta, \underset{\sim}{R})\, d\underset{\sim}{\eta}\,. \tag{2.70}$$

The construction of the overlap integral is discussed in detail in section 3.1. We note here that if Ψ_B is represented by the cluster expansion (2.6) we have

$$\chi^{fi}(\xi, \underset{\sim}{R}) = \sum_{n^i n_x^i} C(n^i n_x^i n_A^f)\, \phi_x^{n_x^i}(\xi)\, \bar{\chi}_{xA}^{n_i}(\underset{\sim}{R}) \tag{2.71}$$

$$\mathcal{O}^{fi}(\underset{\sim}{R}) = \sum_{n^i} C(n^i n_x^f n_A^f)\, \bar{\chi}_{xA}^{n^i}(\underset{\sim}{R}) \tag{2.72}$$

Thus the reaction picks out the term or terms in the cluster expansion in which nucleus B has x nucleons correlated as in the free composite particle x in state n_x^f with nucleus A in state n_A^f. The plane wave form of g becomes

$$g^{PW} \propto \int \bar{\chi}_{xA}^{n}(\underset{\sim}{R})\, \exp[i(\gamma \underset{\sim}{k}_{bB} - \underset{\sim}{k}_{bA}) \cdot \underset{\sim}{R}]\, d\underset{\sim}{R} \tag{2.73}$$

$$\propto \int \bar{\chi}_{xA}^{n}(\underset{\sim}{R})\, \exp[-i \underset{\sim}{p}_x \cdot \underset{\sim}{R}]\, d\underset{\sim}{R} \tag{2.74}$$

which is the momentum distribution of the cluster x in nucleus B; this explains the name given to the integral g containing the distorted waves.

It is now clear that the factorized matrix element (2.64) is composed of a factor describing the interaction of the projectile with the cluster and a nuclear structure factor describing the state of the cluster in the target nucleus. This prompts an approximate treatment of the two-body interaction in a knock-out reaction in which we replace the square of the half-off-shell matrix element, summed over projection quantum numbers, by the cross-section for free scattering of projectile b from cluster x, i.e.

$$\sum' |\langle \underset{\sim}{P}_{on} | t_{bx} | \underset{\sim}{P}_{off} \rangle|^2 \rightarrow \frac{d\sigma}{d\Omega} bx(E', \bar{\theta}) \tag{2.75}$$

where the scattering angle in the bx centre of mass system is defined by

$$P_{off} P_{on} \cos \bar{\theta} = \underset{\sim}{P}_{off} \cdot \underset{\sim}{P}_{on}. \tag{2.76}$$

This may be called the quasi-free or peripheral model for the interaction. In the case of quasi-elastic transfer, photodisintegration, pion capture, etc., we replace the square of the matrix element by the cross-section for the appropriate reaction with the free composite particle x.

There is no unambiguous procedure for determining the relevant lab energy E′ at which the cross-section should be evaluated. Two possible prescriptions are the initial energy prescription (IEP) which relates E′ to the energy in the initial state and the final energy prescription (FEP) which relates E′ to the energy in the final state. Figure 9 shows a comparison [Ro77] of the cross-section derived from the half-off-shell matrix element

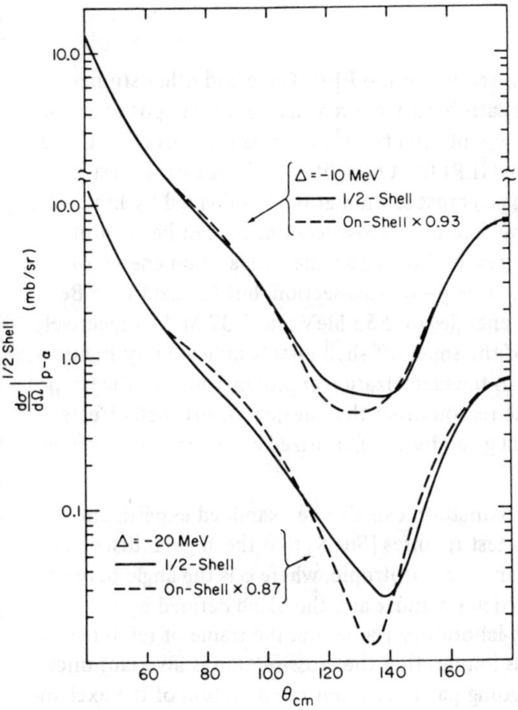

Fig. 9
Half-off-shell p–α cross-sections compared with the on-shell cross-sections calculated in the FEP and multiplied by the ratio of the final to initial momenta in the two-body centre-of-mass system [Ro77].

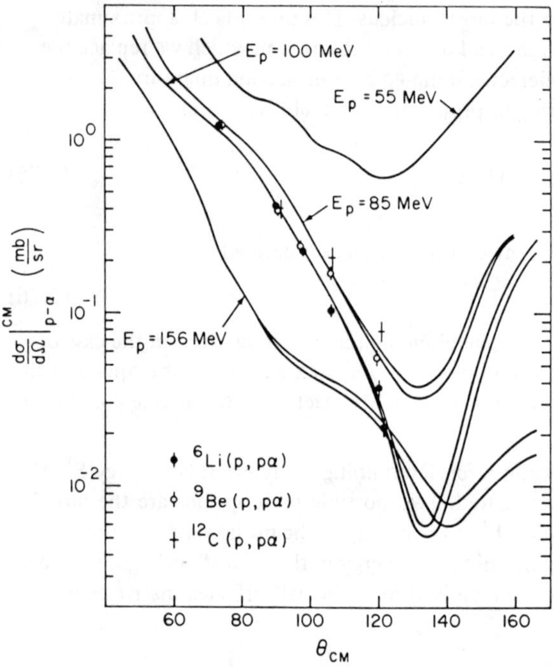

Fig. 10
Angular distributions for L = 0 (p, pα) reactions, normalized near to 90°, compared with free p–α cross-sections at various energies. The double curves indicate uncertainties in the p–α cross-sections [Ro 77].

for p–α scattering with the on-shell cross-section in the FEP. These and other studies [Ro 69 a, Ch 77] suggest that the FEP is a satisfactory approximation for (p, pα) reactions at ∿ 100 MeV but at 75 MeV the (p, pd) (p, pt) and (p, p^{3}He) cross-sections on ^{12}C and ^{16}0 require the initial energy prescription (IEP) for a best fit [Gr 77]. A further test has been made [Ro 77] by comparing the (p, pα) cross-section at p_x ∿ 0 divided by kinematic factors relevant to that reaction with the free p–α cross-section. As can be seen in Figure 10, the ^{6}Li(p, pα) data, which correspond to the smallest separation energy of 1.47 MeV, show good agreement with the free p–α cross-section, but the data for ^{9}Be and ^{12}C which correspond to separation energies of 2.53 MeV and 7.37 MeV respectively show increasing discrepancies. Because of the small off-shell effects indicated by Figure 9, Roos *et al* speculate that the breakdown in the factorization approximation for the (p, pα) reaction in the latter cases arises for some reason other than neglect of off-shell effects. Nevertheless, the agreement with the data given by the factorized cross-section is impressive, as can be seen from Figure 11.

The validity of the peripheral approximation can also be examined experimentally by means of the Trieman-Yang test. This test requires [Sh 65] that the angular distribution with respect to the Triemann-Yang angle ϕ is isotropic, where ϕ is the angle between the plane defined by the momenta of particles b and x and the plane defined by the momenta of particles B and A in the anti-laboratory frame, i.e. the frame of reference in which the incident particle is at rest. This implies that the cross-section is invariant under rotation of the plane containing the outgoing particles about the direction of the exchan-

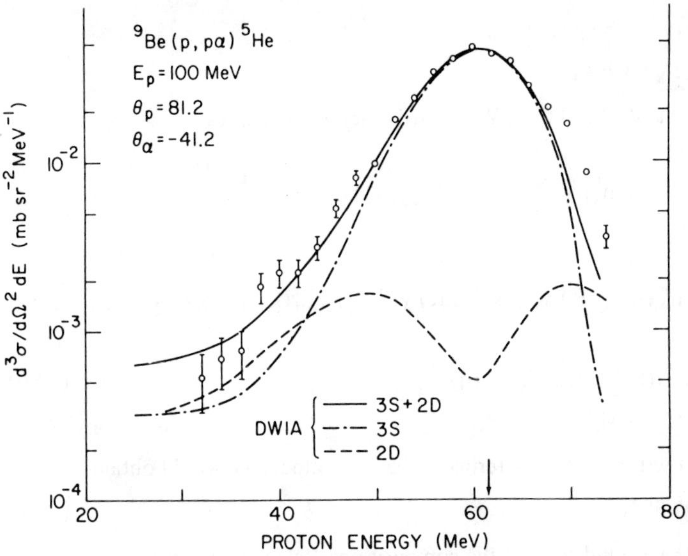

Fig. 11 DWIA calculations for L = 0 and L = 2 knock-out in the $^9Be(p, p\alpha)\,^5He$ reaction compared with the experimental data [Ro 77].

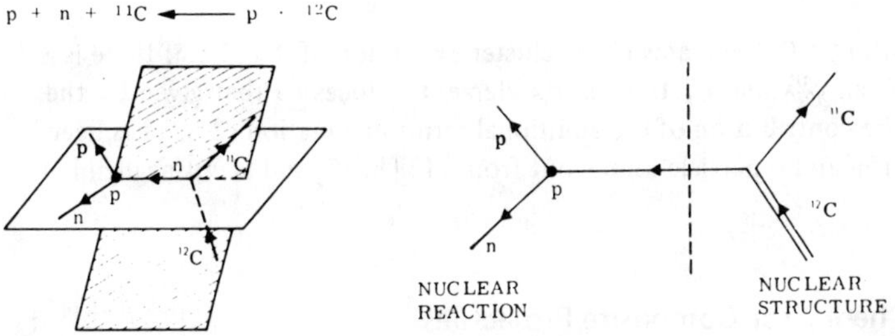

Fig. 12 The anti-laboratory frame of reference for the Trieman-Yang test illustrated for the knock-out of a neutron from ^{12}C in the (p, pn) reaction.

ged particle, as shown in Figure 12. This result is strictly valid if the exchanged particle has spin zero or $\frac{1}{2}$, but has been applied [Li73] to the $^6Li(p, pd)$ reaction to show that the peripheral mechanism, i.e. the pole graph, dominates at low recoil momentum.

The peripheral approximation is somewhat extreme in the sense that it implies that there exists in the target nucleus a group of nucleons which resembles very closely the free composite particle, i.e. it assumes the result which we wish to verify. The internal structure of the cluster may be investigated by the choosing a microscopic model for the interaction between the projectile and the cluster, i.e.

$$\tau_{bx} = \sum_{k=1}^{x} \tau_{bk}(\underset{\sim}{u}_k) \tag{2.77}$$

where

$$\underset{\sim}{u}_k = \underset{\sim}{R}_{b'} - \underset{\sim}{R}_x - \underset{\sim}{\xi}_k = \underset{\sim}{s} - \underset{\sim}{\xi}_k. \tag{2.78}$$

Using equations (2.62) and (2.63) the DWIA matrix element becomes

$$T_{fi}(\text{DWIA}) = \sum_{n^i n_x^i} C(n^i n_x^i n_A^f) \sum_k \int e^{i\underset{\sim}{q} \cdot \underset{\sim}{u}_k} \tau_{bk}(u_k)\, d\underset{\sim}{u} \int e^{i\underset{\sim}{q} \cdot \underset{\sim}{\xi}_k} \phi_x^{n_x^{f*}}(\xi)\, \phi^{n_x^i}(\xi)\, d\underset{\sim}{\xi}$$

$$\times \int \chi_{bA}^{-*}(\underset{\sim}{k}_{bA}, \underset{\sim}{R})\, \chi_{xA}^{-*}(\underset{\sim}{k}_{xA}, \underset{\sim}{R})\, \chi_{xA}^{-n^i}(\underset{\sim}{R})\, \chi_{bB}^{+}(\underset{\sim}{k}_{bB}, \gamma\underset{\sim}{R})\, d\underset{\sim}{R} \tag{2.79}$$

$$= \tau_{bx}(q) \sum_{n^i n_x^i} g_{xA}^{n^i} F_x^{n_x^f n_x^i}(q) \tag{2.80}$$

where $\underset{\sim}{q} = \underset{\sim}{p}_b - \underset{\sim}{q}_b$. Now for elastic scattering in the same model we would obtain

$$T_{bx}(q) = \tau_{bx}(q)\, F_x^{00}(q) \tag{2.81}$$

where ϕ_x^0 represents the ground state of the composite particle x, and hence

$$T_{fi}(\text{DWIA}) = T_{bx}(q) \left[\sum_{n^i} g_{xA}^{n^i} + \sum_{n^i n_x^i \neq 0} g_{xA}^{n^i} F_x^{on_x^i}(q)/F_x^{00}(q) \right] \tag{2.82}$$

Thus if the term with $n_x^i = 0$ dominates in the cluster expansion of Ψ_A, i.e. if there is a strong overlap between $\phi_x^{n_x^i}$ and ϕ_x^0 this matrix element reduces to that given by the quasi-free model. The contribution of the additional terms in equation (2.82) has been evaluated for deuteron and α-particle knock-out from ^{6}Li [Ja65] and is not negligible in this case.

2.3 Scattering Theory for Composite Projectiles

In this section we note some results obtained using the projection operator formalism introduced by Feshbach [Fe58, Le66, Fe73a] and express these results in a form suitable for application to composite projectiles.

We consider a free particle x interacting with a nucleus A so that the Hamiltonian for the system is given by equation (2.4), and introduce the projection operators P and Q, such that PΨ represents all open channels with a free particle x in its ground state at infinity and Q = 1 − P. Here Ψ is the complete wavefunction for the x + A system but we neglect exchange of nucleons between x and A. The Schrödinger equation

$$(E - H)\,\Psi = 0 \tag{2.83}$$

reduces to

$$(E - H_{PP})\, P\Psi = H_{PQ}\, Q\Psi \tag{2.84a}$$

$$(E - H_{QQ})\, Q\Psi = H_{QP}\, P\Psi \tag{2.84b}$$

where $H_{PP} = PHP$, etc. Equation (2.84b) may be formally solved to give

$$Q\Psi = \frac{1}{E - H_{QQ}} H_{QP} P\Psi. \tag{2.85}$$

In this expression it may be necessary to insert $+i\epsilon$ into the denominator if Q includes any open channels. Equation (2.84a) can now be written in the form

$$(E - H^{eff}) P\Psi = 0 \tag{2.86}$$

where the effective, energy-dependent Hamiltonian is given by

$$H^{eff} = H_{PP} + H_{PQ} \frac{1}{E - H_{QQ}} H_{QP}. \tag{2.87}$$

If P is now expressed as a sum over M open channels we have

$$P = \sum_{m'=0}^{M} |\Phi^{m'}_{A+x})(\Phi^{m'}_{A+x}| \equiv \sum_{m'=0}^{M} |\phi^{n'_x}_{x} \Phi^{n'_A}_{A})(\phi^{n'_x}_{x} \Phi^{m'_A}_{A}| \tag{2.88}$$

where the round brackets indicate integration over the internal coordinates of the target and projectile, but not over the relative coordinate. Hence

$$H_{PP} = \sum_{m'm} |\Phi^{m}_{A+x})\,[(H_x + H_A + T_{xA})\,\delta_{mm'} + U^{mm'}_{xA}\,]\,(\Phi^{m'}_{A+x}| \tag{2.89a}$$

$$H^{eff} = H_{PP} + \sum_{m'm} |\Phi^{m}_{A+x})(\Phi^{m}_{A+x}| \quad \hat{V}Q \frac{1}{E - H_{QQ}} Q\hat{V} |\Phi^{m'}_{A+x})(\Phi^{m'}_{A+x}| \tag{2.89b}$$

where $\hat{V} = V_{xA} + V^{C}_{xA}$ and $U^{mm'}$ is the matrix element of this operator defined by equation (2.10). The usual procedure is to separate out those eigen-states of H_{QQ} which are important in the vicinity of energy E. If Q' projects on to these states and Q'' projects on to the other states of H_{QQ}, so that $Q = Q' + Q''$, it is possible to define a generalized optical potential by writing

$$V^{mm'}_{xA} = U^{mm'}_{xA} + (\Phi^{m}_{A+x}|\hat{V}Q'' \frac{1}{E - H_{QQ}} Q''\hat{V}|\Phi^{m'}_{A+x})$$

$$+ (\Phi^{m}_{A+x}|\hat{V}Q' \frac{1}{E - H_{QQ}} Q'\hat{V}|\Phi^{m'}_{A+x}) \tag{2.90a}$$

$$= U^{mm'}_{xA}(\text{opt}) + (\Phi^{m}_{A+x}|\hat{V}Q' \frac{1}{E - H_{QQ}} Q'\hat{V}|\Phi^{m'}_{A+x}) \tag{2.90b}$$

where U(opt) is a slowly varying energy-dependent potential which includes the effect of distant resonances. From now on we will refer to $U^{mm'}$ as the folding model potential (or folded potential) and $U^{mm'}(\text{opt})$ as the generalized optical potential.

Equation (2.86) can be rewritten in the form

$$E-P\,[H_x+H_A+T_{xA}+\hat{U}(opt)]\,P(P\Psi)=PVQ'\frac{1}{E-H_{QQ}}\,Q'V\,P(P\Psi).$$

where $\hat{U}(opt)$ is an operator whose matrix elements are the $U^{mm'}(opt)$. The homogeneous scattering problem is represented by the equation

$$(E'-H'_{PP})\,\Psi^+=0 \tag{2.91a}$$

$$H'_{PP}=P[H_x+H_A+T_{xA}+\hat{U}_{opt}]P \tag{2.91b}$$

which corresponds to a set of coupled equations describing the direct reactions of the projectile plus target from which the distorted wave matrix elements can be derived when the coupling between channels is weak [Le 66, Le 73]. The solutions of this equation can be written in the form

$$P\Psi^\pm=\sum_m{}'\,\Psi^{m\pm}=\sum_m{}'\,\phi_x^{n_x}\,\Phi_A^{n_A}\,X_{xA}^{n\pm} \tag{2.92}$$

where $m=(n_x n_A n)$ and the functions $X^\pm=(\phi_x\Phi_A\,|\Psi^\pm)$ are solutions of the coupled equations obtained from inserting $\Psi^\pm$ into equation (2.91). When the coupling terms are neglected, the $X^\pm$ reduce to the usual distorted wavefunctions $\chi^\pm$ with incoming and outgoing boundary conditions which obey equations of the form indicated by (2.26a, b) with the U_{aA}, etc., interpreted as phenomenological optical potentials.

We may also introduce a wavefunction ψ^+ which satisfies the equation

$$(E-H_{PP})\psi^+=0 \tag{2.93}$$

and scattering functions $\chi^\pm=(\phi_x\Phi_A\,|\psi^+)$ which obey the coupled equations

$$(\epsilon_{xA}^m-T_{xA}-U_{xA}^{mm})\,\chi_{xA}^n=-\sum_{m'\neq m}{}'\,U_{xA}^{mm'}\,\chi_{xA}^{n'}, \tag{2.94}$$

i.e. they are derived from the folded potentials $U^{mm'}$. Thus these functions $\chi^\pm$ obey exactly the same equations as the bound state functions $\bar{\chi}$ introduced in section 2.1.

Eliminating $P\Psi$ from equations (2.84) and using equation (2.93) we find

$$(E-H_{QQ}-H_{QP}\frac{1}{E-H_{PP}}H_{PQ})\,Q\Psi=H_{QP}\,\psi^+ \tag{2.95}$$

where

$$\frac{1}{E-H_{PP}}=\frac{\mathscr{P}}{E-H_{PP}}-i\pi\delta\,(E-H_{PP}). \tag{2.96}$$

For the special case of spin zero projectiles and a single channel, Feshbach [Fe 73a] has shown that

$$<\psi^+|H_{PQ}|Q\Psi>=\frac{<K>}{1+i\pi<K>} \tag{2.97}$$

where K is the reactance matrix

$$<K> = <\psi^+|H_{PQ} \frac{1}{E-\epsilon_{QQ}} H_{QP}|\psi^+> \tag{2.98}$$

$$E-\epsilon_{QQ} = E - H_{QQ} - H_{QP} \frac{\mathscr{P}}{E-H_{PP}} H_{PQ}. \tag{2.99}$$

Feshbach also introduces a set of eigenfunctions of the operator ϵ_{QQ} so that

$$\frac{1}{E-\epsilon_{QQ}} = \sum_\lambda \frac{1}{E-E_\lambda} |\Psi_\lambda><\Psi_\lambda| \tag{2.100}$$

where the E_λ are real, and hence

$$<K> = \sum_\lambda \frac{|<\psi^+|H_{PQ}|\Psi_\lambda>|^2}{E-E_\lambda} \tag{2.101}$$

Thus K is related to the R-matrix defined as

$$R = \sum_\lambda \frac{\gamma_\lambda^2}{E-E_\lambda} \tag{2.102}$$

where γ_λ^2 is a reduced width.

The corresponding transition matrix element is

$$T_{fi} = T_{fi}^{pot} + e^{2i\delta} \frac{<K>}{1+i\pi<K>} \tag{2.103a}$$

$$= T_{fi}^{pot} + e^{2i\delta} <\psi^+|H_{PQ}|Q\Psi> = T_{fi}^{pot} + <\psi^-|H_{PQ}|Q\Psi> \tag{2.103b}$$

$$= T_{fi}^{pot} + e^{2i\delta} <\psi^+|H_{PQ} \frac{1}{E-H_{QQ}-W_{QQ}} H_{QP}|\psi^+> \tag{2.103c}$$

where T_{fi}^{pot} is the potential amplitude arising from equation (2.93) i.e.

$$T_{fi}^{pot} = <\underset{\sim}{k}_f \Phi_A^f \phi_x^f|\hat{V}|\psi^+> \tag{2.104}$$

and we have put

$$W_{QQ} = H_{QP} \frac{1}{E-H_{PP}} H_{PQ}. \tag{2.105}$$

If we assume that the space defined by Q contains a single isolated resonance we have

$$Q = |\Psi_s><\Psi_s| \tag{2.106}$$

$$\begin{aligned} <\Psi_s|W_{QQ}|\Psi_s> &= <\Psi_s|H_{QP} \frac{\mathscr{P}}{E-H_{PP}} H_{PQ}|\Psi_s> \\ &\quad - i\pi <\Psi_s|H_{QP}\, \delta(E-H_{PP}) H_{QP}|\Psi_s> \\ &= \Delta - \frac{1}{2} i\Gamma \end{aligned} \tag{2.107}$$

and the transition amplitude becomes

$$T_{fi} = T_{fi}^{pot} + \frac{<\psi^-|H_{QP}|\Psi_s><\Psi_s|H_{QP}|\psi^+>}{E - E_s - \Delta + \frac{1}{2}\, i\,\Gamma} \tag{2.108}$$

where the second term is the typical Breit-Wigner form for an isolated resonance. Finally, $\psi^{\pm}$ can be expanded in a complete set of eigenstates $\chi^{\pm}$ to give expressions for the total width and the partial width for decay through channel m. This gives

$$\Gamma_s^m = 2\pi |<\Psi_s|H_{QP}|\phi_x^{n_x}\Phi_A^{n_x}\, \chi_{xA}^{n+}>|^2 \tag{2.109}$$

$$\Gamma_s = \sum_m \Gamma_s^m. \tag{2.110}$$

The transition amplitude can also be written in the form

$$T_{fi} = T_{fi}^{opt} + \frac{<\Psi_f^-|\hat{V}Q'|\Psi_s><\Psi_s|Q'\hat{V}|\Psi_i^+>}{E - E_s - \Delta - \frac{1}{2}\, i\,\Gamma} \tag{2.111}$$

where, using equation (2.92)

$$T_{fi}^{opt} = <\underset{\sim}{k}_f, \Phi_A^f \phi_A^f \phi_x^f|\hat{U}(opt)|\Psi_i^+>$$

$$\Gamma_s = 2\pi \sum_m |<\Psi_s|Q'V|\Psi_i^{m+}>.$$

The importance of these formulae is that they provide a unified description of scattering, reactions and resonances without any reference to the arbitrary channel radius which appears in the R-matrix theory of reactions [Wi47, Te52]. In addition, we have clear definitions of the potentials arising in the theory. The role of Q Ψ as an intermediate state or time-delaying component is illustrated in Figure 13 which is due to Feshbach [Fe73a]. It should be noted that no particular choice of model for the target nucleus has been implied.

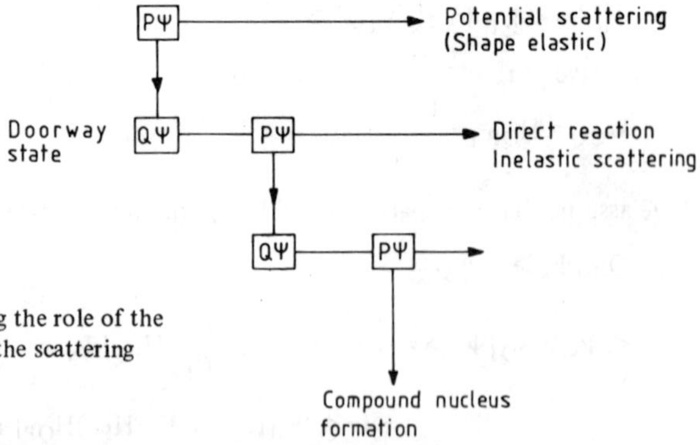

Fig. 13
Schematic diagram indicating the role of the components PΨ and QΨ of the scattering wavefunction [Fe73a].

The folding model yields an expression for the microscopic potential for elastic scattering of the form

$$U^{00}_{xA}(\underset{\sim}{R}) = U^{N}_{xA}(\underset{\sim}{R}) + U^{C}_{xA}(\underset{\sim}{R}) \qquad (2.112)$$

where, using equation (2.5), the nuclear part of the potential is given by

$$U^{N}_{xA}(\underset{\sim}{R}) = \sum_{k=1}^{x} \sum_{i=1}^{A} \int \Phi^{*}_{A}\phi^{*}_{x} V_{nn}(\underset{\sim}{\xi_k} - \underset{\sim}{\eta_i} + \underset{\sim}{R})\, \phi_x \Phi_A \, d\underset{\sim}{\eta}\, d\underset{\sim}{\xi}\,. \qquad (2.113)$$

and we assume that the dependence of the Coulomb interaction on the internal coordinates can be neglected. Equation (2.113) can be evaluated using a suitable nucleon-nucleon interaction or an effective interaction. This procedure is usually called the double folding model. Alternatively the interaction V_{xn} between the projectile x and a nucleon in the target nucleus may be introduced. This interaction can be constructed from the relation

$$V_{xn}(\underset{\sim}{R} - \eta_i) = \int \phi^{*}_{x}(\xi)\, V_{nn}(\xi_k - \eta_i + \underset{\sim}{R})\, \phi_x(\xi)\, d\xi \qquad (2.114)$$

or may be treated phenomenologically. The resulting nuclear potential is given by

$$U^{N}_{xA}(\underset{\sim}{R}) = A \int \rho_A(\underset{\sim}{r})\, V_{xn}(\underset{\sim}{R} - \underset{\sim}{r})\, d\underset{\sim}{r} \qquad (2.115)$$

where $\rho_A(\underset{\sim}{r})$ is the nuclear ground state density, normalized to unity, i.e.

$$\rho_A(\underset{\sim}{r}) = \frac{1}{A}\sum_{i=1}^{A} \int \Phi^{*}_{A}(\eta)\, \delta(\underset{\sim}{\eta_i} - \underset{\sim}{r})\, \Phi_A(\eta)\, d\underset{\sim}{\eta}\,. \qquad (2.116)$$

This is usually called the single folding model. The single and double folding models are usually used when the structure or density distribution of the target nucleus is of particular interest. When the interaction V_{xn} is represented by a phenomenological local interaction this interaction is likely to be energy-dependent. Then in order to make the total centre-of-mass energy the same in the free nucleon-x scattering as in the x-nucleon interaction in the nucleus A, the free n-x interaction must be taken at a lab energy E^{lab}_{n} which is $1/x$ of the lab energy of the projectile x incident on nucleus A.

In yet another approach the nucleon-target interaction is folded into the ground state density of the projectile, i.e.

$$U^{N}_{xA}(\underset{\sim}{R}) = \int \phi^{*}_{x}(\xi)\, V_{nA}(\xi_i + \underset{\sim}{R})\, \phi_x(\xi)\, d\xi. \qquad (2.117)$$

This is called the Watanabe model [Wa58], and is usually preferred when the structure of the projectile is of greatest interest. It has been applied particularly to deuteron scattering [Jo70, 72], to helion scattering [Me72], and also to scattering of ^{6}Li ions [Wa72].

The Watanabe model has been used to examine non-locality and energy-dependence of the microscopic potential [Ja74]. The interaction V_{xn} is taken to be an energy-independent non-local potential of Perey-Buck form so that its matrix elements in coordinate space are given by [Pe62]

$$<\underset{\sim}{r}|V_{nA}|\underset{\sim}{r}'> = H_{nA}(\underset{\sim}{r}-\underset{\sim}{r}', \beta_{nA})\, U^{N}_{nA}[\tfrac{1}{2}(\underset{\sim}{r}+\underset{\sim}{r}')] \tag{2.118}$$

where β_{nA} is the non-locality parameter for the interaction of a nucleon with the target nucleus A and

$$H(s, \beta) = C\,\beta^{-3} e^{-s^2/\beta^2}, \int H(s)\, d^3_s = 1. \tag{2.119}$$

The mixed density function for the projectile can be written as

$$\rho_x(\underset{\sim}{s}, \underset{\sim}{t}) = \int \phi^*_x(\underset{\sim}{s}, \xi_2 \ldots \underset{\sim}{\xi}_x)\, \phi_x(\underset{\sim}{t}, \xi_2 \ldots \underset{\sim}{\xi}_x)\, d\underset{\sim}{\xi}_2 \ldots d\underset{\sim}{\xi}_x \tag{2.120}$$

where $\rho_x(\underset{\sim}{s}, \underset{\sim}{s}) = x\rho_x(\underset{\sim}{s})$. If it can be assumed that the wavefunction ϕ_x is smooth over the range of the non-locality so that the variation of $\rho_x(\underset{\sim}{s}, \underset{\sim}{t})$ over this range can be neglected, the potential U^N_{xA} defined by equation (2.117) is a non-local potential whose matrix elements are

$$<\underset{\sim}{R}|V_{xA}|\underset{\sim}{R}'> = H_{xA}(\underset{\sim}{R}-\underset{\sim}{R}', \beta_{xA})\, U^{n}_{xA}[\tfrac{1}{2}(\underset{\sim}{R}+\underset{\sim}{R}')] \tag{2.121}$$

where

$$U_{xA}[\tfrac{1}{2}(\underset{\sim}{R}+\underset{\sim}{R}')] = x \int d^3 s\, \rho_x(\underset{\sim}{s})\, U^N_{xn}\, [\gamma \underset{\sim}{s} - \tfrac{1}{2}(\underset{\sim}{R}+\underset{\sim}{R}')] \tag{2.122}$$

$$\gamma = (x-1)/x \tag{2.123}$$

$$\beta_{xA} = \frac{x+A}{x(A+1)}\, \beta_{nA}. \tag{2.124}$$

An improved approximation can be obtained by representing ϕ_x by a gaussian function so that

$$\rho_x(\underset{\sim}{s}, \underset{\sim}{t}) \propto \exp[-(s^2+t^2)/2\alpha^2]$$

and

$$<\underset{\sim}{R}|V_{xA}|\underset{\sim}{R}'> = H_{xA}(\underset{\sim}{R}-\underset{\sim}{R}', \beta_{xA})\, (\widetilde{\beta}_{xA}/\beta_{xA})^3\, U^N_{xA}[\tfrac{1}{2}(\underset{\sim}{R}+\underset{\sim}{R}')] \tag{2.125}$$

$$(\widetilde{\beta}_{xA})^{-2} = (\beta_{xA})^{-2} + (\widetilde{\alpha})^{-2} \tag{2.126}$$

$$\widetilde{\alpha} = 2(A+x)\,\alpha/xA. \tag{2.127}$$

Thus the finite size of the projectile x leads to a modification of the range of the non-locality and a renormalization of the depth of the potential.

Because of the factor $(\widetilde{\beta}_{xA}/\beta_{xA})^3$ in equation (2.125) the depth of the non-local projectile-nucleus potential is always less than x times the depth of the nucleon-target potential. The reduction has been estimated to be about 6 % for incident deuterons [Jo70,

72] and about 15 % for incident α-particles [Ja74]; the reductions in the phenomenological potentials necessary to fit elastic scattering are somewhat larger than these predictions [Ge74, Pe67].

The depth of an energy-dependent equivalent local potential is given in the effective mass approximation by

$$V_0(E) = V_0(1 - \mu_{xA}\, \beta_{xA}^2\, E_{xA}^{cm}/2\hbar^2) \tag{2.118}$$

where μ_{xA} is the reduced mass and E_{xA}^{cm} is the total energy in the centre-of-mass frame. Using equation (2.124) and assuming that $A \gg 1$ we find

$$V_0(E) = V_0(1 - \frac{1}{x}\beta_{nA}^2 \frac{M}{2\hbar^2} E_x^{lab}) \tag{2.129}$$

$$\cong V_0(1 - \frac{1}{x} a_{nA}\, E_x^{lab}), \quad a_{nA} = \beta_{nA}^2\, M/2\hbar^2, \tag{2.130}$$

where M is the nucleon mass, E_x^{lab} is the energy in the lab frame of the projectile x incident on nucleus A, and a_{nA} determines the energy dependence of the nucleon-nucleus potential. In this model the energy dependence of the projectile-nucleus interaction arises from the non-locality of the nucleon-nucleus interaction.

The single-folding model and the Watanabe model yield very similar results for the energy dependence of the real part of the potential for α-particle scattering [Ja74] and these predictions are in good agreement with the energy dependence of the phenomenological potential required to fit data above 60 MeV. (At lower energies the shape of the potential is also energy-dependent [Pu79, Gu79].) Since equation (2.124) implies that $\beta_{xA} = 2\beta_{nA}/(A+1)$ if $x = A$ and $\beta_{xA} \cong \beta_{nA}/x$ if $A \gg x$, we predict that the real part of the potential for heavy ion scattering is almost energy independent. There is some evidence in support of this result [Cr76, Vi77], but other work [Si72, Ma72a] shows an increase of real depth with incident energy in the low energy region just above the Coulomb barrier.

Both single and double folding have been used extensively to construct the real potential for α-particle scattering with some success [Ba77, Pu79]. Since $V_{\alpha n}$ is real over most of the energy range of interest, the imaginary part of the optical potential arises in this formalism from the second term in equation (2.90a). Attempts are now being made to calculate the imaginary part. In many cases, however, the imaginary part is either treated phenomenologically or assumed to be of the same shape as the folded real potential. This procedure introduces ambiguities because different real potentials can yield agreement with the data if the imaginary potential is allowed to vary sufficiently [Ja76]. These folding models have also been used to construct potentials for heavy ion scattering [Lo79]. The double-folding model has been used [Ta78a 79a] to calculate the real parts of the coupling terms in equation (2.17) for heavy ion scattering just above the Coulomb barrier. The real generalized optical potential then has a dependence on $J(J+1)$ where J is the total angular momentum of the colliding system, and this can be represented by a linear energy-dependence which increases with increasing incident energy.

The Watanabe model can be used to predict the real and imaginary central potential and the spin-orbit potential. For example, it predicts the correct order of magnitude for the spin-orbit potential for mass 3 projectiles but the predicted linear decrease of the volume integral of the real potential with energy is not in agreement with the experimental results above 40 MeV [Na72b]. This model has been used most extensively as a starting point for studies of the deuteron-nucleus interaction [Jo70, Jo72, Kn75, Ka76] and yields a real spin-dependent potential with a diffuseness and shape having some resemblance to phenomenological potentials deduced from fits to data. The most significant feature of this projectile, however, is its small binding energy; the calculation of corrections to the Watanabe model must therefore use a formulation for deuteron scattering and reactions which takes this explicitly into account [Jo70, Ra75, Au78, Io76, 78].

The energy dependence of the optical potential can be of considerable importance in direct cluster reactions. For example, it can be seen from equations (2.52) – (2.54) that the energy of the outgoing particles in a knock-out reaction depends very strongly on the scattering angle and hence a correct description of the cross-section as a function of angle or energy of one of the outgoing particles or as a function of energy difference of the two particles will depend quite critically on a correct representation of the energy dependence of the scattering of the outgoing particles. The energy dependence of the potentials for scattering of composite projectiles can be related to that of nucleon-nucleus scattering, as discussed above, but the energy dependence of the nucleon potential is not precisely known, particularly for light target nuclei [Ab79, Le78] and at intermediate energies [Na79, In79].

The use of an energy dependent local potential in place of a non-local potential has other effects which may be significant. The Schrödinger equation for a non-local potential has the form

$$(\nabla^2 + \frac{2\mu}{\hbar^2} E^{cm})\, \psi(\underset{\sim}{R}) = \frac{2\mu}{\hbar^2} \int <\underset{\sim}{R}|V|\underset{\sim}{R}'> \psi(\underset{\sim}{R}')\, d\underset{\sim}{R}' \tag{2.132}$$

and the radial wave equation becomes

$$\left[\frac{d^2}{dR^2} + k^2 - \frac{\ell(\ell+1)}{R^2}\right] u_\ell(kR) = \frac{2\mu}{\hbar^2} \int_0^\infty V_\ell(R, R')\, u_\ell(kR')\, dR' \tag{2.133}$$

where $k^2 = 2\mu E^{cm}/\hbar^2$ and

$$V_\ell(R, R') = 2\pi R R' \int_{-1}^{+1} <\underset{\sim}{R}|V|\underset{\sim}{R}'> P_\ell(\cos\theta)\, d(\cos\theta). \tag{2.134}$$

Thus non-local interactions introduce ℓ-dependence. There are a number of different methods for deriving an equivalent local potential [Fr67]. The effective mass approximation yields a relation between the non-local potential U^N introduced in equation (2.118) and an equivalent energy-dependent local potential $U^L(R, E^{cm})$ of the form

$$U^N(R) = U^L(R, E) \exp\left\{\frac{\beta^2}{4}\frac{2\mu}{\hbar^2}[E^{cm} - U^L(R, E)]\right\} \tag{2.135}$$

where β is the range of the non-locality as defined in equations (2.118) and (2.119). Equation (2.128) can be derived from this expression by assuming that the energy dependence is small and with some restrictive assumptions about the imaginary part of the potential. Although the potentials U^N and U^L are chosen to give the same scattering the eigenfunctions ψ^N of the non-local potential U^N are smaller in the nuclear interior than the eigenfunctions of the local potential U^L [Pe 63]. This is known as the Perey effect and affects the magnitude of distorted wavefunctions and bound state wavefunctions in the nuclear interior. The relation between the two wavefunctions is given by [Fe 66]

$$\psi_L(\underset{\sim}{R}) = \psi_N(\underset{\sim}{R}) \exp\left[-\frac{\beta^2}{8}\frac{2\mu}{\hbar^2} U^L(R, E)\right] \tag{2.136}$$

which is in accord with the result deduced empirically by Perey [Pe 63],

$$\psi_L(\underset{\sim}{R}) \simeq \psi_N(\underset{\sim}{R}) \left\{1 + \frac{\beta^2}{4}\frac{2\mu}{\hbar^2} U^L(R, E)\right\}^{-\frac{1}{2}}. \tag{2.137}$$

From our earlier discussion of the relation between scattering of projectile x from nucleus A and scattering of a nucleon from A we can write for $A \gg x$

$$\begin{aligned} \beta_{xA}^2\, \mu_{xA}\, U_{xA}^L &\simeq \frac{1}{x^2}\, \beta_{nA}^2\, x\, \mu_{nA}\, x\, U_{nA}^L \\ &\simeq \beta_{nA}^2\, \mu_{nA}\, U_{nA}^L. \end{aligned} \tag{2.138}$$

For nucleon energies in the range 3–27 MeV the reduction in the scattering wavefunction is 15–20 %. Equation (2.138) implies that the same magnitude of reduction should arise for scattering of a composite projectile at least up to 27x MeV. Together with a reduction in a normalized bound state wavefunction, such reductions in the magnitude of distorted wavefunctions in the nuclear interior could have a substantial effect on distorted wave calculations unless the reaction is localized in the extreme surface of the nucleus beyond the range of the optical potential.

2.4 Antisymmetrization

Formulae representing the effect of antisymmetrization on the distorted wave matrix elements for direct reactions have been available for a considerable time [To 61, Au 70, Le 73]. For example, the transition amplitude for inelastic scattering consists of the sum of amplitudes representing direct inelastic scattering, the knock-on process, and target stripping. The contribution from knock-on terms has been extensively evaluated in DWBA using various effective nucleon-nucleon interactions [Am 67, At 70, Sc 69]; it increases with increasing multipolarity and decreasing incident energy. When impulse approximation is used the direct inelastic and knock-on terms can be combined to give

$$\begin{aligned} T_{\text{inel}} &\simeq A \langle \chi^-(\underset{\sim}{k}_f, \underset{\sim}{r}_0)\, \Phi_A(\underset{\sim}{r}_1, \underset{\sim}{r}_2 \ldots \underset{\sim}{r}_A) - \chi^-(\underset{\sim}{k}_f, \underset{\sim}{r}_1)\, \Phi_A(\underset{\sim}{r}_0, \underset{\sim}{r}_2 \ldots \underset{\sim}{r}_A)| \\ &\qquad \tau(0,1)|\Phi_A(\underset{\sim}{r}_1, \ldots \underset{\sim}{r}_A)\, \chi^+(\underset{\sim}{k}_i, \underset{\sim}{r}_0)\rangle \\ &\simeq A \langle \bar{\chi}(\underset{\sim}{k}_f, \underset{\sim}{r}_0)\, \Phi_A(\underset{\sim}{r}_1, \underset{\sim}{r}_2 \ldots \underset{\sim}{r}_A)|(1 - P_{01})\, \tau(0,1)|\Phi_A(r_1 \ldots r_A)\, \chi^+(\underset{\sim}{k}_1, \underset{\sim}{r}_0)\rangle \end{aligned} \tag{2.139}$$

where the operator P_{01} interchanges particles 0 and 1. Thus when $\tau(0, 1)$ is taken from fits to free two-body scattering it must be chosen to be the two-body amplitude corresponding to properly antisymmetrized wavefunctions.

Similarly, the transition amplitude for transfer reactions consists of terms corresponding to direct transfer, knock-on and heavy-particle stripping processes. When the latter two contributions are neglected, the multiplicative factors modifying the direct transition matrix element can be derived in the following way [Au70, To77a]. The required matrix element for the reaction

$$a + A \rightarrow b + B$$

has the form

$$T_{fi} = \int \chi^{-*}(\underset{\sim}{k}_f, \underset{\sim}{r}_f) < \Psi_{bB} | W | \Psi_{aA} > \chi^{+}(\underset{\sim}{k}_i, \underset{\sim}{r}_i)\, d\underset{\sim}{r}_i\, d\underset{\sim}{r}_f \tag{2.140}$$

where Ψ_{aA}, Ψ_{bB} are properly antisymmetrized wavefunctions for the initial and final states. Each of the wavefunctions Ψ_{aA} can be replaced by a linear combination of product wavefunctions of the form $\phi_a\, \Phi_A$ where ϕ_a and Φ_A are separately antisymmetrized. There are

$$N_i = \frac{(A+a)!}{a!\, A!} = \binom{A+a}{a}$$ ways of doing this. Similarly there are

$$N_f = \frac{(B+b)!}{b!\, B!} = \binom{B+b}{b}$$ ways of writing the final state wavefunction in

product form. Hence Ψ_{aA} contains N_i mutually orthogonal product functions each normalized by $N_i^{-\frac{1}{2}}$ and each term overlaps with one term in Ψ_{bB} giving N_i equal contributions to the matrix element. Hence

$$\begin{aligned} W_{fi} &= < \Psi_{bB} | W | \Psi_{aA} > \\ &= N_f^{-\frac{1}{2}}\, N_i^{-\frac{1}{2}}\, N_i < \phi_b\, \Phi_B | W \mathscr{A} | \phi_a\, \Phi_A > \end{aligned} \tag{2.141}$$

where $\mathscr{A}$ is the antisymmetrizer between the two systems. Each term in $\phi_a\, \Phi_A$ can be converted to a term in $\phi_b\, \Phi_B$ by transferring x nucleons from a to A. The number of ways this can be done is

$$X = \frac{a!}{(a-x)!\, x!} = \binom{a}{x} \tag{2.142}$$

so that

$$\begin{aligned} W_{fi} &= \binom{A+a}{a}^{\frac{1}{2}} \binom{B+b}{b}^{-\frac{1}{2}} \binom{a}{x} < \phi_b\, \Phi_B | W | \phi_i\, \Phi_A > \\ &= \binom{B}{x}^{\frac{1}{2}} \binom{a}{x}^{\frac{1}{2}} < \phi_b\, \Phi_B | W | \phi_a\, \Phi_A >. \end{aligned} \tag{2.143}$$

In this discussion it has been assumed that protons and neutrons are indistinguishable, i.e. that the isospin formalism has been used; if this is not the case the protons and neutrons must be separately antisymmetrized. Equation (2.143) is the basis for definitions of the spectroscopic factor and the essential feature is that $\phi_a\ \phi_b\ \Phi_A\ \Phi_B$ are fully antisymmetrized wavefunctions.

The effect of antisymmetrization in nuclear scattering is displayed very clearly within the framework of the Resonating Group Method (RGM) [Ta78, Wi77]. The scattering wavefunction ψ^+ of equation (2.93) is now written in the form

$$\psi^+ = \mathscr{A} \sum \phi_x^{n_x}\ \Phi_A^{n_A}\ \chi_{xA}^{n\pm}\,. \tag{2.144}$$

Starting from the Hill-Wheeler equation, it is easy to show that identical equations to (2.9) – (2.15) are obtained except that the bound state function $\bar{\chi}$ is everywhere replaced by χ^+ and that $\bar{U}$ is replaced by the folded potential appropriate to scattering at energy E. The single channel case represented by equation (2.19) can be derived in the notation used by Buck *et al* [Bu77a] who write

$$\Psi_{A+x} = \int d\underset{\sim}{R}\,|\Phi_A(\eta)\,\phi_x(\xi)\,\delta(\underset{\sim}{r}-\underset{\sim}{R})>\chi_{xA}(\underset{\sim}{R}) \tag{2.145}$$

so that

$$\int d\underset{\sim}{R}'\, H_{xA}(\underset{\sim}{R},\underset{\sim}{R}')\,\chi_{xA}(\underset{\sim}{R}') = \epsilon_{x\ell}\int d\underset{\sim}{R}'\, A(\underset{\sim}{R},\underset{\sim}{R}')\,\chi_{xA}(\underset{\sim}{R}') \tag{2.146a}$$

or

$$H|\chi> = \epsilon\,\hat{A}|\chi> \tag{2.146b}$$

where $\hat{A}$ is the operator for antisymmetrization,

$$H_{xA}(\underset{\sim}{R},\underset{\sim}{R}') = <\phi_x\,\Phi_A\,\delta(\underset{\sim}{r}-\underset{\sim}{R})|(T_{xA}+V_{xA}+V_{xA}^C)\,\hat{A}|\phi_x\,\Phi_A\,\delta(\underset{\sim}{r}-\underset{\sim}{R}')> \tag{2.147}$$

$$A(\underset{\sim}{R},\underset{\sim}{R}') = <\phi_x\,\Phi_A\,\delta(\underset{\sim}{r}-\underset{\sim}{R})|\hat{A}|\phi_x\,\Phi_A\,\delta(\underset{\sim}{r}-\underset{\sim}{R}')> \tag{2.148}$$

$$= \delta(\underset{\sim}{R}-\underset{\sim}{R}') - K(\underset{\sim}{R},\underset{\sim}{R}'). \tag{2.148b}$$

From these equations it follows the normalization of Ψ for a scattering state is given by

$$<\Psi(\epsilon)|\Psi(\epsilon')> = <\chi(\epsilon)|1-K|\chi(\epsilon')> = \delta(\epsilon-\epsilon'). \tag{2.149}$$

The exchange operator K is hermitian and positive definite. The operator (1 + K) is also positive definite [Fe73a]. The exchange operator can be expanded in terms of its eigenstates and eigenvalues in the form

$$K(\underset{\sim}{R},\underset{\sim}{R}') = \sum_{n\ell m} |g_{n\ell m}^{\lambda}>\lambda_{n\ell}<g_{n\ell m}^{\lambda}|\,. \tag{2.150}$$

When the exchange terms are evaluated using oscillator functions the eigenfunctions $g^m_{n\ell}(\underset{\sim}{R}) = \langle \underset{\sim}{R} | g_{n\ell m} \rangle$ are also oscillator functions whose eiengvalues depend on the oscillator phonon number $Q = 2(n-1)+\ell$, where we take the lowest value of n to be $n = 1$. Values of $1-\lambda_{no}$ are given in Table 1 for the $^{16}O + {}^{16}O$ system [To77].
In the special case when x is a single nucleon and Φ_A is represented by a Slater determinant of orthonormal single-particle functions ϕ_i, we have

$$K(\underset{\sim}{R}, \underset{\sim}{R}') = \sum_{i=1}^{A}{}' \phi_i^*(\underset{\sim}{R}')\, \phi_i(\underset{\sim}{R}), \tag{2.151}$$

i.e. all the eigenvalues are unity. When the eigenvalues are unity it follows that

$$\langle g^1 | 1 - K | g^1 \rangle = 0 \tag{2.152}$$

and that

$$\int d\underset{\sim}{R}\, \hat{A} | \phi_x \Phi_A\, \delta(\underset{\sim}{r} - \underset{\sim}{R}) \rangle g^1(\underset{\sim}{R}) = 0 \tag{2.153}$$

Thus the states with $\lambda = 1$ are the redundant states of the RGM and are forbidden by the Pauli principle.

Table 1: Values of the expectation values of $1-K$ for the $^{16}O + {}^{16}O$ system [To 77]

$2(n-1)$	$1-\lambda_{no}$
26	0.015
30	0.091
34	0.206
38	0.357
42	0.509
46	0.640

States with $1-\lambda = 0$ occur only when the cluster and the core are described by identical length parameters. As soon as the length parameters differ the eigenvalues $1-\lambda$ no longer vanish although some of them remain very small. When realistic length parameters are used in the case of a large core and a small cluster, e.g. $\alpha + {}^{208}Pb$, there are a large number of partially forbidden states. The eigenvalues $1-\lambda_{n\ell}$ for the $\alpha + {}^{208}Pb$ and $\alpha + {}^{16}O$ systems calculated by Tonozuka and Arima [To79] are shown in Figure 14; the sharper increase in $1-\lambda$ for $\alpha + {}^{16}O$ compared to $\alpha + {}^{208}Pb$ reflects the smaller difference in oscillator parameters used in the former case.

In the Orthogonality Condition Model (OCM) the redundant states are removed using a projection operator [Sa69, 73a]

$$\Lambda = 1 - \sum_{\lambda=1}{}' |g^\lambda\rangle\langle g^\lambda|. \tag{2.154}$$

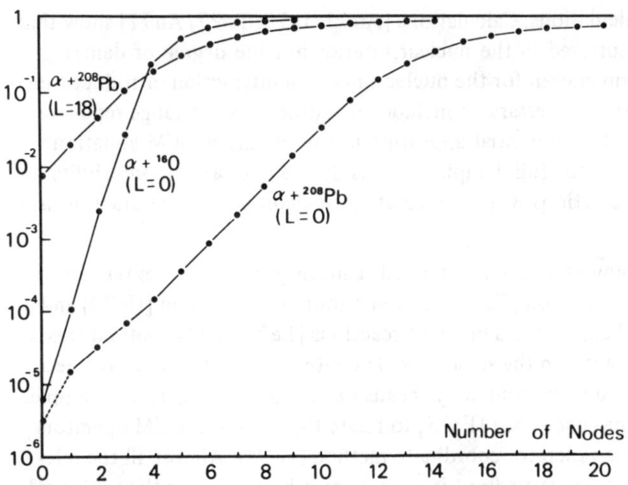

Fig. 14 Eigenvalues of 1 – K for the $\alpha + {}^{208}$Pb and $\alpha + {}^{16}$O systems [To 79].

This yields the equation

$$\Lambda[T + \hat{V} - \epsilon] | \omega \rangle = 0, \quad | \omega \rangle = \Lambda | \chi \rangle \tag{2.155}$$

which is an equation for the modified function ω which has no redundant components. Only the local parts of the RGM equation are retained, i.e. it is not necessary to calculate the exchange potential operator G but the equation is still non-local because of the presence of the operator Λ. This procedure is widely used [Ne 71, Ku 75, Bu 75a, Su 77] and in some cases it appears to be permissible to satisfy the orthogonality condition approximately by solving only the local wave equation and discarding the solutions which approximate to the redundant solutions. This approach appears to be justified for light systems [Sa 69, Ne 71, Ku 75] or where the interior region of the nucleus is important, for example in the calculation of energy levels [Bu 75, 77a, 77b, Pi 79].

Buck *et al* [Bu 77a] have presented arguments to show that another equation of the same form as the OCM equation may be a good approximation to the full RGM equation. This yields

$$\Lambda[T + \hat{V}]\, \Lambda | \omega \rangle = \epsilon | \omega \rangle$$

$$\text{or} \quad \Lambda[T + \hat{V}] | \omega \rangle = \epsilon | \omega \rangle, \quad | \omega \rangle = A^{\frac{1}{2}} | \chi \rangle \tag{2.156}$$

$$\text{or} \quad \Lambda[T + \hat{V} - \epsilon] | \omega \rangle = 0$$

where it is assumed that

$$\Lambda = \Lambda^{\frac{1}{2}} = A^{\frac{1}{2}} = A. \tag{2.157}$$

In this case $|\omega\rangle$ must be interpreted as $A^{\frac{1}{2}}|\chi\rangle$, i.e. $A^{\frac{1}{2}}$ times the RGM function. It is expected [Bu 77a] that these functions will differ only for small separations of the clusters; in particular, they have the same asymptotic behaviour, and hence give the same phase shifts in scattering calculations. Calculations [An75, Ja78, To77, An77] show that the functions $|\omega>$ are suppressed in the nuclear interior but the degree of damping depends strongly on the form chosen for the nucleon-nucleon interaction introduced in equation (2.5). It appears to be necessary to include an additional short-range repulsive term in the local cluster-nucleus potential appearing in the modified OCM equation [Bu77a] in order to reproduce the full damping effects due to nuclear compressibility which are seen in RGM calculations when a saturating nucleon-nucleon interaction is used.

The RGM is a very powerful technique which can, in principle, be extended to include a single three cluster function [Ta78] or even a four-cluster system [He79] and can be used to examine exchange effects in direct reactions [Le79a]. The method treats scattering states and bound states in the same way. The internal wavefunctions of the interacting systems need not be represented by means of the cluster model; for example, the two-centre shell model has been used [Fr75] to relate the non-local RGM operators to the matrix elements of the generator coordinate method (GCM). Almost all the relevant RGM calculations have been carried out in an oscillator basis because the analytical behaviour and the SU(3) symmetry properties of oscillator functions can be used to great advantage [Wi66, He79]. This is unfortunate from our point of view but these studies may nevertheless provide valuable insight into the behaviour of exchange terms.

Detailed examination of the exchange operator $K(\underset{\sim}{R}, \underset{\sim}{R}')$ and the exchange potential $G(\underset{\sim}{R}, \underset{\sim}{R}')$ [Ta78, 79, Le79b] indicate that the one-nucleon exchange term and the core-exchange term are most important of all the exchange terms. The core-exchange term is important only when the difference between the mass numbers of the interacting systems is small. The one-nucleon exchange term is important over a wide range of energies. The effective local potentials corresponding to these one-exchange and core-exchange terms have, respectively, essentially a Wigner-exchange or Majorana-exchange character. Hence the effective potential will have an odd-even ℓ-dependence. Comparisons of predictions derived from such effective potentials with full resonating group calculations have shown that the odd-even effect is important for the systems $p + {}^3He$, $p + \alpha$, ${}^3He + \alpha$, and ${}^{12}C + {}^{13}C$ [Vo74, Bu77a, Oe70, Ko74] thus confirming the significance of a small difference in mass number. The odd-even effect appears to be influenced also by the structure of the interacting systems. For example, from studies of the $n + {}^6Li$ system [St78] and the $\alpha + {}^6Li$ system [Su79] it appears that the presence of the nucleons in the unfilled p-shell in ${}^6Li$ introduces a blocking effect on the exchange terms. For $\alpha + {}^6Li$ scattering, core exchange is not as important as in the case of $d + \alpha$ scattering which has the same difference in mass number. Two-nucleon exchange may be neglected in the former case but three- and four-nucleon exchange must be considered together unless there is strong absorption which suppresses the short-range three-nucleon exchange term.

A study of the scattering of p, ${}^3He$ and α-particles from ${}^{16}O$ up to 25 MeV has been carried out with a local equation and an exchange potential of the form [Le77]

$$V_{ex} = V_a(R) + (-1)^{\ell}\, V_b(R) \tag{2.158}$$

$$V_a(R) = C_a\, e^{-\beta_a R^2}\, V_N(R) \tag{2.159a}$$

$$V_b(R) = C_b\, e^{-\beta_b R^2}\, V_N(R) \tag{2.159b}$$

Table 2: Parameters of the exchange potential defined by equations (2.158) and (2.159) [Le77]

Interacting System	C_a	β_a (fm^{-2})	Rms radius of $V_N + V_a$	Rms radius of V_N
$\alpha + {}^{16}O$	0.40	0.07–0.10	3.55–3.56	3.44
${}^{3}He + {}^{16}O$	0.40	0.02–0.05	3.62–3.66	3.55
$p + {}^{16}O$	0.45	0–0.05	3.25–3.34	3.25

where V_N represents the direct nuclear potential given by the folding model, V_a is thought to represent the effect of knock-on terms while V_b represents heavy-particle pick-up. Satisfactory agreement with full RGM calculations can be obtained with $C_b = 0$. The values of the other parameters are given in Table 2. It can be seen that V_a is of shorter range than V_N; nevertheless, the rms radius of $V_N + V_a$ is greater than the rms radius of V_N alone and hence considerable caution should be exercised when deductions are made about nuclear sizes directly from folding model predictions. For perfect agreement with RGM calculations it is necessary to add an additional term to the exchange potential and this has usually been taken to be a repulsive potential extending out to the region where the density distributions of the interacting systems begin to overlap significantly [Pa76, Pa75, Ba77a, Sw67, Ba77b, Da77].

Other attempts have been made to examine the effects of antisymmetrization on the equivalent local potential. Majka [Ma78b] has calculated the real part of the $n-\alpha$ and $n-{}^{6}Li$ potentials using Green's strong density-dependent interaction based on the Kallio-Kolltveit force and including the term arising from exchange of the interacting pair of nucleons. The energy dependence of the resulting potential can be represented by (2.128) and comes mainly from the density dependence of the interaction. As can be seen from Table 3 the rms radius of the potential is increased by the density dependence and then reduced by antisymmetrization. For composite particles there is a problem with double counting in the single-folding and Watanabe models because the phenomenological two-body interaction will implicitly include some exchange (and other) effects. This may explain why single-folding models have been more successful in fitting data than double-folding models [Gi79]. Majka *et al* [Ma78c] have calculated the same one-nucleon exchange contribution to the real part of the optical potentials for $\alpha + {}^{40}Ca$ and ${}^{6}Li + {}^{40}Ca$ scattering again using Green's strong density-dependent interaction. Table 4 shows that different results are obtained with exchange in the nucleon-target system or exchange in the nucleon-projectile system. Again, the main part of the effect arises from the density-dependence rather than exchange. Fliessbach [Fl79] has examined the real α-nucleus potential using the energy dependence of the nucleon-nucleus potential in the framework of the Watanabe model. He finds that at $E_\alpha \gtrsim 200$ MeV the Pauli correction is disappearing so the energy-dependence of the α-nucleus potential arises solely from the energy-dependence of the nucleon-nucleon potential; this is in agreement with the analysis by Jackson and Johnson [Ja74] which led to equations (2.124) – (2.130) although both Fliessbach and Perkin *et al* [Pe75] include the fermi momentum of the bound nucleons. At lower energies $E_\alpha \lesssim 50$ MeV, however, the Pauli effect dominates and the potential decreases rapidly with decreasing energy. A similar result is found for the volume integral of the real part of the ^{3}He potential [Pe 75]. It should be noted that most of these calculations treat only one term (albeit probably the most important term) and do not discuss the other non-local contributions to the scattering equation; therefore the solution of a local equation containing only this modification to the potential may not necessarily reveal the true effect of antisymmetrization on the cross-section.

The role of single-nucleon exchange has been explored for the real part of the heavy ion potential [Lo 79, Go 76, Sa 75, Sa 76]. Early calculations which used a pure even-state nucleon-nucleon interaction with a rather long range led to an over-estimate of the potential at the strong absorption radius [Ei 77, Sa 75]. It is necessary to include an odd-state interaction to cancel the OPEP contribution; forces which have this character yield rather

Table 3: Parameters of the $n-\alpha$ and n-^{6}Li potentials for $E_n = 10$ MeV. The symbol D indicates the direct folding potential, DD the potential including density-dependence and DD + E the total potential including density-dependence and exchange [Ma 78b]

System		J_N (MeV fm^3)	$<r^2>^{\frac{1}{2}}$ (fm)
$n+\alpha$	D	345.1	2.309
	DD	353.1	2.483
	DD + E	450.1	2.360
$n+{}^6$Li	D	345.5	3.013
	DD	407.9	3.139
	DD + E	582.5	3.087

Table 4: Parameters of the real part of the potential for $\alpha + {}^{40}$C and ^{6}Li + ^{40}Ca scattering. The symbol D indicates the direct folding potential, P indicates that antisymmetrization and density dependent effects are included in the nucleon-projectile system, and T indicates that antisymmetrization and density-dependence are included in the nucleon-target system [Ma 78c]

Projectile		U(R=0) (MeV)	J_V (MeV fm^3)	$<r^2>^{\frac{1}{2}}$ (fm)
α	D	230.3	346.9	4.084
	P	237.6	400.4	4.015
	T	186.2	393.9	4.246
^{6}Li	D	245.5	344.1	4.472
	P	370.6	483.4	4.248
	T	242.8	390.7	4.616

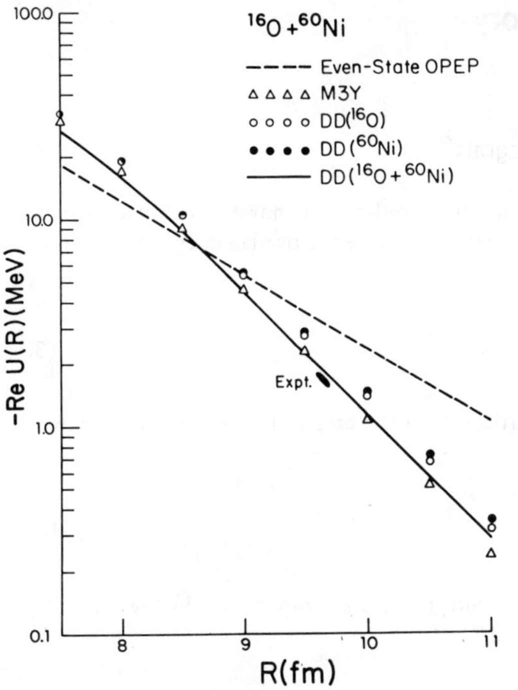

Fig. 15 The potential for the $^{16}O + ^{60}Ni$ system obtained from the double-folding model with various interactions. DD denotes the density-dependent G-matrix, M3Y denotes a density-independent G-matrix. The dashed curve indicates that part of the even state interaction attributed to OPEP [Lo 79].

similar results as can be seen from Figure 15. Other methods have been developed for calculation of the heavy ion potential [Si 79, Mo 78, Ng 75, Br 75], some of which yield shallow attractive potentials with short-range repulsive cores.

Remarkably little is known about the energy-dependence of the imaginary potential, which must arise from several sources, although methods of calculation are being developed [Br 79c]. A phenomenological study [Pu 79] of α-particle scattering from ^{90}Zr shows that below $E_\alpha \sim 80$ MeV the imaginary strength decreases rapidly with decreasing energy, presumably due to Pauli effects and reduction of open reaction channels. Between 80–140 MeV the energy-dependence is linear and of the form

$$W(E) = W_0(1 + \alpha_W E_\alpha) \tag{2.160}$$

$$W_0 = 19.2 \pm 1.4 \text{ MeV}, \alpha_W = 0.0003 \pm 0.0006 \text{ MeV}^{-1} .;$$

while for scattering from ^{40}Ca the volume integral of the imaginary part increases smoothly with incident energy in the range $E_\alpha \sim 22$–142 MeV [Mi 77]. The imaginary part of the heavy ion potential is strongly energy dependent [Vi 77].

3 Nuclear Structure Theory

3.1 Construction of Overlap Integrals

The description of direct cluster reactions leading to removal of a cluster x from nucleus B in state i leaving nucleus A in state f involves the overlap integral (see equations 2.38b, 2.70)

$$X^{fi}(\xi, \underset{\sim}{R}) = \int \Phi_A^{f*}(\eta)\, \Psi_B^i(\xi, \eta, \underset{\sim}{R})\, d\eta. \tag{3.1}$$

When the cluster appears as a free particle, as in the case of knock-out reactions, we obtain

$$\mathcal{O}^{fi}(\underset{\sim}{R}) = \int \phi_x^{f*}(\xi)\, X^{fi}(\xi, \underset{\sim}{R})\, d\xi. \tag{3.2}$$

We will now call this the reduced overlap integral. If Ψ_B is represented by the cluster expansion (2.6) we have (see equations 2.71, 2.72)

$$X^{fi}(\xi, \underset{\sim}{R}) = \sum_{n^i n_x^i} C(n^i\, n_x^i\, n_A^f)\, \phi_x^{n_x^i}(\xi)\, \bar{\chi}_{xA}^{n^i}(\underset{\sim}{R}) \tag{3.3}$$

$$\mathcal{O}^{fi}(\underset{\sim}{R}) = \sum_{n^i} C(n^i\, n_x^f\, n_A^f)\, \bar{\chi}_{xA}^{n^i}(\underset{\sim}{R}) \tag{3.4}$$

where $\bar{\chi}$ obeys equation (2.9). In section (3.2) we consider the content of C and means of calculating this constant. In the case of transfer reactions the reduced overlap integral is more complicated because the cluster does not appear as a free particle but is attached to a different system. If we may also use a cluster expansion to describe the projectile in a stripping reaction, for example, we obtain an additional overlap integral of the form (see equation 2.38c)

$$Y^{fi}(\xi, \underset{\sim}{r}) = \int \phi_b^{f*}(\lambda)\, \phi_a^i(\lambda, \xi, \underset{\sim}{r})\, d\lambda \tag{3.5}$$

$$= \sum_{s^i s_x^i} C(s^i\, s_x^i\, s_b^f)\, \phi_x^{s_x^i}(\xi)\, \bar{\chi}_{xb}^{s^i}(\underset{\sim}{r}) \tag{3.6}$$

where $\bar{\chi}_{xb}$ obeys an equation similar to (2.9), and the reduced overlap integral becomes

$$\mathcal{O}^{fi}(\underset{\sim}{R}) = \int X^{fi*}(\xi, \underset{\sim}{R})\, Y^{fi}(\xi, \underset{\sim}{r})\, d\xi. \tag{3.7}$$

In this section we examine methods of representing the radial behaviour of the overlap integrals. For reactions which we expect to be localized in the surface region the most important piece of information we have is the asymptotic behaviour of $\bar{\chi}_{xA}$ which is given by equation (2.21). The function $\bar{\chi}_{xb}$ must obey a similar equation containing the separation energy $S_{xb} = -\epsilon_{xb}$. In the separation energy procedure (SEP) or well-depth (WD) method the function $\bar{\chi}$ is generated as an eigenstate of a one-body potential whose parameters are adjusted to give the required cluster binding energy for each transition, i.e. we solve equation (2.16). Of the quantum numbers represented by n^i, the possible values of the angular momentum quantum numbers LJ are determined by the selection rules for the reaction. The choice of the principal quantum number N is more difficult. When x ist a single nucleon N is usually taken to be the same as the principal quantum number of the active nucleons in the sub-shell LJ, but for clusters the value of N is usually chosen according to the oscillator rule [Wi66]

$$2(N-1)+L+2(n_x-1)+\ell_x = \sum_{i=1}^{x} [2(n_i-1)+\ell_i] \tag{3.8}$$

where n_i ℓ_i are the quantum numbers of a set of single nucleons presumed to form the cluster and $2(n_x-1)+\ell_x$ represents the quanta of internal excitation of the cluster x. In the spirit of the OCM it is believed that this procedure excludes those states of relative motion which are forbidden by the Pauli principle.

The parameters of the one-body potential may be chosen on some rather arbitrary grounds, possibly by reference to optical potential parameters, or may be obtained by the folding model. This allows for a good deal of variation in the choice of parameters. For example, for the description of the ground state of ^{16}O as $\alpha + {}^{12}C$(gs), equation (3.8) gives N = 3, L = 0. Chant *et al* [Ch77] used a potential obtained by folding an $n-\alpha$ interaction into the ^{12}C ground state density distribution and fitting the resulting potential with the Saxon-Woods parameters

$$a = 0.7825 \text{ fm}, \quad R_0 = 1.09(12^{\frac{1}{3}}) = 2.50 \text{ fm}. \tag{3.9}$$

Parameters chosen for phenomenological bound state potentials include

$$a = 0.65 \text{ fm}, \qquad R_0 = 1.25(12^{\frac{1}{3}}) \text{ fm} = 2.86 \text{ fm [Da76]} \tag{3.10}$$

$$a = 0.65 \text{ fm}, \qquad R_0 = 2.0(12^{\frac{1}{3}}) \text{ fm} = 4.58 \text{ fm [Wo76]} \tag{3.11}$$

$$a = 0.65 \text{ fm}, \qquad R = 1.2(12^{\frac{1}{3}} + 4^{\frac{1}{3}}) \text{ fm} = 4.66 \text{ fm [Yo73]} \tag{3.12}$$

$$a = 0.65 \text{ fm}, \qquad R_0 = 1.25(12^{\frac{1}{3}} + 4^{\frac{1}{3}}) \text{ fm} = 4.85 \text{ fm [De73]} \tag{3.13}$$

$$a = 0.7825 \text{ fm}, \quad R_0 = 2.52(12^{\frac{1}{3}}) = 5.77 \text{ fm. [Ch77].} \tag{3.14}$$

The 3s functions $u_{NL}(R) = R\,\bar{\chi}_{NL}(R)$ corresponding to some of these potentials are plotted in Figure 16. The variations in the magnitude of these functions in the surface

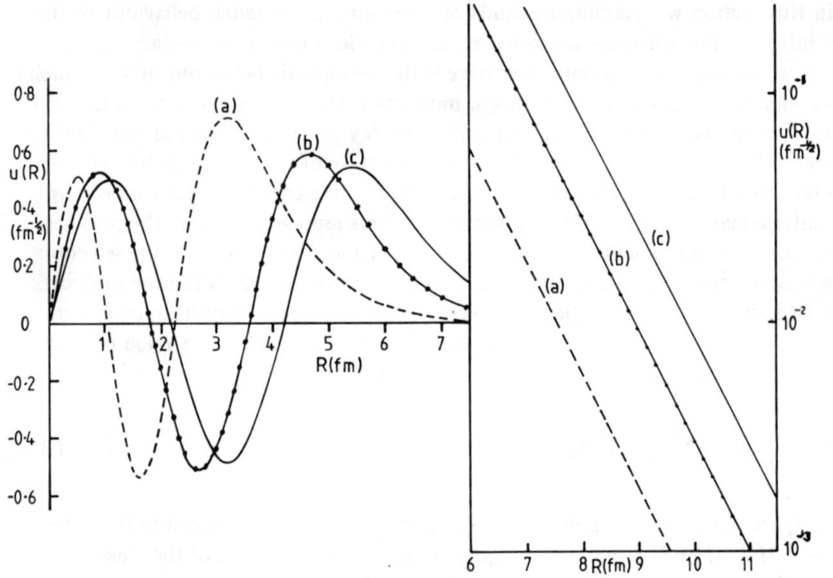

Fig. 16 The function $u_{NL}(R)$ for the 3s state of the $\alpha + {}^{12}C$ system generated in a Saxon-Woods potential with the parameters. (a) $R_0 = 1.09\ A_c^{1/3}$ fm, a = 0.7825 fm, (b) $R_0 = 1.25\ (A_c^{1/3} + 4^{1/3})$fm, a = 0.65 fm, and (c) $R = 2.52\ A_c^{1/3}$ fm, a = 0.7825 fm. (Note change of scale on the right hand side of the diagram.)

region produce large changes in the magnitude of the predicted cross-sections for various reactions.

It may be argued that the phenomenological bound state potentials take implicit account in some way of the effects of antisymmetrization and coupling to other channels, but when a folded potential is used it is important to attempt to take these effects into account explicitly. Several attempts have been made to do this for the overlap integral appropriate to single nucleon transfer, by solving the set of coupled equations [Pi65, Ro67, St66] or by an iterative treatment of the inhomogeneous coupling term [Pr69, Ph68, Ka67, Ig69]. Some of these techniques have been applied to evaluate the overlap integral for two-nucleon transfer reactions, usually for two interacting neutrons outside a closed shell spherical core, e.g. ^{18}O, ^{42}Ca. In this case the overlap integral may be written as

$$X^{fi}(\underset{\sim}{r}_{A+1}, \underset{\sim}{r}_{A+2}) = \int \Phi_A^{f*}(\eta)\, \Psi_B^i(\underset{\sim}{r}_{A+1}, \underset{\sim}{r}_{A+2}, \eta)\, d\eta \tag{3.15a}$$

and it obeys the equation

$$[\epsilon - T_{A+1} - T_{A+2} - V(r_{A+1}) - V(r_{A+2}) - W(\underset{\sim}{r}_{A+1}, \underset{\sim}{r}_{A+2})]\, X^{fi}(\underset{\sim}{r}_{A+1}, \underset{\sim}{r}_{A+2}) = 0 \tag{3.15b}$$

where $\epsilon = E_B^i - E_A^f$, $V(r_{A+1})$ $V(r_{A+=})$ are the potentials which bind each nucleon to the core and $W(r_{A+1}, r_{A+2})$ is the residual interaction between the two nucleons. The most common early procedure was to omit the residual interaction but bind each neutron at a separation energy of $\frac{1}{2}\,\epsilon$; this implies that the effect of the residual interaction on the energy is included in the potentials $V(r_{A+!})$ and $V(r_{A+2})$. Alternatively each neutron may be bound at the separation energy ϵ_n for a nucleus with one neutron outside the core [Dr66]; this implies that $\langle W(r_{A+1}, r_{A+2})\rangle = \epsilon - 2\epsilon_n$ which is the pairing energy of two neutrons. Extensive configuration mixing calculations have been made using bound orbitals generated in a Saxon-Woods potential [Ja69] or including continuum contributions [Ib70]. Very large configuration mixing calculations (the extended basis shell model) are usually carried out in an oscillator basis [Ib75, Va76, Ia75, Fe73] which makes it difficult to obtain the correct fall-off of the overlap integral far beyond the nuclear radius [Ib75, Ba78a]. A modified Sturm-Liouville expansion method [Ba76a, Va79] reduces the computational effort compared with the oscillator expansions but is still considered to be impractical when there are many active valence nucleons. The Pauli principle is taken into account in these calculations by omitting those Sturm-Liouville functions which have quantum numbers corresponding to closed shells.

In most calculations on the knock-out of a deuteron the overlap integral has been constructed from shell model wavefunctions using fractional parentage techniques and transforming two single-particle wavefunctions of oscillator form into a product of a function of the relative coordinate $\mathbf{r} = \mathbf{r}_{A+1} - \mathbf{r}_{A+2}$ and a function of the centre-of-mass coordinate $\overline{\mathbf{R}} = \frac{1}{2}(\mathbf{r}_{A+1} + \mathbf{r}_{A+2})$. Both of these functions are of oscillator form so neither have the correct asymptotic behaviour. The use of a shell model coordinate system is unsatisfactory for two reasons. Firstly, the coordinate $\overline{\mathbf{R}}$ is not equal to $\mathbf{R} = \mathbf{R}_x - \mathbf{R}_A$, where x represents a deuteron in this case, but is given by

$$\overline{\mathbf{R}} = A\,\mathbf{R}/(A+2) + \mathbf{R}_B = \mathbf{R} + \mathbf{R}_A \tag{3.16}$$

where $\mathbf{R}_B$ is the coordinate of the centre of mass of the initial nucleus B = A + x and $\mathbf{R}_A$ is the centre of mass of the residual nucleus A. This is a consequence of the lack of translation invariance of the shell model Hamiltonian and implies that the wavefunction is not free from spurious centre of mass motion [El55]. Secondly, in plane wave approximation the momentum distribution seen in a deuteron knock-out reaction becomes

$$g = \int \phi_d^*(\mathbf{r}_1 - \mathbf{r}_2)\, X^{fi}(\mathbf{r}_1, \mathbf{r}_2)\, \exp\left[i(\mathbf{p}_b - \mathbf{q}_b - \mathbf{q}_d)\cdot \frac{1}{2}(\mathbf{r}_{A+1} + \mathbf{r}_{A+2})\right] d\mathbf{r}_{A+1}\, d\mathbf{r}_{A+2} \tag{3.17a}$$

where the overlap integral is given by

$$X(\mathbf{r}_{A+1}, \mathbf{r}_{A+2}) = \int \Phi_A^{f*}(\mathbf{r}_1 \ldots \mathbf{r}_A)\, \Phi_B^{i}(\mathbf{r}_1 \ldots \mathbf{r}_{A+1}, \mathbf{r}_{A+2})\, \delta(\mathbf{r}_1 + \mathbf{r}_2 + \ldots \mathbf{r}_{A+1} + \mathbf{r}_{A+2}) \times$$

$$\times \exp\left[-i\mathbf{q}_A \cdot (\mathbf{r}_1 + \mathbf{r}_2 + \ldots + \mathbf{r}_A)/A\right] d\mathbf{r}_1 \ldots d\mathbf{r}_A \tag{3.17b}$$

and the δ-function has been included to remove centre of mass motion. This expression illustrates the point made by Berggren, Brown and Jacob [Be62] that the overlap integral is independent of q_A only for appropriate choices of the relative coordinates.

Centre-of-mass corrections to the overlap integral for transfer reactions have also been examined [Pi76, 77, Vi73]. A recent study of two-nucleon transfer reactions

[Pi 79 a] starting from equation (3.15) yields enhancements of the cross-sections on light nuclei; for example the (t, p) reactions on ^{16}O and ^{40}Ca are increased by factors of 2.1 and 1.6 respectively.

The cluster model provides an appropriate choice of coordinates for the description of knock-out reactions. Unfortunately almost all cluster model calculations for the (e, ed) and (p, pd) reactions have truncated the cluster expansion to a single term, and in so doing lose all spectroscopic information, and have also used gaussian functions which means that the asymptotic behaviour of both the relative motion of the clusters and the internal cluster wavefunction are incorrect. For a cluster as loosely bound as the deuteron the error introduced in the internal wavefunction may be as serious as the error in the relative motion, particularly in a reaction such as (γ, np) when the deuteron cluster is emitted in a continuum state [Ja 67 a]. A study of exchange effects for the break-up process $^6Li \rightarrow \alpha + d$ has shown [Ja 69 a, 70 a] that correct estimation of exchange effects depends quite critically on a correct description of the overlap integral; with gaussian functions the exchange effects seem large but for a relative function with the correct asymptotic behaviour the magnitude of the exchange terms is very much reduced.

In most studies of α-knock-out and α-transfer reactions (see section 4.1 and 4.3) the overlap integral is generated in a phenomenological bound state potential and so has the correct asymptotic behaviour. In a few cases, for example in calculations on α-decay [To 79, Ar 78] a relative function derived from oscillator functions is matched to a suitable Coulomb or Hankel function at some arbitrarily chosen channel radius. A study of the effect of this procedure in the (p, 2p) reaction [Ho 78] has shown that this can be a very unsatisfactory method of determining the single nucleon overlap integral for a surface reaction and there is no reason to suppose that it will be any better for the cluster overlap integral.

The reduced overlap integral for the system $^{212}Po(gs) \rightarrow \alpha + {}^{208}Pb(gs)$ has been calculated by Rhoades-Brown [Rh 78, Ja 78] using the modified OCM equation (2.156) and assuming that the resonant state can be replaced by a suitably chosen bound state (see section 4.2). In this work the wavefunction of the α-particle was taken to be of gaussian form and so was the $n-\alpha$ interaction but the core nucleus ^{208}Pb was represented by a Slater determinant of single-particle wavefunctions generated in a Saxon-Woods potential. Thus each single-particle wavefunction had the correct asymptotic behaviour and the parameters of the Saxon-Woods potential were chosen so that the proton density distribution gave good agreement with elastic electron scattering and the matter distribution gave good agreement with pion reaction cross-sections and other data sensitive to the neutron distribution. The folded real optical potential for α-particle scattering obtained from this matter distribution and the chosen $n-\alpha$ interaction gave good agreement with α-particle scattering data. Using the density matrix expansion of Negele and Vautherin [Ne 72] the one-nucleon exchange term and all higher exchange terms can be expressed in terms involving only the nuclear density distribution and its derivatives. The validity of this procedure has been examined by Sprung *et al* [Sp 75]. The structure of the modified OCM equation ensures that the reduced overlap integral has the correct asymptotic behaviour and the correct normalization. The diagonal elements of the exchange operator $K(\underset{\sim}{R}, \underset{\sim}{R}')$ are shown in Figure 17. It can be seen that the one-particle exchange term has

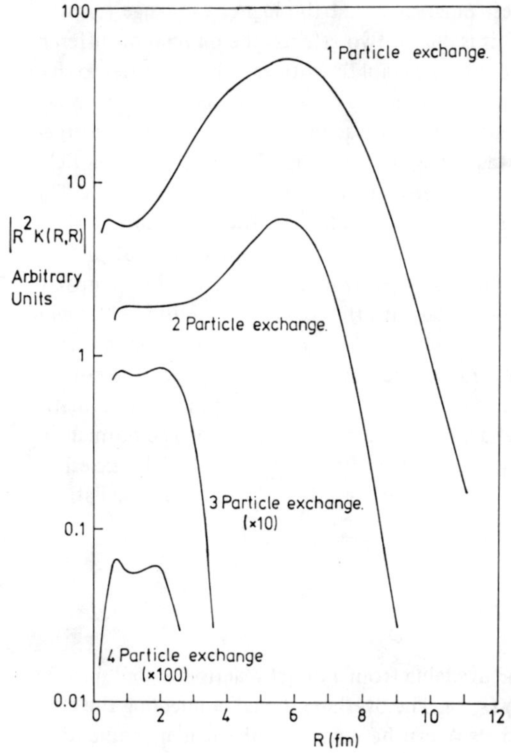

Fig. 17
The behaviour of $|R^2K(R,R)|$ and its components for the system α + ^{208}Pb [Ja78].

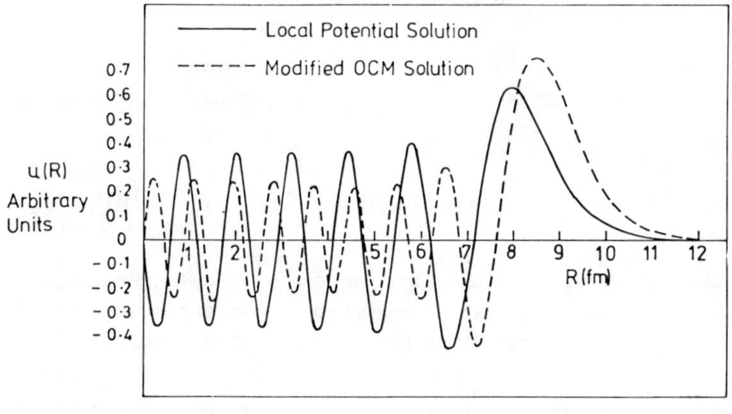

Fig. 18 The function $u_{NL}(R)$ for the 12s state of the α + ^{208}Pb system generated in a local α-nucleus potential (full line) and the corresponding solution of the modified OCM equation (broken line) [Ja78].

by far the largest magnitude and longest range and that the higher exchange terms become progressively less important. This is due to two effects: the number of different ways each exchange process can occur diminishes rapidly with each higher order exchange component and the higher order exchange components depend on ever increasing powers of the local density matrices. In these calculations it was found that the surface correction in the density matrix expansion was negligible when applied to the two particle exchange term and it was neglected for the three-and four-particle exchange terms. The behaviour of the eigenvalues $1-\lambda_{n0}$ behave similarly to those shown in Figure 14 although the actual magnitudes are different because of the different treatment of the nuclear density. The normalized overlap integral is shown in Figure 18 and is compared with the function derived from a local potential with the oscillator condition (3.8) which requires a 12s state. It can be seen that the main effect of antisymmetrization in this calculation is to reduce the amplitude of the reduced overlap integral inside the potential in accordance with the Perey effect. The damping of the interior region is not dramatic because of the simple non-saturating behaviour of the $n-\alpha$ interaction. The normalization condition leads to an enhancement of the reduced overlap integral of the exterior region and an outward shift of the last maximum in accordance with other calculations [An75, To77].

3.2 Spectroscopic Factors

The nuclear structure information available from a direct reaction is contained in the two overlap integrals $X(\xi, \underset{\sim}{R})$ and $Y(\xi, \underset{\sim}{r})$. The overlap integral connecting the target nucleus $B = A + x$ and the residual nucleus A can be written with angular momentum labels as

$$X^{fi}(\xi, \underset{\sim}{R}) = \binom{B}{x}^{\frac{1}{2}} \int \Phi_A^{J_A M_A^*}(\eta)\, \Psi_B^{J_B M_B}(\eta, \xi, \underset{\sim}{R})\, d\eta. \tag{3.18}$$

For convenience we omit isospin labels. It must in principle be possible to make an expansion of Ψ_B in terms of a complete set of states Φ_A. Such a fractional parentage expansion is given by

$$\Psi_B^{J_B M_B} = \sum_{JMJ_PM_P} (J_P M_P\, JM | J_B M_B)\, I_{AB}(J_P J_B J)\, \Phi_A^{J_P M_P}\, [\phi_x^{jm}(\xi)\, \bar{\chi}_{xA}^{NL\Lambda}(\underset{\sim}{R})]_{JM} \tag{3.19}$$

where I is the fractional parentage coefficient and the square brackets imply angular momentum coupling to the resultant values JM, so that j and L are coupled to J while J_p and J are coupled to J_B. Substituting expansion (3.19) in equation (3.18) we obtain

$$X^{fi}(\xi, \underset{\sim}{R}) = \binom{B}{x}^{\frac{1}{2}} \sum_{JM} (J_A M_A\, JM | J_B M_B)\, I_{AB}(J_A J_B J)\, [\phi_x^{jm}(\xi)\, \bar{\chi}_{xA}^{NL\Lambda}(\underset{\sim}{R})]_{JM}. \tag{3.20}$$

In this case when x is a single nucleon ϕ^{jm} is simply a spin function and $\bar{\chi}$ is a normalized form factor which may be expressed in terms of its angular and radial functions, i.e.

$$[\phi^{jm}(\xi)\,\bar{\chi}^{NL\Lambda}(\underset{\sim}{R})]_{JM} = \sum_{L\Lambda m} (jmL\Lambda|JM)\,\phi^{jm}(\xi)\,R^{NLJ}(\underset{\sim}{R})\,i^L\,Y_L^\Lambda(\hat{\underset{\sim}{R}}) \tag{3.21}$$

with $j=\frac{1}{2}$ and $\underset{\sim}{L}+\underset{\sim}{j}=\underset{\sim}{J}$. In all these expressions N is strictly an additional label but it will be interpreted in general as a principal quantum number so that the radial function R^{NLJ} has $2(N-1)+L$ nodes.

It is usual to define the spectroscopic factor as

$$S^{\frac{1}{2}}(J_A J_B J) = \binom{B}{x}^{\frac{1}{2}} I_{AB}(J_A J_B J). \tag{3.22}$$

Several authors [Ic73, Ku74, To77] define a spectroscopic amplitude for multi-particle processes as

$$A_{AB}(J_A J_B J) = \binom{B}{x}^{\frac{1}{2}} \int [\Phi_A^{J_A M_A}[\phi_x^{jm}\,\bar{\chi}^{NL}]_{JM}]^*_{J_B M_B}\,\Psi_B^{J_B M_B}\,d\eta\,d\xi\,d\underset{\sim}{R} \tag{3.23}$$

$$= S_x^{\frac{1}{2}}(J_A J_B J). \tag{3.24}$$

It is not possible to proceed any further in the discussion of spectroscopic factors or spectroscopic amplitudes without involving a nuclear model. Comparing the cluster model expansion (2.6) with the fractional parentage expansion (3.19) we have

$$C(n\,n_A\,n_x)\binom{B}{x}^{-\frac{1}{2}} = (J_P M_P J M|J_B M_B)\,(jmL\Lambda|JM)\,I_{AB}(J_P J_B J)$$

$$C(n\,n_A\,n_x) = S_x^{\frac{1}{2}}(J_P J_B J)\,(J_P M_P J M|J_B M_B)\,(jmL\Lambda|JM). \tag{3.25}$$

Use of the cluster model is particularly appropriate with a quasi-free or peripheral model of the direct reaction and when the interaction V_{bx} (see equation 2.34a) for a transfer reaction is replaced by a δ-function. *Unfortunately there is no simple way of determining the coefficient,* C, *except when the cluster expansion is restricted to one term and there is as yet no standard cluster model formalism for the construction of the spectroscopic factors.* The only approach to this problem of which we are aware is the work of Noble and Coelho [No71, Co73] for the systems $^{16}O \to \alpha + {}^{12}C$, $^{12}C \to \alpha + {}^{8}Be$. Their values have been by Jackson and Brenner [Ja79, Br79] in a calculation of stopped pion capture in ^{16}O and lead to the results shown in Figure 5.

For a microscopic approach to cluster reactions it is appropriate to use an independent-particle model for the spectroscopic factors and it is of particular interest to relate multi-particle spectroscopic amplitudes to amplitudes for a sub-set of the cluster nucleons, for example to relate α-particle spectroscopic amplitudes to two-proton and two-neutron amplitudes. *Unfortunately there is at present no method for evaluating spectroscopic amplitudes for multi-particle processes within the traditional shell model approach involving fractional parentage coefficients and angular momentum recoupling which does not make some use of oscillator functions.* This traditional shell model approach proceeds with the following steps [Ic73, To77a]. The shell model coordinates $\underset{\sim}{r}_i$ are converted to a set of internal coordinates $\underset{\sim}{s}_i$. The shell model wavefunction can then be expressed in the form

$$\Psi_B^{J_B}(\underset{\sim}{r}_B) = \phi^{n\ell}(\underset{\sim}{R}_B)\,\Psi^{J_B}(\underset{\sim}{s}_B)$$

and similarly for $\Phi_A(\underset{\sim}{r}_A)$ and $\phi_x(\underset{\sim}{r}_x)$, where $\phi^{n\ell}$ is in the lowest allowed state $n=1, \ell=0$. It is assumed that each nucleus A, B and x can be adequately described by the same oscillator length parameter $\nu_0 = 1/a^2 = \mu\omega/\hbar$ where μ is the reduced mass. Using the generalized Talmi transformation the functions $\phi^{n\ell}$ and ϕ^{jm} can be rewritten as a bilinear sum of functions of the coordinate $\underset{\sim}{R}_B$ and the relative coordinate $\bar{\underset{\sim}{R}} = \underset{\sim}{R} + \underset{\sim}{R}_A$ (see equation 3.16). The Brody-Moshinsky bracket yields the factor $(B/A)^{N+\frac{1}{2}L-1}$ and a fractional parentage expansion is made for Ψ_B to give

$$\left(\frac{B}{x}\right)^{\frac{1}{2}} \Psi_B^{J_B}(\underset{\sim}{r}_B) = \sum_{J_P J_x \Gamma} <\Psi_B^{J_B}||\chi^{J_x\Gamma\dagger}||\Phi_A^{J_P}>[\Phi_A^{J_P}(\underset{\sim}{r}_A)\,\phi^{J_x\Gamma}(\underset{\sim}{r}_x)]_{J_B} \qquad (3.26)$$

where the first factor on the right hand side is the fractional parentage amplitude expressed as a reduced matrix element of a four-nucleon creation operator with angular momentum J_x and label Γ specifying the shell model character of the four nucleons. By convention this amplitude includes the factor $\left(\frac{B}{x}\right)^{\frac{1}{2}}$. The spectroscopic amplitude finally becomes

$$A_{AB}(J_A J_B J) = \left(\frac{B}{A}\right)^{N+\frac{1}{2}L-1} \sum_\Gamma <\Psi_B^{J_B}||\chi^{J\Gamma\dagger}||\Phi_A^{J_A}> G(J\Gamma, j, NL) \qquad (3.27)$$

where

$$G(J\Gamma, j, NL) = \int [\phi^j(\underset{\sim}{s}_x)\,\phi^{NL}(\underset{\sim}{R}_x)]_J^*\,\phi^{J\Gamma}(\underset{\sim}{r}_x)\,d\underset{\sim}{r}_x. \qquad (3.28)$$

In these expressions the oscillator length parameter appearing in the function ϕ^{NL} is now ν where [Ic73]

$$\nu = x\,\nu_0(A/B). \qquad (3.29)$$

For heavy nuclei the assumption of equal length parameters is very unsatisfactory [Wa79]. In a study of the effects of different length parameters in relatively light systems, Ichimura *et al* [Ic73] assume that

$$\nu_{0x}/\nu_0 = [x/B]^{-\frac{1}{2}} \qquad (3.30)$$

and hence

$$\nu = \nu_{0x}\,A[x/b]^{4/3}. \qquad (3.31)$$

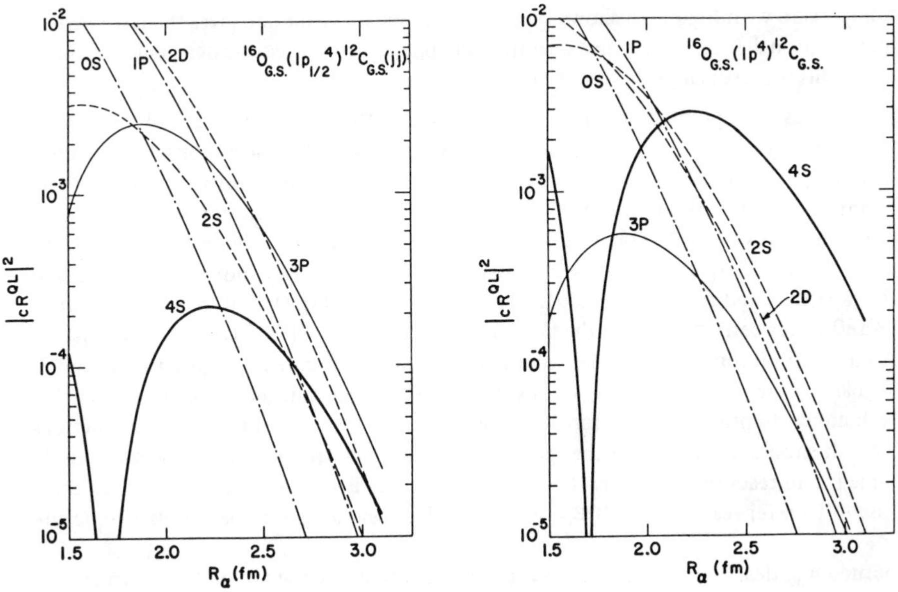

Fig. 19 The square of radial functions for the relative motion of various clusters of a $(lp)^4$ group of nucleons transferred between ^{16}O and ^{12}C ground states. (a) Results obtained from an intermediate wavefunction for ^{12}C. (b) Results obtained in the jj coupling limit. The curves are labelled by QL where $Q = 2(N - 1) + L$ [Ku74].

The integral G has been evaluated for two-nucleon transfer [Gl65, 75] in a slightly different form and for four-nucleon removal [Sm61, Ro68, 69, Ic73, Ku74]. The four-nucleon parentage amplitudes and spectroscopic amplitudes have been calculated for the lp-shell and the sd shell [He75, Ba65, Ro69, Ku73, Ch78, Dr75]. Kurath and Towner [Ku74] have shown how the α-particle amplitudes may be reduced to two-proton and two-neutron amplitudes and the latter have been calculated in various models [Cl68, Ar63, Yo62]. The amplitudes for np removal, needed for example for deuteron knock-out reactions, have also been calculated by various authors [Be65, Ba65, Ro68].

Figure 19 shows the results of a calculation [Ku74] on the transfer of a $(lp)^4$ group of nucleons between the ground state of ^{16}O and ^{12}C. The quantity plotted is the square of the radial parts of ϕ^{NL} weighted by their intensity in the $(lp)^4$ cluster and summed over all internal states with the same relative function. The functions are labelled by $Q = 2(N-1)+L$ and L. The oscillator rule (3.8) gives $N = 3$ $L = 0$ and $Q = 4$ if the cluster is in the lowest allowed state with no internal excitation; smaller values of N and Q correspond to internal excitation of the cluster. It can be seen from the Figure that the state of maximum Q and N is dominant in the surface region when an intermediate coupling wavefunction is used for ^{12}C, and hence for a reaction localized in the surface it would be meaningful to describe the process as removal of an α-cluster with no internal excitation. When the nuclear wavefunctions are taken in the jj coupling limit, so that the transferred cluster is $(lp_{\frac{1}{2}})^4$, this state of maximum N is very much reduced in magnitude.

Comparison with Figure 16 shows that the shell model calculation gives the last maximum at a smaller separation distance than the phenomenological cluster model calculations; this is presumably the result of using oscillator functions.

It has been pointed out [Fl75, Wa80a] that in the definition (3.23) of the spectroscopic amplitude the final state is not properly normalized for the reasons discussed in section 2.4. When the normalization is carried out correctly, or approximately so, a very significant increase is obtained in the spectroscopic factor for α-cluster removal from a heavy nucleus within the framework of the shell model and a further increase is obtained when the use of equal oscillator length parameters is replaced by more suitable values for the α-particle and the heavy residual nucleus [Fl76, Wa80a]. One of these calculations [Wa80a] was carried out with the Glasgow shell model code [Wh71, 72] which departs from the traditional shell model techniques of fractional parentage coefficients and angular momentum recoupling and exploits second quantization and computational techniques. In principle, therefore, this approach is not restricted to oscillator functions.

It should be noted that the spectroscopic factors introduced in this section are the same for all reactions involving the overlap of nucleus $B = A + x$ with nucleus A. In the case of transfer reactions $b + B \rightleftarrows A + a$ there will appear an additional spectroscopic amplitude A_{ab} describing the process $a \rightleftarrows b + x$ and containing the factor $\binom{a}{x}^{\frac{1}{2}}$ which appears in equation (2.143). There are other ways of describing the spectroscopic information in nuclear reactions [Me71, Br72, Fl78] but they lead to an undesirable energy dependence or time dependence of the spectroscopic factors. The clean separation provided by the conventional theory between the reaction mechanism and the nuclear structure content has proved extremely powerful.

The spectroscopic factors defined in this section can readily be related to the reduced widths γ^2 which appear in the R-matrix theory of resonance reactions. The relation for the process $B \to A + x$ is [La58]

$$\gamma^2_{xA}(R_c) = \frac{\hbar^2 R_c}{2\mu_{AB}} S_x |R^{NL}(R_c)|^2 \tag{3.32}$$

where R_c is the channel radius of the R-matrix theory, μ_{AB} is the reduced mass and R^{NL} is the normalized radial part of ϕ^{NL}. A dimensionless quantity θ^2 is also used and is defined as

$$\begin{aligned}\theta^2_{xA}(R_c) &= (2\mu_{AB} R_c^2/3\hbar^2)\, \gamma^2_{xA}(R_c) \\ &= \frac{1}{3} R_c^3 S_x |R^{NL}(R_c)|^2 .\end{aligned} \tag{3.33}$$

3.3 Investigation of Nuclear Models

The study of single-nucleon transfer and knock-out reactions has proved extremely effective for comparison of the spectroscopic predictions of various nuclear models and for studying trends through excited states in the same system and through similar states in neighbouring nuclei. Such information is also of great interest in cluster reactions. It would be particularly satisfying to obtain a conclusive answer to the question – does the shell model in its extended form provide an adequate description of the way in which nuclei fragment? If the answer to this question is negative then we would also like to know whether the cluster model offers an improved description. It is therefore appropriate to examine the formalism of sections 3.1 and 3.2 in order to establish to what extent it allows us to make a clear distinction between the use of the shell model and the cluster model.

It is evident that there is a well-defined theory of shell model spectroscopic amplitudes which is constrained to oscillator functions through the need to remove the centre-of-mass motion and recouple the functions into internal and relative functions. The construction of overlap integrals with the correct radial behaviour then involves either the somewhat dangerous procedure of matching the relative function for the cluster to a function with the correct asymptotic behaviour and overlooking the presence of the incorrect coordinate $\overline{R}$ defined by equation (3.16) or of taking single-particle wavefunctions with the correct asymptotic behaviour and assuming that their centre-of-mass motion can be neglected or described approximately in the same manner as for oscillator functions. In the latter case it may be necessary to take a very large number of single-particle configurations to yield the correct relative function. A more simple approach is to use equation (3.8) which describes the redistribution of the quanta of excitation associated with single nucleons in an oscillator potential. There is nothing to prevent the representation of the relative function in terms of a summation over allowed values of N and L but in this case we need the weighting factors associated with the different states of internal excitation of the cluster and these factors are model-dependent, as noted in the discussion of Figure 19.

For light nuclei and reactions localized in the nuclear surface it may prove to be acceptable to take the values of L determined by the selection rules and the corresponding highest allowed values of N given by equation (3.8) for the dominant single-particle configuration. Configuration mixing of higher oscillator shells increases N by one for each pair of particles promoted to a higher major shell and again the weighting factors for each excited configuration are required. For heavy nuclei, important configuration mixing can take place within an oscillator major shell which is not reflected in any change in N. This reflects the high degeneracy of the oscillator model which may produce unrealistic results for heavy nuclei.

Using the folding model it is possible to construct a potential which describes the interaction of the cluster x and the residual nucleus A in the bound state and the formalism allows for a shell model description of x and A. However, in this model it is still essential to take account of exchange of nucleons between x and A by solving equation (2.15) in some level of approximation, i.e. *the use of the oscillator rule (3.8) with a folded potential does not adequately take account of antisymmetrization.* Apart from this practical problem there is a complete shell model theory which may be used with a microscopic reaction theory.

Because of the computational difficulties associated with a finite-range, microscopic reaction theory the quasi-free or peripheral model is frequently used with a zero-range or factorized interaction. In the latter case the cluster model is the natural choice for the nuclear structure aspect of the problem. The overlap integral is clearly defined and again requires solution of equation (2.15), but there have been only very limited attempts to evaluate spectroscopic amplitudes in the cluster model.

An alternative approach to the nuclear structure has been adopted by Tomoda and Arima [To 78] who write the wavefunction of states in ^{20}Ne as a linear combination of shell model states constructed from a ^{16}O core plus $(2s1d)^4$ configurations and cluster model states constructed from a ^{16}O core plus α-cluster states. All functions are of oscillator form and so some care must be exercised over representations which appear in both the shell model and cluster model states. This procedure can be used to introduce strong α-cluster correlations without an excessive amount of configuration mixing in the shell model states. Spectroscopic amplitudes for the α-decay of excited states of ^{20}Ne are given.

A relation between the overlap integral and the nuclear density matrix or transition density has been derived [Be 65, El 67]. If we consider first a nucleus $B = A + 1$ the density matrix or mixed density function (see equation 2.120) is given by

$$\rho_B^{if}(\underset{\sim}{R}, \underset{\sim}{R}') = \int \Psi_B^{f*}(\underset{\sim}{R}, \underset{\sim}{\eta}_1 \dots \underset{\sim}{\eta}_A)\, \Psi_B^{i}(\underset{\sim}{R}', \underset{\sim}{\eta}_1, \dots \underset{\sim}{\eta}_A)\, d\underset{\sim}{\eta}_1 \dots d\underset{\sim}{\eta}_A$$

$$= \sum_p \langle \Psi_B^f | \underset{\sim}{R}\, \Phi_A^p \rangle \langle \underset{\sim}{R}'\, \Phi_A^p | \Psi_B^i \rangle$$

$$= \sum_p \mathcal{O}_{BA}^{fp*}(\underset{\sim}{R})\, \mathcal{O}_{BA}^{pi}(\underset{\sim}{R}') \tag{3.34}$$

where p represents the sum over all parent states in the A system. Similarly the ground state density, normalized as in equation (2.116) can be written as

$$\rho_B^{ii}(\underset{\sim}{R}) = \frac{1}{B} \sum_p |\mathcal{O}_{BA}^{pi}(\underset{\sim}{R})|^2. \tag{3.35}$$

Using the fractional parentage expansion (2.6) for Φ_B and hence for $\mathcal{O}$, as in equation (3.4), we obtain

$$\rho_B^{ii}(\underset{\sim}{R}) = \frac{1}{B} \sum_p |\sum_{n^i} C(n^i\, n_x^p\, n_A^p)\, \overline{\chi}^{n^i}(\underset{\sim}{R})|^2 \tag{3.36}$$

where $\overline{\chi}(\underset{\sim}{R})$ is the wavefunction of the single nucleon relative to the centre-of-mass of the residual nucleus A in the nucleus B in state i. Equation (3.25) gives the connection between the coefficient C and the spectroscopic factor. In contrast the shell model gives

$$\rho_B(\overline{\underset{\sim}{R}}) = \frac{1}{B} \sum_p |\phi_B^p(\overline{\underset{\sim}{R}})|^2 \tag{3.37}$$

where the ϕ_B are single-particle wavefunctions. If configuration mixing is excluded the functions ϕ are normalized to unity; otherwise these functions include the appropriate fractional parentage coefficient, as in equation (3.36). The coordinates are related by

$$\bar{\underset{\sim}{R}} = (A-1)\,\underset{\sim}{R}/A + \underset{\sim}{R}_B$$

although for a heavy nucleus we can neglect centre-of-mass motion and the difference between $\underset{\sim}{R}$ and $\bar{\underset{\sim}{R}}$. Equations (3.36) and (3.37) give the distribution of nucleon centres of mass in nucleus B. If we require the charge distribution of nucleus B we must include the nucleon charge distributions in the usual convolution integral.

For a system B = A + x, where x is a cluster of nucleons, the distribution of centres of mass of clusters x in the ground state of nucleus B is given by equations (3.35) and (3.36) where $\bar{\chi}(\underset{\sim}{R})$ is now the relative cluster function as defined in equation (2.6) and following equations. The distribution of nucleon centres of mass is now given by

$$\rho_B^{ii}(\underset{\sim}{s}) = \sum_{k=1}^{x} \int \sum_{p} |\chi^{pi}(\underset{\sim}{\xi}_1 \ldots \underset{\sim}{\xi}_x, \underset{\sim}{R})|^2\, \delta(\underset{\sim}{s} - \underset{\sim}{R} - \underset{\sim}{\xi}_k)\, d\underset{\sim}{\xi}_1 \ldots d\underset{\sim}{\xi}_x. \tag{3.38}$$

If the system is close to true clustering so that the terms which correspond to the free particle x dominate, the expression (3.38) reduces to

$$\rho_B^{ii}(\underset{\sim}{s}) \simeq \sum_{k=1}^{x} \int \sum_{p} |\mathcal{O}^{pi}(\underset{\sim}{R})|^2\, \rho_x(\underset{\sim}{\xi}_1 \ldots \underset{\sim}{\xi}_x)\, \delta(\underset{\sim}{s} - \underset{\sim}{R} - \underset{\sim}{\xi}_k)\, d\underset{\sim}{\xi}_1 \ldots d\underset{\sim}{\xi}_x \tag{3.39}$$

where ρ_x is the density distribution of the free particle x in its ground state normalized to unity. This is a typical convolution integral.

The formalism given above shows that the nuclear ground state density and transition densities for nucleus B can be expressed in terms of a complete set of overlap integrals and this may serve as a further test of the nuclear model. In practice, unless a few parentage coefficients are very large in comparison to all the others or the density is evaluated only at a large distance from the residual nucleus so that all but a few of the relative functions have died away, construction of a density distribution from experimentally determined cluster overlap integrals would be a tedious business since it is necessary to include all the possible states of relative motion and also all the possible states of internal excitation of the cluster associated with each parent state of the residual nucleus.

4 Application to Selected Reactions

4.1 Alpha-particle Knock-out Reactions

In quasi-free knock-out reactions two outgoing particles are detected in coincidence, and this provides considerable freedom of choice in what aspects of the reaction mechanism are examined. For spectroscopic studies it is usual to choose the experimental geometry to emphasise a range of values of the cluster fermi momentum p_X (see equation 2.66c) about zero and hence to emphasise low momentum components of the function describing the relative motion of the cluster and the residual nucleus in the target nucleus (see equation 2.74). As the momentum distribution has a maximum at $p_X = 0$ for a relative s-state and a minimum for other angular momentum states, this choice assists in the separation of distributions corresponding to transitions to different final states. This emphasis on low momentum components of the relative wavefunction implies an emphasis on its long-range radial behaviour.

As discussed in section 2.2, direct knock-out reactions can be analysed to a good degree of accuracy using the DWIA formula given by equation (2.67) where the proportionality sign hides a simple kinematic factor. The distorted momentum distribution, defined by equation (2.68), contains the spectroscopic information through the overlap integral.

In this section we examine in detail those knock-out reactions which involve an interaction with an α-particle cluster in the nucleus. The (p, pα) reaction has been studied over an energy range for the incident proton of 40–600 MeV, while the (α, 2α) reaction has been studied over the energy range 25–850 MeV and a few measurements on the (e, eα) reaction have been made at electron energies around 500 MeV.

Extensive studies of the (p, pα) reaction have been made on 1p-shell and 2s1d-shell nuclei but rather few studies have been made with heavier targets. We concentrate here on the measurements made by the Maryland group at 60 and 100 MeV and by the Orsay group at 157 MeV. At 157 MeV, Bachelier *et al* [Ba 73, 76] observe that for the 2s1d-shell nuclei ^{24}Mg, ^{28}Si and ^{40}Ca the excitation of the 0^+ ground state in the residual nucleus dominates in the 0–100 MeV/c region of the spectrum. Excitation of the first 2^+ state in the residual nucleus is important for ^{24}Mg and ^{40}Ca and becomes comparable to the 0^+ excitation for all the three targets in the 100–200 MeV/c region. As expected the cross-sections show maxima at $p_\alpha = -q_A \sim 0$ for $0^+ \rightarrow 0^+$ transitions while transitions to excited 2^+ and 4^+ states show minima at this point. Using DWIA the authors extract both absolute and relative spectroscopic factors as the ratio of the experimental cross-section to the theoretical cross-section and these are compared with the predictions of SU(3) in Table 5. The experimental uncertainties given in the Table do not include the effect of variation of the potential parameters in the DWIA calculation. The absolute values show considerable dependence on the potential parameters; for example the distorted momentum distribution shows a 50 % variation in magnitude though little variation in shape when the parameters of the optical potential for the outgoing α-particle are varied over a quite reasonable range. Because of the importance of distortion, the distorted momentum distribution shows a very substantial change in magnitude – by a factor of 4 – when the parameters of the bound state potential are varied, but for the same change in the relative cluster function the plane wave momentum distribution changes by only 20 %. Thus, it is very difficult to disentangle uncertainties in the optical potentials from uncertainties in the relative cluster function. With the choice of potential parameters taken by Bachelier *et al* the (p, pα) reaction is localized in the region of the target nucleus where the density has fallen to 3 % of its central value.

The Maryland group obtained a spectroscopic factor of 0.25 for the ^{24}Mg (p, pα) ^{20}Ne(gs) transition [Ch75] and also found that the main contribution to the matrix element comes from the 3 % density region [Ch77]. Their results [Ro77] for the spectroscopic factors for the (p, pα) reaction on lp-shell nuclei are given in Table 6, where they are compared with shell model predictions [Ku73, Sm61]. In these calculations the relative cluster function is generated in a folding model potential without any additional consideration of antisymmetrization except through use of the oscillator rule. Sensitivity to variation in the radius of this potential is small for ^{6}Li but increases as the separation energy increases, probably because the reactions with higher separation energy are more likely to be localized within the range of the optical potentials. This suggests a need for some consistency of treatment of the bound state and optical potentials. No energy dependence is included for the optical potential for the outgoing α-particle. From our discussion of the energy-dependence of this potential in sections 2.3 and 2.4 this would seem to be a weak point in the analysis although, as the authors point out, it is very difficult to deduce reliable parameters for α-particle scattering from light nuclei over a range of energies. Results for the (p, pα) reaction at 600 MeV [De78] suggest that greater sensitivity to the detailed radial behaviour of the overlap integral may occur at this energy.

Several studies have concentrated on the (p, pα) reaction on ^{6}Li [Ja70, To76, Bo77a, Ro76]. Bachelier *et al* [Ba73] note that the (p, pα) cross-section on ^{6}Li is two orders of magnitude larger than the cross-section for 2sld-shell targets. Roos *et al* [Ro76] have compared the ^{6}Li (p, pα) and ^{6}Li (p, p^{3}He) reactions and find that the α+d and t+^{3}He parentages in the ground state of ^{6}Li are quite comparable in magnitude. These

Table 5: Spectroscopic factors for the transitions (i) ^{24}Mg $\rightarrow \alpha$ + ^{20}Ne, (ii) ^{28}Si $\rightarrow \alpha$ + ^{24}Mg and (iii) ^{40}Ca $\rightarrow \alpha$ + ^{36}Ar as determined from the (p, pα) reaction at 157 MeV [Ba76] and compared with theoretical values deduced from the SU(3) model [Dr75, He75]

	Excitation of 0⁺ g. s.			Excitation of 2⁺ g. s.		
	(i)	(ii)	(iii)	(i)	(ii)	(iii)
Absolute S_α (exp)	0.23 ± 0.07	0.24 ± 0.05	0.50 ± 0.07	0.4 ± 0.1	0.4 ± 0.1	0.9 ± 0.4
Absolute S_α (theory)	0.08	0.09	0.09	0.01	0.11	0.45
Realtive S_α (exp)	1.0 ± 0.4	1	$2.1^{+0.6}_{-0.3}$	$0.9^{+0.5}_{-0.4}$	$1.5^{+0.6}_{-0.4}$	$3.8^{+2.2}_{-1.8}$
Relative S_α (theory)	0.9	1	1.0	0.1	1.2	5.0

Table 6: Spectroscopic factors determined from the (p, pα) reactions at 100 MeV on 1p-shell nuclei leading to the ground state of the residual nucleus [Ro 77] compared with shell model predictions [Ku 73, Sm 61]

	S_α (exp)	S_α (theory)
^6Li (p, pα)^2H	0.58 ± 0.02	
^7Li (p, pα)^3H	0.94 ± 0.05	1.12
^9Be (p, pα)^5He	0.45 ± 0.02	0.56
^{12}C (p, pα)^8Be	0.59 ± 0.05	0.56

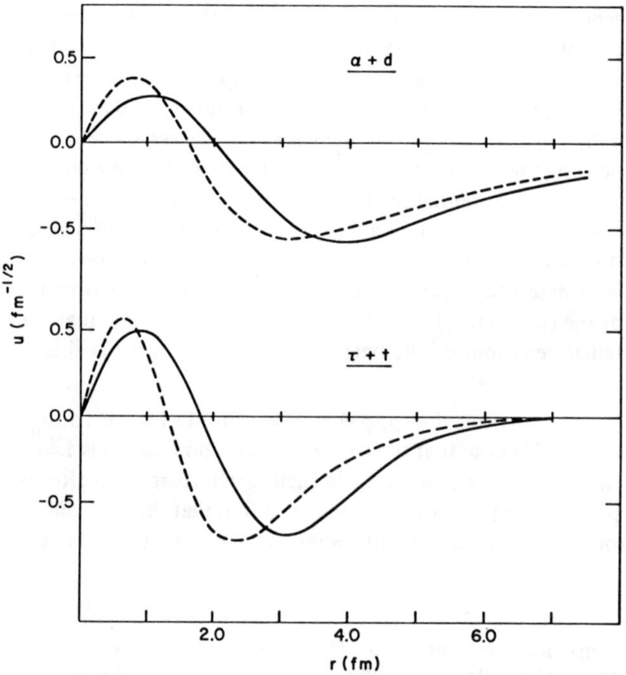

Fig. 20 Radial functions $u_{NL}(R)$ for the α + d and t + ^3He systems in the ^6Li ground state. The solid lines represent cluster model wavefunctions and the dashed lines correspond to functions generated in a Saxon-Woods potential [Ro 76].

data were analysed using a relative cluster function generated in a Saxon-Woods potential whose parameters are chosen to reproduce some features of low energy elastic scattering and also using a properly antisymmetrized cluster model function modified to have the correct asymptotic behaviour. (It is quite possible that this latter procedure is inconsistent because the function with a longer tail would give different results for the exchange components.) These functions are compared in Figure 20. The distorted momentum distributions calculated with the two functions differ essentially in magnitude. The study of the ^{6}Li (p, pα) np system at 40 MeV [Bo77a] shows that the np pair can be regarded as a cluster and suggests that there is a significant component of the α+np (unbound system) in the ground state of ^{6}Li. This again demonstrates that it is incorrect to truncate the cluster expansion of this nucleus at one term representing the α+d system.

Experiments on the (α, 2α) reaction on ^{6}Li and ^{7}Li have been carried out in order to study the reaction mechanism. The free $\alpha-\alpha$ cross-section varies quite rapidly with incident energy and centre of mass angle [Da65], as shown in Figure 21. Hence we expect much greater sensitivity to the choice of the prescription for choosing the equivalent on-shell energy than in the case of the (p, pα) reaction. Jain *et al* [Ja70b] have shown that this is indeed the case and that the extent of the effect depends on the geometry of the experiment; they conclude that the (p, pα) reaction is to be preferred to the (α, 2α) reaction for spectroscopic studies. Studies of the ^{6}Li (α, 2α) reaction over the energy range 50–80 MeV [Pu69, Wa71] have shown that the final energy prescription (FEP) is to be preferred but does not give equally good agreement with the data over the whole energy range. Watson *et al* [Wa71] have given a particularly careful treatment of the reduced overlap integral by constructing a cluster wavefunction for the α+d system which has the correct separation energy and rms radius. The function for the relative

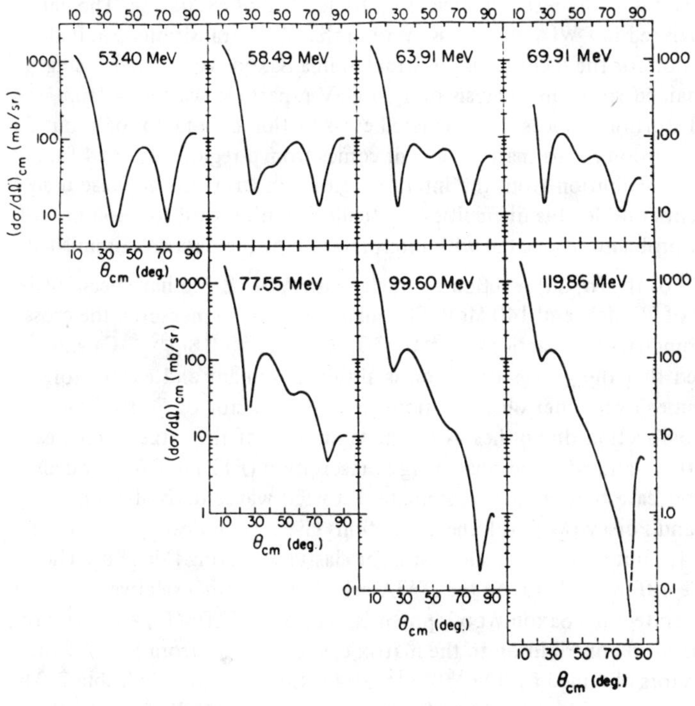

Fig. 21 The differential cross-sections for $\alpha-\alpha$ elastic scattering measured at various lab energies [Wa71].

motion was generated in a Saxon-Woods potential with $a = 0.7$ fm, $R_0 = 1.1(6^{1/3}) = 2.0$ fm and a repulsive core at $R_c = 1.25$ fm. This potential had been adjusted to give agreement with the low energy $d + \alpha$ phase shifts. In this way, antisymmetrization may be taken into account in a phenomenological manner. Their calculations are then carried out in PWIA with a radial cut-off.

Data on the ^{6}Li (α, 2α) reaction at 43–55 MeV [Ga 70, Pi 69] have been analysed in DWIA with the final energy prescription [Ja 73]. The relative motion of the $\alpha + d$ system is described by a Hankel function with the correct asymptotic behaviour and an interior cut-off at $R_0 = 1.5(6^{1/3}) = 2.72$ fm, or alternatively by a gaussian function matched to a Hankel function tail at 0.3 fm. Single and double nucleon exchange terms were considered. The exchange terms give a small contribution to the cross-section unless the real part of the optical potential is large. The contributions to the matrix element from the interior region $R \lesssim 3$ fm are small and constant for the relevant range of the cluster fermi momentum p_α.

Some studies of the (α, 2α) reaction have been carried at very high energies where the $\alpha-\alpha$ cross-section is slowly varying and hence uncertainties due to off-shell effects are less than a few percent. At 700 MeV [Do 79], the pole approximation with PWIA is adequate to describe the data because only a small range of cluster momenta around zero was covered. The (α, 2α) reaction on ^{16}0 and ^{28}Si has been studied at 650 and 850 MeV [Ch 76]. At these energies, resolution of the final states presents difficulties and these data include excited states of the residual nucleus up to 30 MeV of excitation. The data for ^{16}0 have been analysed in DWIA on the assumption that $L = 0$ transitions dominate and the relative function for the α-cluster is constructed in a Saxon Woods well. Using optical potential obtained from the analysis of 1.37 GeV α-particle scattering from ^{12}C it appears that distortion reduces the calculated cross-section by a factor of about 5. The bulk of the contribution to the matrix element comes from a region around 4 fm but there is a significant contribution from the interior region. Experiments at these high energies clearly present considerable difficulties, particularly with regard to resolution of finalstates, but other problems associated with interpretation of the data are diminished.

Detailed studies of the (α, 2α) reaction on nuclei with $A \sim 12$–66 have been made for incident energies of 90 MeV and 140 MeV. Sherman *et al* [Sh 76] measured the cross-section in various geometries for reactions on ^{12}C, ^{16}0, ^{24,26}Mg, ^{28,30}Si, ^{40,44}Ca and ^{66}Zn. It was observed that the gs$\rightarrow$gs transition is always dominant and except for ^{12}C and ^{40}Ca dominates over other observed transitions by a factor of 5–10. The magnitude of the cross-section diminishes as the mass number of the target increases. They have analysed their data with the final energy prescription (FEP) in PWIA and also in DWIA. In the latter case they used approximate distorted waves derived from the model of McCarthy and Pursey [McC 61]; these functions give a good description of inelastic scattering [Ja 74a] but may be less successful for elastic scattering [Ma 78d]. They find surprisingly little difference between the DWIA calculations with a relative oscillator function and one generated in a Saxon-Woods potential with $a = 0.73$ fm, $R_0 = 1.3A^{1/3}$ fm, and also find that the main contribution to the matrix element comes from $R > 2.5$ fm. The spectroscopic factors obtained for the $^{16}0 \rightarrow {}^{12}$C transitions are given in Table 7. The same data have been reanalysed by Chant and Roos [Ch 75] using the full DWIA method.

Table 7: Spectroscopic factors for the process $^{16}O \to \alpha + {}^{12}C$ deduced from studies of the $(\alpha, 2\alpha)$ reactions. All calculations are in DWIA except the one case indicated where the PWIA results are normalized to the DWIA calculation by the same authors

Source	Reference	$S_\alpha (0^+ \to 0^+)$	$S_\alpha (0^+ \to 2^+)$	Ratio $(2^+/0^+)$
Shell model	[Ku 73]	0.23	1.31	5.7
90 MeV-PWIA	[Sh 76]	2.9	0.70	0.24
90 MeV	[Sh 76]	2.9		
52.5 MeV	[Ch 77]	150		
90 MeV	[Ch 77]	15	1.20	8
140 MeV	[Ch 78]	40.3		
850 MeV	[Ch 77]	< 1.8		
90 MeV	[Ja 79d]	0.24 – 0.32		
90 MeV	[Ja 79e]	1.1		

Although their results, given in Table 7, restore the expected relative importance of the transitions to the ground state and the first 2^+ state in ^{12}C the absolute spectroscopic factors are much too large and the fits to the data are not good.

Further experimental data for the $(\alpha, 2\alpha)$ reaction on ^9Be, ^{12}C, ^{16}O and ^{20}Ne have been obtained by the Maryland group [Ch 78, Wa 79] at 140 MeV. A DWIA analysis of the ^{16}O data with what appear to be reasonable parameters again yields a very large spectroscopic factor as noted in Table 7. The same authors [Ch 77] have reanalysed data at 52.5 MeV [Gu 71] and 850 MeV [Ch 76] and find that the discrepancy between the shell model spectroscopic factor and that deduced from DWIA calculations increases rapidly with decreasing incident energy. In these calculations the relative function for the α-cluster in ^{16}O is generated in a Saxon-Woods potential whose parameters are derived by fitting a folded potential. The Saxon-Woods parameters are given by equation (3.9) and the corresponding radial function is shown in Figure 16. The effect of antisymmetrization is not included except through the oscillator rule (3.8). No energy-dependence is included for the parameters of the α-particle optical potential but it is observed [Ch 77] that the extreme localization of the reaction limits sensitivity to changes in these parameters.

Several attempts have been made to interpret and improve the theoretical description of the $(\alpha, 2\alpha)$ reaction on light nuclei, and particularly on ^{16}O. Chant, Roos and Wang [Ch 78] find that they can fit the data at 140 MeV using the shell model spectroscopic factor of 0.23 if they greatly increase the radius of the Saxon-Woods well in which the relative cluster function is generated. The increased radius is given in equation (3.14) and the new relative function is shown in Figure 16. The increased magnitude of the function in the extreme surface is sufficient to increase the DWIA cross-section and the reaction is localized in the region of 5.8 fm. B.K. Jain [Ja 79d] has used a surface localization model to describe the effect of distortion at 90 MeV and excludes all contributions to the matrix element from the region for which $R_c \leqslant 1.15\,(A^{1/3} + 4^{1/3})$ fm where A is the mass number of the target. For ^{16}O, this cut-off radius is $R_c = 4.73$ fm. He then uses a

Saxon-Woods potential with radius $R_0 = 1.3\,(A-4)^{1/3}$ fm to generate the relative cluster function and obtains spectroscopic factors for ^{16}O in excellent agreement with the shell model value and for the other nuclei studied by Sherman *et al* he obtains good agreement with SU(3) spectroscopic factors. In contrast, A.K. Jain and Sarma [Ja 79e] have used the Saxon-Woods potential for the relative cluster function derived by Chant and Roos [Ch 77] but have shown that the optical potentials used by these authors do not give good agreement to the elastic scattering data. Instead Jain and Sarma use an optical potential with an imaginary part which is smaller for $R < 3$ fm and larger outside this distance. The real potentials show a similar difference. Hence they obtain a larger DWIA cross-section and a smaller spectroscopic factor (see Table 7). In their calculation the main contribution to the matrix element comes from the region where $R \sim 2.5-5.0$ fm; the contribution from $R > 5$ fm falls sharply with increasing p_α while contributions from $R < 3.5$ fm remain roughly constant.

From our discussion of the properties of the overlap integral in section 3.1 and the discussion of antisymmetrization in section 2.4, it is evident that the use of a folded potential without further consideration of antisymmetrization, as has been done by Chant and Roos [Ch 77] and by A.K. Jain and Sarma [Ja 79e], is not correct. The potential must either be modified phenomenologically, by inserting a hard or soft core or by extending the radius of the potential, or the exchange terms must be included explicitly through an RGM or OCM calculation. These procedures produce a relative function which is enhanced at and beyond the last maximum, as can be seen from Figure 16 and Figure 18, and hence will lead to a larger DWIA cross-section and smaller spectroscopic factor. We may also expect that the region of localization observed by Jain and Sarma would be pushed out by $\sim 0.5-1.5$ fm. However, our discussion in sections 2.3 and 2.4 of the energy-dependence of the α-particle optical potential and the lack of knowledge of the potentials for scattering from light targets makes it clear that Jain and Sarma are right to challenge this aspect of earlier calculations and, in this respect, their calculation is probably the most realistic. The combination of these two effects implies that the region of localization chosen by B.K. Jain [Ja 79d] is probably most reasonable even though the process may not be as strongly dominated by strong absorption as he has assumed.

The effects discussed above will not be so significant in the case of the $(p, p\alpha)$ reaction as there is only one outgoing α-particle although we have already noted in the case of the $(p, p\alpha)$ reaction that there is a close connection between the effect of the choice of relative function and the choice of the optical potentials.

4.2 Alpha-Decay

A qualitative explanation of the phenomenon of α-decay was given by Gamow [Ga 28] in the earliest days of the quantum theory. Subsequently, it proved possible to give a reasonably satisfactory description of relative decay rates [Ma 64, Ra 65] but substantial difficulties remained over the production of absolute decay rates for heavy nuclei. This problem has recently received considerable attention and a number of different developments of the theory have been given [Ha 65, Ka 70a, Fl 75, 76, Ar 74, Ja 77c, 78, To 76, 78].

The prediction of absolute decay rates has two aspects. The nuclear structure part of the problem is concerned with the extent to which the α-particle can be regarded as preformed in the parent nucleus. This is represented by the spectroscopic factor which in this case is determined experimentally as the ratio of the experimentally observed width Γ to the one-body width Γ_{ob}, i.e.

$$\Gamma = S_\alpha \Gamma_{ob}, \quad \Gamma = 0.693\ h/T_{\frac{1}{2}} \tag{4.1}$$

where $T_{\frac{1}{2}}$ is the half-life of the decay. Thus, α-decay provides another means of determining α-particle spectroscopic factors which is particularly useful in the study of unstable parent states. The nuclear reaction part of the problem is concerned with the escape of the α-particle through the potential barrier and so involves a consideration of the α-nucleus interaction at low energies. The treatment of the two aspects of the problem should be consistent, of course, although this has not been easy to achieve in practice.

Theoretical expressions for the decay width can be developed using the perturbation theory of decaying states [Go64] or resonance theory for the scattering of the α-particle by the residual nucleus [To78] or R-matrix theory [Ma64, Th54]. In R-matrix theory the configuration space is separated into two regions: in the external region defined by $R > R_c$, the total wavefunction can be expressed as a sum of channel functions of simple product form while the wavefunction in the internal region is replaced by a complete set of eigenfunctions X_λ with real eigenvalues E_λ. In a single channel approximation, the width of the narrow resonance is given by

$$\Gamma = 2P\gamma_a^2 \tag{4.2}$$

where γ_α^2 is the reduced width determined by the overlap of X_λ and the channel function at the channel surface, i.e.

$$\gamma = \sqrt{\frac{\hbar^2}{2\mu R_c}} \int\limits_{R=R_c} X_\gamma^* \psi_c \, ds_c \tag{4.3}$$

where μ is the reduced mass of the α-particle. This reduces to equation (3.32) if the overlap function is normalized. The penetrability P is given by

$$P = \frac{kR_c}{\overline{F}^2(R_c) + \overline{G}^2(R_c)} \simeq \frac{kR_c}{\overline{G}^2(R_c)} \tag{4.4}$$

where the functions $\overline{F}$ and $\overline{G}$ are those solutions of the one-body Schrödinger equation containing the Coulomb potential and the nuclear optical potential whose asymptotic behaviour corresponds to the usual Coulomb functions F and G. The R-matrix is

$$R = \sum_\lambda \frac{\gamma_\lambda^2}{E - E_\lambda}$$

as in equation (2.102). The extension of these formulae to broad resonances has been discussed by Arima and Yoshida [Ar74].

In Gamow's theory the penetrability is calculated in a WKB approximation assuming that the α-particle is in an s-state relative to the residual nucleus, and it is then given by

$$P = \exp\left\{-\frac{2\sqrt{2\mu}}{\hbar}\int_{r_i}^{r_t} [U(R) - T_\alpha]^{\frac{1}{2}}\, dR\right\} \tag{4.5}$$

where T_α is the measured kinetic energy of the emitted α-particle, U(R) is the total real potential, and r_i, r_t are the inner and outer turning points.

The channel radius R_c is not defined within the R-matrix theory except for the rather general requirement that it should be sufficiently large to exclude coupling between various exit channels. Even so, values of R_c well inside the inner turning point have been used [Ra63, Ra67]. It is well established [Sh68, Ra67, Ja77c] that a small change in R_c can change the penetrability by an order of magnitude.

A further source of uncertainty in these calculations lies in the ambiguities in the α-nucleus optical potential determined from low energy scattering data. However, several studies [Go70, Ba71, Ba74, Ba78b, Ja76] show that the position r_b of the peak of the barrier formed by the Coulomb potential plus the real nuclear potential and the maximum height $U(r_b)$ of this barrier are relatively well-determined. Sets of values of r_b and $U(r_b)$ obtained from phenomenological potentials and folded potentials all of which fit the data for low energy α-particle scattering from ^{208}Pb are given in Table 8. These potentials yield penetrabilities which differ by a factor of as much as 2.5 [Sh68, Ja77c].

Several authors [Ha68, Ja77c, To76, Sa78] have used the unified reaction theory of Feshbach [Fe58, 73] and MacDonald [Ma64] to develop a new formalism for α-decay. In principle, this method avoids the use of arbitrary channel radii and ambiguous potentials. Such an approach yields a first-order expression for the width of the form

$$\Gamma = 2\pi \int d\rho(E) |\langle \Psi | (H-K) | \psi^+ \rangle|^2 \tag{4.6}$$

where $\rho(E)$ is the density of states, H is the total Hamiltonian

$$H = H_\alpha + H_A + TR + V_{\alpha A} + V_{\alpha A}^c \tag{4.7}$$

and $V_{\alpha A}$ is the total nuclear interaction as defined in equation (2.5). The initial state Ψ is an eigenstate of the Hamiltonian K and the final state ψ^+ is taken to be an eigenstate of a Hamiltonian K′ which may or may not be the same as K. If $K \neq K'$ the initial and final states will not be orthogonal and at worst may correspond to completely different nuclear models. In the work of Jackson and Rhoades-Brown [Ja77c, 77d, Ja78] the initial Hamiltonian is taken to be

$$K = H_\alpha + H_A + T_R + U_R \tag{4.8}$$

where U_R is a real potential chosen so that the initial state is a bound state and not a decaying state. The wavefunction of the final state is taken to be

$$\psi^+ = \phi_\alpha \Phi_A^{nA} \chi_{\alpha A}^{n+} \tag{4.9}$$

Table 8: Values of the barrier radius r_b and barrier height $U(r_b)$ for $\alpha + {}^{208}Pb$ calculated from phenomenological Saxon-Woods potentials and microscopic folded potentials all of which fit low energy elastic scattering data

Potential	Reference	r_b (fm)	$U(r_b)$ (fm)
Saxon-Woods	[Go 70]	10.94	20.42
	[Ba 74 (set A, BL)]	10.9	20.49
	[Ba 78b (a = 0.52)]	11.14	20.21
	[Ba 78b (a = 0.58)]	10.96	20.42
	[Ba 78b (a = 0.62)]	10.83	20.57
Folded	[Ja 77c (H)]	10.8	20.60
	[Ja 77c (H')]	11.0	20.33
	[Ja 77c (M)]	11.0	20.56

where χ^+ is a non-resonant scattering state generated in the potential

$$\left.\begin{aligned} U(R) &= U_R(R) && R \leqslant r_b \\ &= U^N_{xA}(R) + U^c_{xA} && R > r_b \end{aligned}\right\} \tag{4.10}$$

where the nuclear potential is generated in the folding model as given by equations (2.112)–(2.115). This choice is called by Tobocman [To 78] the "alternative" version of the Feshbach-MacDonald formalism since it involves an off-resonance continuum wavefunction. This is in accord with the more recent approach of Wildermuth *et al* [Wi 79]. Also, in accord with these authors, exchange of nucleons between the free α-particle and the residual nucleus in the final state is neglected on the grounds that the overlap of the two systems is very weak owing to the Coulomb barrier.

The difficulties, and variation between authors, arise in the treatment of the initial state. It is generally agreed that a single configuration, whether within the framework of the shell model or the cluster model, is inadequate to describe both the absolute decay rates and the branching ratios of heavy α-emitters. For the decay of ${}^{212}Po$ Tonozuka and Arima [To 79] have carried out an extended shell model calculation to show the importance of many higher configurations, while several authors restrict the expansion to a small number of shell model states but couple these to a small number of scattering states of cluster form [Wi 79, Sa 78, Ro 79]. Jackson and Rhoades-Brown use a cluster model expansion of the initial state [Ja 77c], i.e. they start with an initial wavefunction as in equation (2.6). In the last case, there appears the now familiar relative function $\overline{\chi}_{\alpha A}$ which is generated as a solution of equation (2.18) with the potential

$$\left.\begin{aligned} U_R &= g[U^N_{\alpha A}(R) + U^c_{\alpha A}(R)] + (1-g)\, U_R(r_b && R \leqslant r_b \\ &= U_R(r_b) && R > r_b \end{aligned}\right\} \tag{4.11}$$

Table 9: One body widths Γ_{ob} for the L = 0 decay $^{212}Po(gs) \rightarrow \alpha + ^{208}Pb(gs)$ calculated for a fixed binding energy ϵ_α in the folded potential H

Principal Quantum Number N	Scaling factor g	Γ_{ob} (MeV)
12	0.67	1.02×10^{-14}
13	0.77	1.38×10^{-14}
14	0.87	1.87×10^{-14}
15	0.98	2.48×10^{-14}

Table 10: Calculated values of the one-body width for the L = 0 decay $^{212}Po(gs) \rightarrow ^{208}Pb + \alpha$

Potential		Γ_{ob} (MeV) Correct barrier	Γ_{ob} (MeV) Barrier fixed at 20.4 MeV
Folded potential	H	1.02×10^{-14}	1.27×10^{-14}
	H'	1.46×10^{-14}	1.37×10^{-14}
	M	1.12×10^{-14}	1.32×10^{-14}
Folded potential H with exchange		7.0×10^{-14}	–
Saxon-Woods	BL	1.45×10^{-13}	–
	DLF	3.84×10^{-14}	–

where U^N is the folded potential already defined and g is a parameter which is varied to give a bound state with the required number of nodes and a binding energy of

$$\epsilon_\alpha = T_\alpha - U_R(r_b) - 2\mu L(L+1)/\hbar^2 r_b^2. \tag{4.12}$$

The parameter g can be regarded as renormalizing the total interaction to take account of the use of an equivalent local potential. The unmodified folded potential for $\alpha + ^{208}Pb$ supports a 15s state whereas the oscillator rule (3.8) yields a 12s state for the lowest shell model configuration in ^{212}Po. Table 9 shows the change in the scaling factor g and the corresponding one-body width Γ_{ob} for the decay $^{212}Po(gs) \rightarrow \alpha + ^{208}Pb(gs)$ as the principal quantum number is increased from 12 to 15 for a fixed binding energy given by equation (4.12). These results indicate that configuration mixing of shell model configurations will not produce a significant change in the width in this model.

Further results obtained in this model for the one-body width for the $^{212}Po \rightarrow ^{208}Pb$ transition are given in Table 10. The result obtained for potential H with exchange is the result of the calculation by Rhoades-Brown [Ja77c, Rh77] of $\overline{\chi}$ in the OCM discussed in section 3.1 Apart from the major effect of exchange the difference between the results for the microscopic potentials arises from the variation in the barrier heights noted in Table 8; this is confirmed by the agreement seen when the barrier heights are all adjusted

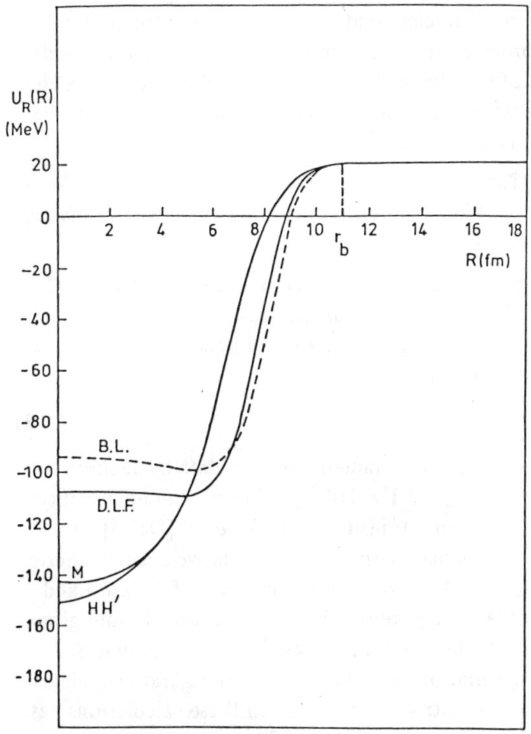

Fig. 22
The bound state potentials used to describe the $\alpha + {}^{208}Pb$ system in the ${}^{212}Po$ ground state. Potentials BL and DLF have Saxon-Woods shape inside r_b and potentials M, H, and H' are calculated from the single-folding model [Ja 77c].

to the same value. The difference between the results for the phenomenological potentials and the microscopic potentials can easily be understood by reference to Figure 22, which shows that the phenomenological potentials have smaller depths and larger half-way radii than the folded potentials and hence the relative α-cluster functions generated in the phenomenological potentials extend further out and yield a larger overlap with the perturbing interaction and the scattering function. The potential DLF fits low energy scattering and gives a 12s state at the correct binding energy [De 76].

The one-body widths given in Table 10 can be used to derive the spectroscopic factor using equation (4.1). For the phenomenological potentials BL and DLF this yields $S_\alpha = 0.010$ and $S_\alpha = 0.040$ respectively. It can be seen from Table 11 that antisymmetrization with the folded potential brings this value into line with the ones given by the phenomenological potentials. Also given in Table 11 is the rms radius of the relative cluster function $\bar{\chi}_{\alpha A}$ and the position R_{max} of its last maximum. It should be noted that in both calculations with the folded potential the rms radius of the nuclear matter distribution of ${}^{208}Pb$ and the barrier height r_b of the potential for the $\alpha + {}^{208}Pb$ system have the correct values. Thus the large values of $\langle r^2 \rangle_{\alpha A}^{\frac{1}{2}}$ and R_{max} are not incompatible with an acceptable rms radius for the residual nucleus [Ja 78a] but arise from the effect of antisymmetrization in suppressing the interior region and increasing R_{max} [Ja 78]. On the other hand, rather small uncertainties in the nuclear matter distribution of the residual nucleus can lead to differences of an order of magnitude in the predicted one-body width for very heavy nuclei [Ja 79f] because of the change in the barrier height. The matrix element for α-decay in heavy and very heavy nuclei receives its main contribution from the region 3–4 fm beyond the barrier radius r_b [Ja79g].

Some additional results given in Table 11 show the effect of increasing the half-way radius parameter r_0 of an arbitrary Saxon-Woods potential. As r_0 increases, S_α decreases; in fact log S_α decreases linearly as r_b increases.

We have already obtained the expression (3.25) which connects the coefficient C in the cluster expansion (2.6) to the conventional theoretical spectroscopic factor S_α. The decay $^{212}Po(gs) \rightarrow \alpha + {}^{208}Pb(gs)$ corresponds to $J_B = 0$, $J_A = 0$ and hence $L = 0$, if we assume that the α-particle is not excited. This gives

$$|C(000)|^2 = S_\alpha \qquad (4.13)$$

and includes all binomial factors. The final result extracted from the data by Jackson and Rhoades-Brown, including exchange, is $S_\alpha = 2.1 \times 10^{-2}$ [Ja78] with an uncertainty of approximately a factor of two. For the same transition, Davies *et al* [Da76] have derived $S_\alpha = 0.73 \times 10^{-2}$ using a modified R-matrix theory and a relative cluster function generated in the DLF potential (see section 4.3) whereas the method of Jackson and Rhoades-Brown gives 4.0×10^{-2} with the same potential. Shell model calculations give a theoretical value of $S_\alpha = 2.0 \times 10^{-5}$ for the lowest configuration while configuration mixing up to 13 $\hbar\omega$, i.e. 89 proton configurations and 83 neutron configurations, yields $S_\alpha = 0.12 \times 10^{-2}$ [Ar78]. The treatment of antisymmetrization in these calculations has been criticised by Fliessbach [Fl76] who obtains $S_\alpha = 0.44 \times 10^{-2}$ and by Watt *et al* [Wa80] who obtain $S_\alpha = 0{,}50 \times 10^{-2}$ for the lowest shell-model configuration for ^{212}Po.

The transitions from the 18^+ isomeric state in ^{212}Po to the gs(0^+), the lowest 3^- state and the lowest 5^-state in ^{208}Pb have been examined by Glendenning and Harada [Gl65] and by Tonozuka and Arima [To79] using R-matrix theory and the shell model with configuration mixing, and by Jackson and Rhoades-Brown [Ja77c] using the cluster model. It is evident from all these calculations that the lowest configuration is not adequate in any model but with sufficient shell or cluster configurations the absolute

Table 11: Spectroscopic factors derived for the L = 0 decay $^{212}Po(gs) \rightarrow \alpha + {}^{208}Pb(gs)$ with various potentials, together with radii of the relative cluster function. The Saxon-Woods potentials have a = 0.65 fm.

Potential		S_α	$\langle r^2 \rangle_{\alpha A}^{\frac{1}{2}}$ (fm)	R_{max} (fm)
Folded potential H		0.149	6.99	8.0
Folded potential H with exchange		0.021	7.80	8.5
Saxon-Woods potential	$r_0 = 1.215$ fm	0.225	6.10	8.0
	$r_0 = 1.265$ fm	0.080	6.29	8.3
	$r_0 = 1.315$ fm	0.041	6.49	8.6
	$r_0 = 1.365$ fm	0.015	6.67	8.9

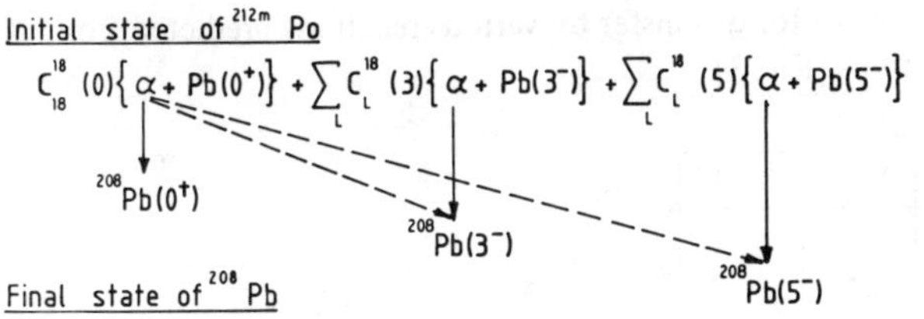

Fig. 23 Schematic diagram illustrating the transitions contributing to the decay of ^{212m}Po [Ja 77c].

values can be approached reasonably closely. In particular the cluster model with a collective description of the ^{208}Pb core appears to be successful in reproducing the branching ratios. The explanation of this result is illustrated in Figure 23. In many calculations only the term in the initial wavefunction corresponding to the core in the ground state of ^{208}Pb is considered and only the transitions to the 3^- and 5^- final states indicated by the dashed lines are included, but it appears to be essential to consider components in the initial state corresponding to excited states of the core and hence to include the transitions indicated by the full lines in the figure in order to obtain a good description of the branching ratio.

These results suggest that the barrier penetration can be very adequately described using the best available optical potentials. The most important requirement is an initial wavefunction which gives a sufficiently realistic description of the nuclear dynamics of the initial state and its overlap with the final state. It is clearly not easy to do this in a correctly antisymmetrized formalism, but the methods which use a cluster expansion or a mixed expansion of cluster and shell model terms appear most promising. It is not obvious, however, that the second approach can easily be used in studies of α-transfer reactions where the comparison with α-decay has proved a useful method of normalizing the distorted wave calculations (see section 4.3).

4.3 Four-Nucleon Transfer Reactions

In recent years extensive studies of multi-nucleon transfer reactions have been made using light and heavy ion beams. In this section we concentrate on the transfer of four nucleons and the comparison of this process with α-particle knock-out reactions and α-decay.

As we have already noted in section 3.2, transfer reactions of the type

$$a + A \rightarrow b + B$$

involve two spectroscopic factors, one for the process $A + x \rightarrow B$ and the other for the process $a \rightarrow b + x$, in the case of a stripping reaction. Mallet-Lemaire [Ma 78a] has tabulated the theoretical spectroscopic factors $S_\alpha(ab)$ given by Kurath [Ku 73] for several possible stripping reactions involving transfer of a 2p2n cluster. These values, given in Table 12, suggest that lithium-induced stripping reactions should be particularly useful.

Table 12: The spectroscopic factors for α-transfer by various reactions predicted from shell model calculations [Ku 73]

Reaction	S_α			
	L = 0	L = 1	L = 2	L = 4
$(^{16}O, ^{12}C)$	0.23			
$(^{15}N, ^{11}B)$			0.41	
$(^{14}C, ^{10}Be)$	0.24			
$(^{14}N, ^{10}B)$				0.69
$(^{12}C, ^{8}BE)$	0.55			
$(^{11}B, ^{7}Li)$	0.26			
$(^{7}Li, t)$		1.19		
$(^{6}Li, d)$	1.13			

Becchetti [Be 78] has discussed other advantages of lithium-induced stripping and pick-up reactions, including the occurrence of J-dependent, oscillatory angular distributions. In contrast, transfer reactions induced by heavier ions, particularly on heavy targets at energies comparable with the Coulomb barrier, show characteristic bell-shaped angular distributions peaked at the grazing angle. At higher energies the angular distributions are forward peaked and, under certain kinematic conditions, they also show oscillatory behaviour [Ma 78a, Sc 78].

Semi-classical ideas have proved particularly useful for the understanding of qualitative features of these reactions [Br 70, An 74]. For a more detailed analysis of the data and extraction of spectroscopic factors the DWBA theory must be used. There are a number of complications involved in using the DWBA formalism as outlined through equations (2.25)–(2.38). By making cluster expansions of the functions Ψ_B and ϕ_a which appear on the right-hand-side of equation (2.35) and neglecting the dependence of $\Sigma\,\Delta V$ on all variables except $\underset{\sim}{r}$, i.e. by neglecting core excitation, the matrix element can be reduced to

$$T_{fi}(DWBA) \propto \langle \chi_b^-(\underset{\sim}{k}_f, \underset{\sim}{r}_f)\, \bar{\chi}_{xA}(\underset{\sim}{R}) | \Delta V(\underset{\sim}{r}) | \chi_a^+(\underset{\sim}{k}_i, \underset{\sim}{r}_i)\, \bar{\chi}_{xb}(\underset{\sim}{r}) \rangle \tag{4.13}$$

The simplest approximation is obtained by the zero-range substitution

$$\Delta V(\underset{\sim}{r})\, \bar{\chi}_{xb}(\underset{\sim}{r}) = D_0\, \delta(\underset{\sim}{r}) \tag{4.14}$$

so that

$$T_{fi}(DWBA\text{-}ZR) \propto D_0 \int \chi_b^{-*}(\underset{\sim}{k}_f, \gamma \underset{\sim}{R})\, \bar{\chi}_{xa}^*(\underset{\sim}{R})\, \chi_a^+(\underset{\sim}{k}_i, \underset{\sim}{R})\, d\underset{\sim}{R}. \tag{4.15}$$

This approximation can be expected to be satisfactory only when the relative function $\bar{\chi}_{xb}$ is an s-state function. Alternatively, equation (4.13) can be rewritten by introducing $\underset{\sim}{r}' = \underset{\sim}{R} - \underset{\sim}{r}$ and becomes

$$T_{fi}(DWBA) \propto \langle \chi_b^-(\underset{\sim}{k}_f, \underset{\sim}{r}' - \{M_x/M_B\}\underset{\sim}{R})\, \bar{\chi}_{xA}(\underset{\sim}{R}) | \Delta V(\underset{\sim}{R} - \underset{\sim}{r}') |$$
$$\times\, \bar{\chi}_{xb}(\underset{\sim}{R} - \underset{\sim}{r}')\, \chi_a^+(\underset{\sim}{k}_i, \underset{\sim}{r}' - \{M_x/M_a\}\underset{\sim}{r}' + \{M_x/M_a\}\underset{\sim}{R}) \rangle. \tag{4.16}$$

If it is permissible to assume that $M_x/M_a \ll 1$ and $M_x/M_B \ll 1$, the no-recoil approximation is obtained [Bu66, 68]

$$T_{fi}(\text{DWBA-NR}) \propto \int \chi_b^{-*}(\underset{\sim}{k}_f, \underset{\sim}{r}')\, G(\underset{\sim}{r}')\, \chi_a^+(\underset{\sim}{k}_i, \; M_b/M_a \; \underset{\sim}{r}')\, d\underset{\sim}{r}' \qquad (4.17)$$

where

$$G(\underset{\sim}{r}') = \int \bar{\chi}_{xA}^{*}(\underset{\sim}{R}) \Delta V(\underset{\sim}{R}-\underset{\sim}{r}')\, \bar{\chi}_{xb}(\underset{\sim}{R}-\underset{\sim}{r}')\, d\underset{\sim}{R}. \qquad (4.18)$$

This approximation is usually evaluated using a spherical Hankel function for one or both of the bound state radial functions so that the integral over $\underset{\sim}{R}$ can be handled rather easily. Other approximate treatments of the DWBA matrix element involve expansion of the distorted wavefunctions to first or higher order in $\underset{\sim}{\nabla}_R$ [Sa 73]. Another approach makes use of the expressions

$$\chi_b^{-*} \simeq e^{i\underset{\sim}{q}_f \cdot \underset{\sim}{r}}\, \chi_b^{-*}(\underset{\sim}{k}_f, \gamma \underset{\sim}{R}), \quad \chi_a^+ \simeq e^{-i\underset{\sim}{q}_i \cdot \delta \underset{\sim}{r}}\, \chi_a^+(\underset{\sim}{k}_i, \underset{\sim}{R}) \qquad (4.19)$$

where $\underset{\sim}{q}_i$ and $\underset{\sim}{q}_f$ may be equal to the asymptotic momenta $\underset{\sim}{k}_i$ and $\underset{\sim}{k}_f$, respectively or they may be replaced by local momenta [Do65, 69, Br74a, b]. This approximation appears rather similar to the approximation used in the DWIA formalism of knock-out reactions which yields equations (2.62) and (2.63). The conditions, however, are very different.

Codes for full finite-range calculation of the matrix element have been constructed. The most commonly used are the code LOLA [De73] and the code SATURN-MARS [Ta74]. Comparisons of the various methods of treating recoil have been considered by Blair *et al* [Bl74] who found that none of the approximate methods gave good agreement with the finite-range results from LOLA for the α-particle transfer reaction ^{40}Ca (^{16}O, ^{12}C) at 42 MeV, although the choice of potential (3.10) for the bound α-particle in ^{16}O gave smaller discrepancies than the potential (3.13) which leads to a more extended α-particle wavefunction. As noted earlier by Dodd and Greider [Do65] there is an interaction between finite-range and recoil effects. Santos [Sa74] has noted that recoil effects are enhanced when there is an increase in mismatch of the critical angular momenta of grazing collisions in the entrance and exit channels as this reduces surface localization.

Extensive studies have been made of lithium-induced reactions. Gutbrod *et al* [Gu71a] examined the (d, ^{6}Li) reaction on light nuclei with 19.5 MeV deuterons. They found that a reasonably satisfactory quantitative description of all transitions to the ground state of the residual nucleus, except for ^{10}B(d, ^{6}Li)^{6}Li, could be obtained with finite-range DWBA and a shell model description of the target and residual nuclei and, in particular, the fall in the maximum cross-section by an order of magnitude from A = 11 to A = 40 could be reproduced. The relative wavefunction of the α-cluster in the initial nucleus was generated in a Saxon-Woods potential with r_0 = 1.25 fm and a = 0.65 fm. The wavefunction of the α-particle in ^{6}Li was taken to be either in a 2s state as predicted by the oscillator rule (3.8) from the shell model configuration or to have the form

$$\phi_\alpha(r) = r^2[\exp(-c_1 r^2) + c_2 \exp(-c_3 r^2)] \qquad (4.20)$$

obtained from a variational calculation. The latter choice gave much better agreement with the data both in shape and magnitude. Considerable sensitivity to the ^{6}Li optical potential was used for each target nucleus.

In more recent work, the (d, ^{6}Li) cross-sections on ^{16}O have been measured at incident energies of 80, 65, 60 and 50 MeV [Oe 78, 79]. Both zero-range and finite-range DWBA calculations have been made to analyse the data at 80 MeV assuming $(1p)^4$ pick-up to all positive parity states and (1s) $(1p)^3$ pick-up to negative parity states. The Talmi-Moshinsky transformation and the oscillator rule were used and the relative α-cluster functions are generated in a Saxon-Woods potential with $R_0 = 1.65\ A^{1/3}$ fm and a = 0.65 fm. The choice of the deuteron optical potential does not seem to be crucial but a strong dependence of the ^{6}Li optical potential is observed in both shape and magnitude. For each optical potential set there is an optimum value of the r_0 parameter of the bound state potential. Results for relative spectroscopic factors are given in Table 13 where they are compared with predictions from the shell model [Ku 73, Ro 73] and the α-chain model [Ic 73]. Further studies of the same reaction over a wider energy range [Oe 79] confirm that the influence of the deuteron optical potential is minor. The effect of energy dependence of the ^{6}Li optical potential has been examined and, although this is not necessary for good fits, it removes the need to vary the bound state parameter r_0 for each final state. The data at 50 MeV could not be fitted well for angles larger than 30°. The bound state functions for the α-cluster were calculated in a Saxon-Woods potential with a = 0.65 fm and $R = 0.97\ (A_c^{1/3} + 4^{1/3})$ fm. The same parameters were found to give good agreement for targets in the 2s1d-shell [Oe 79a] and Cr isotopes [Be 81]. Nevertheless, it is surprising that the same parameters are used for the α + d and α + core systems since the folding model suggests that the corresponding potentials should be significantly different.

Other studies of the (d, ^{6}Li) reaction on 1p and 2s1d shell nuclei have been reported [Be 78a, Oe 78a, Ta 78b]. The quality of fits is rather uneven and comparison of the resulting spectroscopic factors may therefore be misleading; however, Betigeri *et al* [Be 78b] find reasonable agreement with the shell model predictions for $^{16}O \rightarrow {}^{12}C$ while Oelert *et al* [Oe 78a] find better agreement with the α-chain model. For the (d, ^{6}Li) reaction on ^{24}Mg, ^{26}Mg and ^{28}Si Cossairt *et al* [Co 76] find that zero-range DWBA gives good agreement with the data; the extracted spectroscopic factors are in fair agreement

Table 13: Relative spectroscopic factors for the process $^{16}O \rightarrow \alpha + {}^{12}C$ obtained from various transfer reactions

	Incident energy	Reference	S_α (gs → gs) absolute	S_α (gs → 2$^+$) relative	S_α (gs → 4$^+$) relative
^{16}O(d, ^{6}Li)	80 MeV	[Oe 78]	0.76	3.18	1.90
^{16}O(d, ^{6}Li)	19.5 MeV	[Gu 71]	0.333		
^{16}O(^{3}He, ^{7}Be)	30 MeV	[De 69]	0.295		
^{16}O(^{12}C, ^{16}O)	76.8 MeV	[Mo 79]	1.0		
^{12}C(^{16}O, ^{12}C)	80 MeV	[De 73a]	0.55	1.3	
Theory – shell model		[Ku 73]	0.23	5.70	10.16
Theory – α-chain model		[Ic 73]	0.104	3.16	1.97

with the predictions of pure SU(3) but not with the same experiment on ^{24}Mg at 28 MeV. Agreement with the $^{28}Si(^{3}He, ^{7}Be)$ reactions is quite good. The finite range calculations are unsatisfactory, being out of phase with the data, and this seems to be a common problem in calculations on (d, ^{6}Li) and (^{6}Li, d) reactions.

Becchetti [Be 78] has compared spectroscopic factors deduced from (^{6}Li, d) reactions on 1p and 2s1d shell nuclei with theoretical predictions. He notes that general features such as decrease of S_α in mid-shell and blocking in odd-A nuclei by unpaired nucleons are reproduced but there are some sharp discrepancies. For example, the transitions ^{12}C (gs) → ^{16}O (gs) and ^{16}O (gs) → ^{20}Ne (gs) are very much enhanced, and the latter transition is 10–40 times larger than predicted [An 79]. Indeed, the experimental results obtained for states up to 12 MeV excitation in ^{20}Ne [An 79] are in poor agreement with SU(3) predictions. Part of the problem of deducing S_α from the data arises from the relatively poor fits given by zero-range DWBA, but some improvement is obtained by including the p-h admixtures in the ^{16}O ground state proposed by Brown and Green [Br 66, 68]. Similar discrepancies arise in the $^{18}O(^{6}Li, d)$ reaction [An 77a] although the ratio of the experimental gs → gs transitions in ^{16}O and ^{18}O is in good agreement with the predicted value from both the simple SU(3) model and more detailed shell model calculations.

The (^{6}Li, d) reaction on odd A nuclei in the 2s1d shell has been studied by Eswaran *et al* [Es 79]. Spectroscopic factors are deduced from fits to the data with finite-range calculations using the 2s relative wavefunction for the α + d system developed by A.K. Jain *et al* [Ja 73] for analysis of the $^{6}Li(\alpha, 2\alpha)$ and similar reactions. Some results are given in Table 14 and, apart from the $^{17}O \rightarrow {}^{21}Ne$ transition, are in reasonable agreement with the shell model predictions of Chung *et al* [Ch 78a]. The systematic behaviour of (^{6}Li, d) reactions in the 1f2p shell is discussed by Hanson *et al* [Ha 78a] who claim good fits with both zero-range and finite-range calculations. Agreement with shell model predictions is quite good for heavier nuclei [Be 78].

A survey of (d, ^{6}Li) reactions at 35 MeV indicated that the cross-section decreases as A^{-3} approximately [Be 75] but enhancements occur in certain mass regions. One such region corresponds to the rare-earth nuclei and a more detailed study with isotopic targets of Ba, Ce, Nd, Sm and Er has been carried out [Mi 77a]. The data were analysed with zero-range DWBA. The optical potentials for the deuteron and ^{6}Li were fixed; spin-orbit terms were found to have a negligible effect and were omitted. The α-cluster functions were generated in a Saxon-Woods potential with r = 1.30 fm, a = 0.65 fm which are the same as the parameters used by Sherman *et al* [Sh 76] for the (α, 2α) reaction. When the α-cluster in the target is unbound the radial cluster function was matched to the irregular Coulomb function and normalized within a cut-off radius of 20 fm. The oscillator rule (3.8) was used to determine the number of nodes in this function. The zero-range calculations were normalized by requiring that the reduced width deduced from the known lifetime for α-decay of ^{148}Sm using equations (4.1) and (4.2) should be identical to the reduced width deduced from the (d, ^{6}Li) reaction on the same nucleus using the relation (3.32) between the reduced width and the spectroscopic factor. Several different methods were used to calculate the penetrability and all gave the same result for a channel radius $R_c > 10$ fm. However, the channel radius was chosen to be $R_c = 1.7\ A^{1/3}$ fm which gives

Table 14: Spectroscopic factors for gs → gs transitions in odd A nuclei deduced from (^{6}Li, d) reactions assuming L = 2 transfer [Es79].

Transition	Possible L-Transfer	S_α (L = 2) Exp	S_α (L = 2) Theory
^{17}O (5/2$^+$) → ^{21}Ne (3/2$^+$)	2,4	0.34	0.80
^{19}F (1/2$^+$) → ^{23}Na (3/2$^+$)	2	0.02	0.04
^{21}Ne (3/2$^+$) → ^{25}Mg (5/2$^+$)	2,4	0.10	0.08
^{23}Na (3/2$^+$) → ^{27}Al (5/2$^+$)	2,4	0.03	0.07
^{25}Mg (5/2$^+$) → ^{29}Si (1/2$^+$)	2	0.66	0.35
^{27}Al (5/2$^+$) → ^{31}P (1/2$^+$)	2	0.65	0.51
^{29}Si (1/2$^+$) → ^{33}S (3/2$^+$)	2	0.30	0.30
^{31}P (1/2$^+$) → ^{35}Cl (3/2$^+$)	2	0.29	0.34
^{35}Cl (3/2$^+$) → ^{39}K (3/2$^+$)	0.2	0.18	0.61

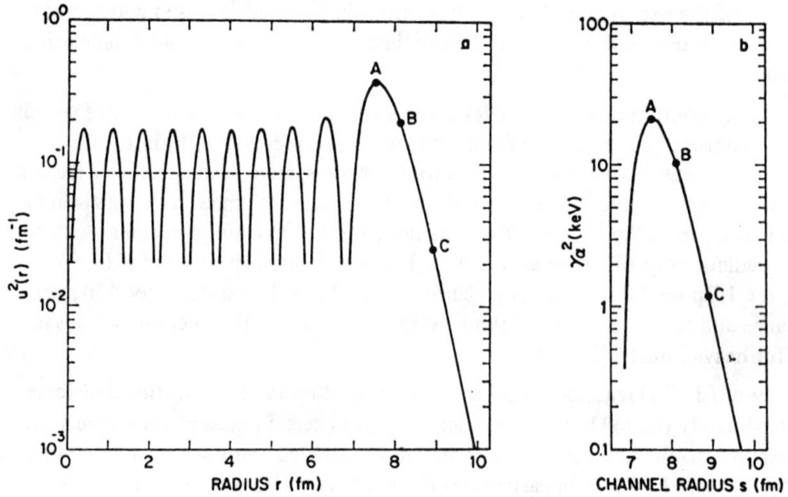

Fig. 24 (a) The square of the radial function $u_{NL}(R)$ for α + ^{144}Nd system, (b) the reduced widths $\gamma_\alpha^2(s)$ as a function of channel radius s. The point A indicates where the logarithmic derivative vanishes, the point B indicates the inner turning point, and the point C corresponds to $s = 1.7\,A^{1/3}$ fm [Mi77a].

R_c = 8.91 fm for ^{148}Sm. The dependence of the reduced width on the channel radius is shown in Figure 24. The sensitivity of the spectroscopic factor and the reduced width to the parameters of the bound state potential is shown in Figure 25.

The normalization coefficient deduced for ^{148}Sm was used for all the rare earth targets. This leads to a fairly accurate prediction of the α-decay lifetime of ^{144}Nd. The very large extrapolation to ^{238}U leads to a discrepancy in the predicted α-decay half-life of an order of magnitude.

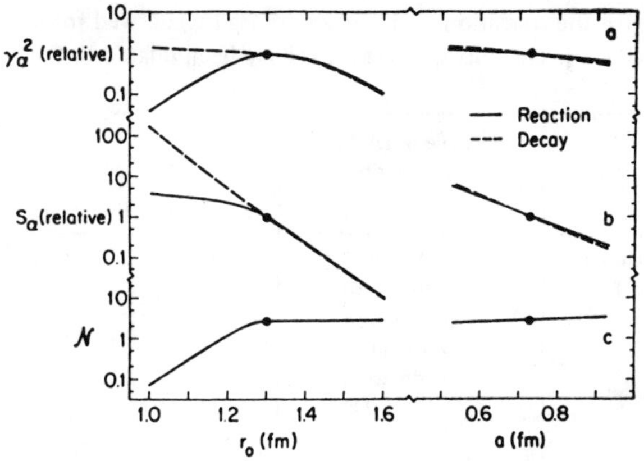

Fig. 25 The relative reduced widths (a), the spectroscopic factors (b) for the reaction ^{148}Sm(d,^{6}Li)^{144}Nd and the α-decay $^{148}\text{Sm} \to \alpha + {}^{144}\text{Nd}$, and (c) the normalization factor needed to obtain the same reduced widths from the reaction and the decay process. The filled circles correspond to results calculated with the bound state potential parameters $r_0 = 1.3$ fm and $a = 0.73$ fm [Mi77a].

The ^{208}Pb (d, ^{6}Li) ^{204}Hg reaction has been further studied at 55 MeV [Be79]. The data have been analysed using both zero-range and finite-range DWBA and various sets of optical potential parameters. In this case the spin-orbit term in the deuteron optical potential appears to have an important effect in damping the angular distribution. For unbound α-cluster states the wavefunctions were calculated as for a bound state with separation energy of 0.5 MeV using the same parameters as Milder *et al* [Mi77a] or $r_0 = 1.20$ fm, $a = 0.65$ fm. The α+d cluster wavefunction was taken to be the 2s wavefunction constructed by Kubo and Hirata [Ku72], which gave results out of phase with the data, or the wavefunction generated with a soft repulsive core by Watson *et al* [Wa71] and used in the analysis of the ^{6}Li (α, 2α) reaction. The latter choice leads to reasonable shape agreement with the data and increases the absolute spectroscopic factor by a factor of ~ 20, as noted in Table 15. The Table also shows the result of normalizing to the ^{148}Sm α-decay. The energy dependence of the deduced spectroscopic factors is disturbing.

Using the spectroscopic factors deduced from this analysis the authors have constructed the distribution of centres of mass of α-clusters in ^{208}Pb by taking the sum

$$\rho(\underset{\sim}{s}) = \sum_{NL} S_\alpha(NL) |\bar{\chi}_{\alpha\text{Hg}}(\underset{\sim}{s})|^2$$

Unfortunately, this quantity has been compared at s = 10 fm directly with the proton distribution without folding in the finite size of the α-cluster which is required to give the correct expressions (3.38) or (3.39) for the nucleon distribution. It can be seen from Figure 24 that at 10 fm the radial part of $\bar{\chi}$ is falling exponentially and hence the effect of folding in the α-particle density with a range of 1.0–1.5 fm will be substantial.[1])

[1]) New results for the (d, ^{6}Li) reaction on heavy nuclei [Ja80] show that inclusion of the finite size of the α-particle increases the deduced values of α-particle clustering probability by an order of magnitude.

Table 15: Spectroscopic factors for the transition $^{208}Pb \rightarrow \alpha + {}^{204}Hg$ (gs) derived from the (d, ^6Li) reaction [Be79]. The normalized results have been related to the α-decay of ^{148}Sm

Parameters of α + Hg potential		Choice of ^6Li wavefunction	Type of DWBA calculation	S_α	
r_0(fm)	a(fm)			E = 55 MeV	E = 35 MeV
1.30	0.73	Kubo and Hirata	Finite-range	0.30	0.10
1.20	0.65	Watson et al	Finite-range	6.0	1.0
1.30	0.73	–	Normalized zero-range	0.014	0.002
1.20	0.65	Watson et at	Normalized finite-range	1.0	0.2

Table 16: Spectroscopic factors for the transition $^{122}Te \rightarrow \alpha + {}^{118}Sn$ (gs) deduced from the (d, ^6Li) reaction [Ja79b]. The normalized results have been normalized to the α-decay of ^{148}Sm.

Parameters of bound state potential r_0(fm)	a(fm)	Choice of ^6Li wavefunction	Type of DWBA Calculation	S_α
1.20	0.65	–	Normalized ZR	0.39
1.30	0.73	–	Normalized ZR	0.022
1.40	0.65	–	Normalized ZR	0.012
1.20	0.65	Kubo and Hirata	Finite-range	1.3
1.30	0.73	Kubo and Hirata	Finite-range	0.34
1.40	0.65	Kubo and Hirata	Finite-range	0.22
1.20	0.65	Watson *et al*	Finite-range	2.97
1.30	0.73	Watson *et al*	Finite-range	0.36
1.40	0.65	Watson *et al*	Finite-range	0.16

The (d, ^6Li) reaction has also been studied at 33 MeV on isotopes of Te, Sn and Cd [Ja79b]. The data were analysed in zero-range DWBA using the normalization constant derived from the α-decay of ^{148}Sm, as discussed above. The oscillator rule (3.8) was used to predict the number of nodes for the relative α-cluster function in the target nucleus. The corresponding bound state potential was taken to have one of three sets of parameters, as listed in Table 16. In some finite-range calculations the $\alpha + d$ relative cluster function was taken to be of the form given by Kubo and Hirata [Ku72] or similar to that used by Watson *et al* [Wa71]. It can be seen from Table 16 that these combinations lead to considerable variations in the extracted values of S_α for a gs $\rightarrow$ gs transition. (The paper of Jänecke *et al* [Ja79b] contains extensive data on excitation of higher states and gives a full discussion of the spectroscopy of these states.)

The zero-range calculations for the rare earth nuclei [Ja 79b] give surprisingly good agreement with the angular distributions. The finite-range calculations give values of S_α which are an order of magnitude larger than the normalized zero-range calculations, as can be seen from Table 16. This suggests either that the finite-range method underestimates the cross-section for some reason or that the treatment of α-decay is too model-dependent to provide a reliable normalization. The use of a correctly antisymmetrized relative cluster function, as described in sections 2.4 and 3.1, would increase the calculated matrix element and decrease the spectroscopic factor deduced by comparison with experiment. Provided that the α-transfer reaction is localized beyond the last maximum of this function, the correction would appear in the same way in α-transfer and α-decay calculations [Ja 78] and hence the results obtained from the two methods would remain consistent.

The reaction (^{16}O, ^{12}C) can be regarded as the inverse of α-decay. Comparisons of DWBA analyses of this transfer reaction with studies of α-decay have been carried out in the lead region [Da 76, Ja 77b] and in the rare earth region [Ru 78]. The approach of Davies *et al* [Da 76] makes use of the R-matrix theory of α-decay and an α-nucleus potential for $\alpha + {}^{208}$Pb derived from a simultaneous fit to the ^{212}Po α-decay lifetime and low energy α-particle scattering from ^{208}Pb [De 76]. Assuming that the α-particle is transferred in a one-step reaction, the finite-range DWBA calculation yields good agreement with the experimental angular distributions for the (^{16}O, ^{12}C) reaction in the lead region, which show the characteristic bell shape. Assuming the value of $S_\alpha = 0.55$ for $^{16}\mathrm{O} \to {}^{12}\mathrm{C} + \alpha$ [De 73], Davies *et al* obtain $S_\alpha = 0.73 \times 10^{-2}$ for $^{212}\mathrm{Po} \to {}^{208}\mathrm{Pb}\ (\mathrm{gs}) + \alpha$. Adjusting the first spectroscopic factor to the theoretical value of 0.23 [Ku 73] gives for the second spectroscopic factor $S_\alpha = 1.75 \times 10^{-2}$; this is in good agreement with the value of $S_\alpha = 2.1 \times 10^{-2}$ derived from the ^{212}Po α-decay by Jackson and Rhoades-Brown [Ja 78], which is discussed at the end of section 4.2. Jackson and Rhoades-Brown have also fitted the same data for the ^{208}Pb (^{16}O, ^{12}C) ^{212}Po (gs) reaction [Ja 77b] taking account of the difference between the exact and the model Hamiltonians which leads to additional terms in the perturbation, as indicated in equations (2.33b) and (2.34b). This approach leads to good shape agreement with the experimental data. However, it is now considered that the treatment of the spectroscopic factors in this work was not consistent with equation (3.25) and that a normalization factor of $N \sim 36$ is needed to yield the same spectroscopic factor from the zero-range calculation of the transfer reaction as is obtained from the description of the ^{212}Po α-decay in the unified theory.

In the case of (^{16}O, ^{12}C) reactions on ^{140}Ce and ^{144}Nd at 88 MeV, finite range DWBA calculations yield good shape agreement with the experimental angular distribution, which again has the bell shape [Ru 78]. However, the normalization to the α-decay lifetime using R-matrix theory yields a cross-section which is too small by a factor of 5. This result could again indicate that the method of normalization to α-decay using R-matrix theory is too model-dependent. For the ^{142}Nd (^{16}O, ^{12}C) reaction the experimental angular distribution has a completely different shape to the other experimental distributions and to the DWBA predictions. No explanation of this discrepancy is available at present.

Systematics of the (^{16}O, ^{12}C) reaction on light and medium mass nuclei have been reviewed by Mallet-Lemaire [Ma 79a] and compared with (^{6}Li, d) and (^{7}Li, t) reactions. The (^{16}O, ^{12}C) cross-sections fall less rapidly with increasing mass number of the target. The mechanism of transfer of 4 nucleons in the (^{16}O, ^{12}C) reaction has been extensively discussed as the existence of α-clustering in ^{16}O is less convincing than in ^{6}Li or ^{7}Li. However, the theoretical results shown in Figure 19 suggest that transfer of a simple α-cluster with no internal excitation can occur if the reaction is localized in the surface of the target nucleus. In other circumstances, additional states of internal excitation of the 2p2n system can contribute. Analysis of the ^{12}C (^{16}O, ^{12}C) ^{16}O reaction [De 73a] suggests that this reaction can be adequately described by finite range DWBA with a simple α-transfer mechanism. The spectroscopic factors for ^{16}O (gs) $\rightarrow$ ^{12}C (gs) + α deduced from this and some other reactions are listed in Table 13 and show some variation due mainly to different choices for the parameters of the bound state potential in which the relative cluster function is generated.

Yoshida [Yo 76] has considered the contribution of the ^{12}C (^{16}O, ^{12}C) ^{16}O transfer reaction to ^{12}C – ^{16}O elastic scattering at energies in the range 24–80 MeV using finite range DWBA. It is necessary to multiply the shell model spectroscopic factor by $N \sim 1.5$ in order to obtain a best fit to the data.

Results for the reaction ^{16}O (^{12}C, ^{16}O) ^{12}C at 76.8 MeV have been presented by Motobayashi *et al* [Mo 79], and the spectroscopic factor deduced by fitting the data is included in Table 13. The same authors have examined several other α-transfer reactions on light nuclei; the products of the spectroscopic factors $S_\alpha(AB)$ $S_\alpha(ab)$ deduced by fitting the experimental data with conventional finite-range DWBA calculations are, with one exception, always substantially larger than shell model predictions. The same authors have repeated their calculations, replacing the overlap integral $S_\alpha^{1/2}$ (AB) $\bar{\chi}_{\alpha A}(\underset{\sim}{R})$ by the function $(1-K)^{1/2}$ $\bar{\chi}_{\alpha A}(\underset{\sim}{R})$ where K is the exchange kernel derived from the OCM (see section 2.4). This procedure is an approximation to the correct procedure of replacing a solution of equation (2.18) by a solution of equation (2.19) or (2.15), but by setting the spectroscopic factor equal to unity the cluster wavefunction of the target nucleus has effectively been truncated after one term which seems a very poor (and unnecessary) approximation.

A variety of α-transfer reactions on light nuclei have been studied experimentally by Bradlow *et al* [Br 79a, b]. They also carried out finite-range DWBA calculations for reactions involving the process α + ^{16}O $\rightarrow$ ^{20}Ne [Br 79a] using a modified version of the LOLA code in which the bound state potential used to generate the α-core relative cluster function is taken to have the form derived by Buck, Dover and Vary [Bu 75]. For unbound states a Gamow prescription is used. The absolute spectroscopic factors are found to be very sensitive to the optical potentials and the relative cluster function chosen for the projectile. Relative spectroscopic factors within a given rotational band are much less sensitive to parameter variation, but theoretical shell model predictions are not reproduced.

The selectivity of the (^{16}O, ^{12}C) and (^{6}Li, d) reactions for different final states appears to be very similar [Ma 79a]. Rather complete studies have been made of the (^{16}O, ^{12}C) reaction on ^{24}Mg and ^{28}Si at 36–53 MeV [Pe 76], on $^{58, 60, 62, 64}$Ni at 46 and 60 MeV [Be 78a], and on $^{54,56,58}Fe$ at 50 MeV [Ha 79]. There is considerable difficulty in reproducing the angular distributions for such targets but very satisfactory results for low-lying states have been obtained using surface transparent optical potentials, i.e. potentials with quite large central values for both the real and imaginary parts but unusually small values of the diffuseness and radius parameters of the imaginary part. These potentials yield less marked localization in the surface, but the very weak excitation of the unnatural parity states implies that the mechanism does not depart significantly from simple α-transfer. Fits to highly excited states are not good. It has been widely observed that DWBA often does not give the Q-dependence of transfer reactions and this has been taken as indicating a need for energy-dependent optical potentials. Nevertheless the spectroscopic factors obtained from recent studies of the (^{16}O, ^{12}C) reaction show good overall agreement with those obtained from the (^6Li, d) reaction [Pe 76, Ma 78a, Be 78a, Ha 79].

4.4 Peripheral Heavy Ion Collisions

It has proved convenient to distinguish between central and peripheral heavy ion collisions [Bo 77, Ge 78, Sc 78]. At low energies, central collisions lead to complete fusion followed by particle evaporation or fission. Peripheral collisions at low energies consist mainly of quasi-elastic processes involving transfer of a small number of nucleons between the colliding systems and deeply-inelastic reactions involving partial equilibration and subsequent decay of the di-nuclear system by particle emission or fission. At high energies, peripheral collisions consist mainly of projectile fragmentation processes in which the target acts as a spectator and little momentum is transferred between the colliding systems.

In this section we consider some aspects of peripheral collisions of ions of mass number A = 4 – 16 on medium and heavy targets and, where appropriate, examine connections with the cluster reactions discussed in sections 4.1 and 4.3.

At energies less than 10 MeV/nucleon, the cross-section for isotope production can be parametrized with the expression [Ar 71 a, b, 73 a, Volkov 76]

$$\sigma(N, Z) = f(Z) \exp(Q_{gg}/T) \tag{4.21}$$

where f(Z) depends only on the nuclear charge of the reaction products and Q_{gg} is the Q-value for two-body transfer to the ground states of the final pair of nuclei. This expression can be derived from a statistical model of partial equilibration of the di-nuclear system at temperature T [Bo 71]. The temperature is connected with the effective energy by the relation

$$a\,T^2 = E_{eff} = E_{cm} - V_c \tag{4.22}$$

where E_{cm} is the energy in the centre-of-mass system, V_c is the Coulomb barrier in the incident channel and a is a constant. Plots of log σ against Q_{gg} yield values of T which are plotted in Figure 26 for the Berkeley data obtained with ^{16}O ions [Ge 78, Sc 78a] and other data for ^{16}O and ^{15}N ions [Vo 76]. At much higher energies the formation of a partially-equilibrated di-nuclear system is unlikely. If the process proceeds by frag-

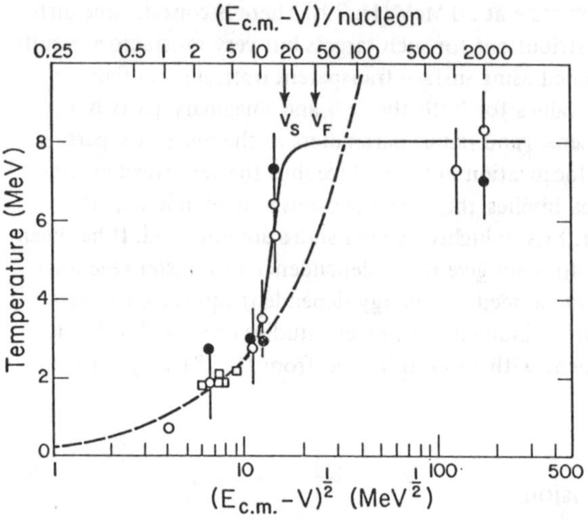

Fig. 26 The variation of temperature with effective energy in the collision of ^{16}O with ^{208}Pb, ^{197}Au, ^{58}Ni and Ta. The open circles correspond to values derived from the widths of the momentum distributions. The filled circles and square blocks correspond to values derived from isotope production systematics. The hatched line is the prediction of a statistical equilibrium model based on the Fermi gas model. The arrows indicate the characteristic velocity of sound v_s and the Fermi velocity v_F in nuclear matter [Sc 78a].

mentation of the projectile the cross-section would be expected to have the form [Lu 75, Ge 78]

$$\sigma(N, Z) = C \sum_i \exp(Q_i/T) \qquad (4.23)$$

where the sum runs over all the break-up channels leading to the observed fragment and Q_i is the corresponding Q-value for projectile fragmentation. Values of T derived using this formula with the data obtained with ^{16}O ions at 20 MeV/nucleon [Bu 76] and for ^{16}O, ^{14}N and ^{12}C at 2.1 GeV/nucleon [Gr 75, He 72] are also shown in Figure 26. The trend of the temperature values suggests that there is a rather sharp transition between partial equilibration and projectile fragmentation at an energy of ~ 19 MeV/nucleon [Sc 78a].

The energy and momentum distributions of the same data have also been studied [Lu 75, Ge 77, Sc 78a]. At both 20 MeV/nucleon and at 2.1 GeV/nucleon, these distributions can be interpreted in terms of the fermi momenta of the nucleons in the projectile. The relevant formulae can be derived using several different models [Fe 73b, Go 74a, Lu 75, Ab 76] all of which involve the assumption of statistical decay of an excited fragment. A further experimental study with ^{16}O ions at 20 MeV/nucleon [Dy 79] has shown that the momentum transferred is substantially smaller than that expected for a transfer reaction but larger than expected for projectile fragmentation. This confirms the suggestion that the energy of ~ 20 MeV/nucleon corresponds to a transition region; however, these authors suggest that an intermediate type of reaction is occurring rather than a mixture of transfer and fragmentation.

The studies of ^{16}O reactions at 20 MeV/nucleon [Bu76] and at 2.1 GeV/nucleon [Gr75, Li75] in the form of measurements of inclusive reactions

$$a + A \rightarrow b + \text{anything}$$

have shown that the relative isotopic yields are independent of the structure of the target [Ge76]. This has been taken to imply that the cross-sections for isotope yield can be factorized in the form

$$\sigma(a + A \rightarrow b) = \Gamma(a, b)\,\delta(a, A) \tag{4.24}$$

and hence

$$R = \frac{\sigma(a+A\rightarrow b)}{\sigma(a+D\rightarrow b)} = \frac{\delta(a, A)}{\delta(a, D)}. \tag{4.25}$$

At relativistic energies this ratio depends only on the size of the target nuclei [Gr75, Li75]. The ratio is also independent of the nature of the observed particle b, as noted in experiments at 20 MeV/nucleon [Bu76] and at 2.1 GeV/nucleon [Gr75, Li75]. This factorization cannot be derived from the Q_{gg} systematics [Ge78] and is not observed at 9 MeV/nucleon [Ge76].

In heavy ion collisions well above the Coulomb barrier, production of fast α-particles is observed [Br61a] with the α-particles emitted predominantly in the forward direction. Angular correlations between fast α-particles and heavy ion fragments have been measured for ^{16}O ions incident on targets of ^{208}Pb and ^{197}Au at 9 MeV/nucleon and 20 MeV/nucleon [Ge77]. Coincident α-particles and heavy ions appear with maximum probability on the *same* side of the beam direction. The angular distributions show a similar shape except that the cross-section for $^{16}O \rightarrow {}^{12}C + \alpha$ has a double peak. At 20 MeV/nucleon the cross-section for $^{16}O \rightarrow {}^{14}N + \alpha$ is almost an order of magnitude smaller than the other reactions, which are comparable, while at 9 MeV/nucleon the cross-section for $^{12}C + \alpha$ is smaller than the cross-section for production of ^{13}C and ^{14}C.

Other coincidence measurements have been made at lower energies. Collisions of ^{16}O ions on ^{58}Ni were studied at 6 MeV/nucleon [Ho77] and α-particles were detected in coincidence with O and C reaction products. Kinematic analysis of the energies gives strong evidence for a sequential process in which the target-like fragments are excited by a deep-inelastic collision and subsequently undergo α-decay. The angular distributions, however, show a pronounced forward peak instead of the expected isotropic distribution. Somewhat similar results were found for the reaction $^{16}O + {}^{27}Al$ at 4 MeV/nucleon [Ha77]. The ^{12}C singles spectra show the broad bell-shaped peak associated with an α-transfer process. The angular correlation of ^{12}C fragments with α-particles is forward peaked along the direction of momentum transfer, indicating that the decay of the intermediate system ^{31}P is rapid.

The break-up of ^{6}Li ions on targets of ^{56}Fe and ^{197}Au has been studied at 12.5 MeV/nucleon [Ca78]. Measurements of singles spectra plus emitted γ-rays indicate that about 35 % of the reaction cross-section involves emission of α-particles and that relatively few of these events are associated with a γ-ray. The ratio of the cross-section for production of any fragment to that for production of an α-particle is independent of the choice of target. (This observation is consistent with the factorization effect (4.24) observed with ^{16}O ions.) The cross-section for production of a given fragment is proportional to $A_{target}^{1/3}$. Coincidence measurements show that α + d coincidences are important, α + n are significant but α + p are much weaker.

Coincidences of α-particles plus heavy ion fragments have been measured [Bh 78,79] for reactions on targets of ^{54}Fe and ^{58}Ni with beams of ^{11}B, ^{13}C, ^{14}N at energies of 8–11 MeV/nucleon. For all projectiles the largest cross-section occurs for production of ^{12}C. The energy spectra for heavy ion fragments in coincidence with α-particles appear insensitive to the α-particle angle and energy. Similarly the energy spectra for coincident α-particles are approximately independent of the heavy ion ejected and the heavy ion angle and are very similar to the α-particle singles spectra except for $Z_{HI} = 6$ shere the singles spectra contain a quasi-elastic component. The angular distributions are symmetric for $Z_{HI} < 6$ and slightly asymmetric for increasing mass of the heavy fragment. Again, there is close similarity to the singles spectra. These results suggest that the cross-section may be factorized in the form

$$\frac{d^4\sigma(HI,\alpha)}{d\Omega_{HI}d\Omega_\alpha\, dE_{HI}dE_\alpha} \propto \frac{d^2\sigma(HI)}{d\Omega_{HI}dE_{HI}}\;\frac{d^2\sigma(\alpha)}{d\Omega_\alpha\, dE_\alpha} \tag{4.26}$$

where $d^2\sigma(HI)$ and $d^2\sigma(\alpha)$ correspond to the singles spectra for heavy ion fragments and α-particles, respectively, at angles θ_{HI} and θ_α. This formula suggests that the emission of the α-particle and the heavy ion fragment are essentially independent. It is proposed [Bh 79] that the α-particle is first emitted by a knock-out mechanism while the remaining projectile fragment and the target form a di-nuclear system which finally breaks up producing the fragments characteristic of a deeply inelastic collision. The latter process must be a fast process because the energies of the ejected heavy ions are well above the Coulomb barrier.

A study of the reaction $^{12}C + ^{160}Gd$ has been made over the energy range 7.5–16.7 MeV/nucleon [Si 79a, Wi 79a]. Cross-sections for inclusive α-particle production have been measured, as well as α–γ concidences [Si 79a] and α–α coincidences [Wi 79a]. The reactions (^{12}C, 2α) and (^{12}C, α) can be interpreted as aristing from capture of an α-cluster and a ^8Be-cluster, respectively, and constitute an important part of the cross-section in the lower energy range. At higher energies, the direct break-up process $^{12}C \rightarrow 3\alpha$ without target excitation is dominant, as can be seen from Table 17. These results lend support to the model of non-equilibrium massive transfer [In 77, Zo 78] in which relatively large fragments of the projectile are transferred into the target. The resulting complex system then decays, mainly by xn reactions. The α-particle spectra arising from the $^{14}N + ^{209}Bi$ reaction have been studied at 6–7 MeV/nucleon [No 78]. The measured cross-sections are given in Table 18. These show that fission is the dominant mode of deexcitation of the compound nucleus, and that the cross-section for α-particle emission is roughly equal to the sum of cross-sections for heavy residual nuclei produced by the (^{14}N, αxn) and (^{14}N, 2αxn) reactions. This suggests the composite system formed after α-particle emission decays by neutron emission and not by fission. Gonthier *et al* [Go 79] have detected α-particles emitted in coincidence with heavy ion fragments or fissionlike residues arising from ^{16}O + Ti collisions at 19.5 MeV/nucleon.

They conclude that the α-particles are emitted prior to fusion of the rest of the projectile with the target in a mechanism consistent with the massive transfer model. At 9 MeV/nucleon and 12 MeV/nucleon the cross-section for α-transfer and ^{8}Be massive-transfer are approximately equal in ^{12}C reactions with ^{100}Mb, ^{128}Te and ^{154}Sm [Zo79].

The production of fast protons, deuterons and tritons arising from collisions of 12 MeV/nucleon ^{14}N ions with $^{152,\,154}$Sm and 9 MeV/nucleon ^{19}F ions on ^{148}Nd have been observed in coincidence with γ-rays [Ya79]. The results are consistent with the massive transfer model and do not resemble ordinary (HI, xn) reactions. The process occurs near to the critical angular momentum for complete fusion of the projectile. The spectra of p, d, t, α arising from 16 MeV/nucleon ^{12}C collisions with ^{56}Fe have also been measured [Ba78,79a]. It is noted [Ba78] that the spectra extend to more than four times the energy per nucleon of the incident projectile. This effect of energy amplification can be explained by addition of two velocities, arising either from excitation of the projectile without change in its laboratory velocity [Ba78] or by transfer of nucleons from target to projectile, or vice versa [Bo80].

Studies have also been made of particle spectra arising from α-particle collisions with various nuclei at 35 MeV/nucleon [Wu78, 79]. Results from coincidence measurements [Ko79] suggest that α-particle fragmentation involves three different peripheral processes – final-state break-up is the process in which the interaction between the α-particle and target leads to excitation of the projectile which subsequently decays by particle emission, while quasi-free break-up leaves the target in a ground or low-lying

Table 17: Cross-sections from ^{12}C + ^{160}Gd collisions at different incident energies [Si79a, Wi79a]

Reaction	Cross-section at 10 MeV/nucleon (mb)	Cross-section at 17 MeV/nucleon (mb)
σ_α (inclusive)	850	2100
σ (^{12}C, α)	177 ± 17	144 ± 23
σ (^{12}C, 2α)	75 ± 9	119 ± 30
σ (^{12}C, 3α)	174 ± 8	573 ± 21

Table 18: Cross-sections from ^{14}N + ^{209}Bi collisions at different incident energies [No78]

Reaction	Cross-section at 6 MeV/nucleon (mb)	Cross-section at 7 MeV/nucleon (mb)
α-particle emission	42 ± 6	63 ± (large uncertainty)
Fission	890 ± 65	1350 ± 100
(^{14}N, xn)	5 ± 1	3 ± 1
(^{14}N, αxn)	31 ± 3	49 ± 8
(^{14}N, 2αxn)	3 ± 1	3 ± 1

state, and in a third process part of the projectile continues undisturbed while the remaining part interacts strongly with the target to give secondary fragments and evaporation. The similarity of the inclusive spectra with the spectra for the same product in coincidence with a low energy particle leads to the conclusion that the third process gives ~ 90 % of the inclusive cross-section. Koontz *et al* [Ko79] call this process "absorptive" break-up and it has clear similarities with the massive transfer process.

Sugihara [Su79a] argues that the high energy spectra of p, d, t and α-particles observed in the massive transfer process are not emitted in an early stage of a process leading towards equilibration but are spectator particles in a direct process in which the remainder of the projectile is tranferred to the target nucleus. If this is correct, pre-equilibrium models should not be applied to the analysis of these data, although there is evidence that they are moderately successful at 20 MeV/nucleon [Sy78].

Attempts to characterize the complex phenomena observed in heavy ion reactions have largely concentrated on macroscopic models of the interacting matter in bulk or in a localized region together a semiclassical approach to the scattering. To complete this section, however, we explore the extent to which a microscopic direct reaction approach together with the cluster model can explain some of the observed phenomena. We start by listing some of the direct processes.

I. Projectile break-up

$$a + A \to b + x + A \tag{4.27}$$

This process may be followed by final state interactions leading to different reaction products.

II. Light ion transfer

$$a + A \to b + B \tag{4.28}$$

$$a = b + x \tag{4.29}$$

$$B = A + x \tag{4.30}$$

III. Massive transfer

$$a + A \to x + C \tag{4.31}$$

$$C = A + b$$

Both transfer processes may be followed by fast or slow decay of the residual nucleus.

IV. Knock-out or knock-on

$$a + B \to a' + x + A \tag{4.33}$$

This reaction may also be followed by final state interactions and either a or c may remain in the compound system.

V. Light ion emission from the target

$$b + B \to x + C \tag{4.34}$$

where B and C have the composite structure indicated by equations (4.30) and (4.32).

This process may be followed by fast or slow decay of the residual nucleus. In reactions I and III the light ion x is removed directly from the projectile, in reactions IV and V it is removed directly from the target nucleus, and in reaction II it comes from the decay of a reaction product. Examples in which particle x is an α-particle are of particular interest.

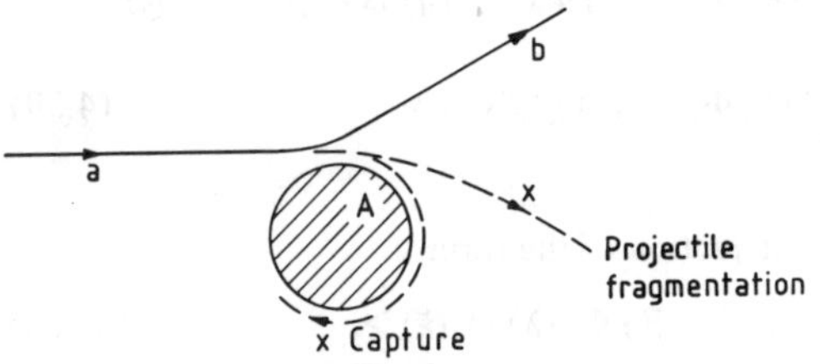

Fig. 27
Schematic diagram illustrating projectile fragmentation and capture of one of the fragments.

We consider first the break-up process represented by equation (4.27). The DWBA formalism has already been applied [Ud 79] rather successfully to describe the break-up of ^{20}Ne ions into ^{16}O + α in collisions with ^{40}Ca. We use a somewhat different formalism in order to derive the factorization effect discussed above. We assume that the heavier particle b continues along the grazing angle trajectory of the incident heavy ion, as shown in Figure 27. We take the initial momentum of the projectile in the laboratory frame to be $\underset{\sim}{p}_a$ and the final momenta to be $\underset{\sim}{p}_b$, $\underset{\sim}{p}_x$ and $\underset{\sim}{p}_A$ where

$$\underset{\sim}{p}_a = \underset{\sim}{p}_b + \underset{\sim}{p}_x + \underset{\sim}{p}_A . \tag{4.35}$$

We use the system of relative coordinates shown in Figure 7 and assume that the target nucleus is very heavy so that the conjugate momenta $\underset{\sim}{k}_a$, etc. are essentially the same as the lab momenta.

We make a cluster expansion of the wavefunction of the projectile in the form

$$\Phi_a^{m_a s_a} = \sum_{sm_s s_b m_b} C(s\, s_a s_b)\,(s m_s s_b m_b \,|\, s_a m_a)\, \Phi_b^{s_b m_b}(\lambda)\, \psi_{xb}^{s m_s}(\underset{\sim}{r}, \xi) \tag{4.36}$$

where C includes the coefficient of fractional parentage and Φ_b is the internal wavefunction for particle B. The function ψ is given by

$$\psi_{xb}^{s m_s}(\underset{\sim}{r}, \xi) = \sum_{\ell m s_x m_x} (\ell m\, s_x m_x \,|\, s\, m_s)\, \overline{\chi}_{xb}^{\ell m}(\underset{\sim}{r})\, \Phi_x^{s_x m_x}(\xi) \tag{4.37}$$

where $\overline{\chi}_{xb}$ is the relative cluster function generated in a bound state potential. We use the prior representation of DWBA with the perturbation W_{aA} given by equation (2.33a, b) but omit the term ΔV^1 and write

$$W_{aA} \equiv \sum_{i=2}^{4} \Delta V_{aA}^{i} \equiv \Delta V_{aA}(\xi, \eta, \lambda, \underset{\sim}{R}). \tag{4.38}$$

The matrix element for break-up of the projectile A leaving the target nucleus in state J'_A and the particles b and x in the same states s_b and s_x is given by

$$T_{if}(J_A \to J'_A) = \sum_{sm_s\ell m} C(s\, s_a\, s_b)\,(s\, m_s\, s_b m_b | s_a m_a)\,(\ell m\, s\, m_x | s m_s) \times$$

$$\times < \chi_x^-(\underset{\sim}{k}_x, \underset{\sim}{R})\,\chi_b^-(\underset{\sim}{k}_b, \underset{\sim}{R} - \underset{\sim}{r})\Phi_b^{s_b m_b}(\lambda)\,\Phi_x^{s_x m_x}(\xi)\,\Phi_A^{J'_A M'_A}(\eta) | \Delta V_{aA}(\xi, \eta, \lambda, \underset{\sim}{R}) \times$$

$$\times | \chi_a^+(\underset{\sim}{k}_a, \underset{\sim}{R} - \delta \underset{\sim}{r})\,\Phi_A^{J_A M_A}(\eta)\,\Phi_b^{s_b m_b}(\lambda)\,\Phi_x^{s_x m_x}(\xi)\,\overline{\chi}_{xb}^{\ell m}(\underset{\sim}{r}) > \tag{4.39}$$

where $\delta = M_b/M_a$.

We now define a projectile-nucleus interaction of the form

$$\Delta V_{aA}(\eta, \underset{\sim}{R}) = < \Phi_b(\lambda)\,\Phi_x(\xi) | \Delta V_{aA}(\xi, \eta, \lambda, \underset{\sim}{R}) | \Phi_b(\lambda)\,\Phi_x(\xi) >. \tag{4.40a}$$

so that the matrix element between the target states becomes

$$< \Phi_A^{J'_A M'_A}(\eta) | \Delta V_{aA}(\eta, \underset{\sim}{R}) | \Phi_A^{J_A M_A}(\eta) > = \sum_{LM} (J_A M_A\, LM | J'_A M'_A)\, W_{aA}^{J_A J'_A LM}(\underset{\sim}{R}) \tag{4.40b}$$

It will be seen that we have neglected spin-dependent terms from the definition (4.40a) of $\Delta V(\eta, \underset{\sim}{R})$. This is exact in the case of $^{16}O \to {}^{12}C$ (gs) + α to which the formalism was first applied [Ja77a] and probably represents a reasonable approximation for other cases. The interaction $W_{aA}(\underset{\sim}{R})$ could be calculated using an appropriate nucleon-nucleon interaction for ΔV^i. Alternatively, if we neglect completely the dependence of ΔV^i on λ, i.e. we treat particle b entirely as a spectator, then $W_{aA} \simeq W_{xA}$ and may be calculated in a single folding model using a parametrized form $\Delta V_{xA}(\eta, \underset{\sim}{R})$ for the interaction of particle x with a nucleon or it may be parametrized using the generalized optical model for particle x with nucleus A.

In terms of $W_{aA}(R)$ as defined above by neglecting the spins of particles b and x, the matrix element (4.39) becomes

$$T_{if}(J_A \to J'_A) = \sum_{LM\, sm_s} C(s\, s_a s_b)\,(s m_s\, s_b m_b | s_a m_a)\,(J_A M_A\, LM | J'_A M'_A) \times$$

$$\times \int \chi_x^{-*}(\underset{\sim}{k}_x, \underset{\sim}{R})\,\chi_b^{-*}(\underset{\sim}{k}_b, \underset{\sim}{R} - \underset{\sim}{r})\, W_{aA}^{J_A J'_A LM}(\underset{\sim}{R})\,\overline{\chi}_{xb}^{sm_s}(\underset{\sim}{r})\,\chi_a^+(\underset{\sim}{k}_a, \underset{\sim}{R} - \delta\underset{\sim}{r})\, d\underset{\sim}{r}\, d\underset{\sim}{R}. \tag{4.41}$$

The cross-section for detecting particles b and x in coincidence is given by

$$\frac{d^3\sigma}{d\Omega_x dE_x d\Omega_b} = \frac{2\pi}{\hbar v_a}\rho_f(E)\frac{1}{2J_A + 1}\frac{1}{2s_a + 1}\sum_{M_A M'_A m_a m_b} |T_{if}|^2$$

where $\rho_f(E)$ is the density of final states. When the sums over Clebsch-Gordan coefficients have been carried out the cross-section becomes

$$\frac{d^3\sigma}{d\Omega_x dE_x d\Omega_b} = \frac{2\pi}{\hbar v_a}\rho_f(E)\frac{1}{(2L+1)(2s+1)}\sum_{LMsm_s} |C(s\, s_a s_b)\, I_{sm_s}^{J_A J'_A LM}(\underset{\sim}{k}_a, \underset{\sim}{k}_b, \underset{\sim}{k}_x)|^2 \tag{4.42}$$

where I is the integral written in full in equation (4.41). If the distorted waves are replaced by plane waves, the integral I factorizes into

$$I_{sm_s}^{J_A J_A' LM}(\underset{\sim}{k}_a, \underset{\sim}{k}_b, \underset{\sim}{k}_x) = {}^{PW}I_1^{J_A J_A' LM}(\underset{\sim}{Q})\, I_2^{sm_s}(\underset{\sim}{P}) \tag{4.43}$$

where

$${}^{PW}I_1^{J_A J_A' LM}(\underset{\sim}{Q}) = \int e^{i\underset{\sim}{Q}.\underset{\sim}{R}}\, W_{aA}^{J_a J_A' LM}(\underset{\sim}{R})\, d\underset{\sim}{R} \tag{4.44}$$

$$I_2^{sm_s}(\underset{\sim}{P}) = \int e^{i\underset{\sim}{P}.\underset{\sim}{r}}\, \bar{\chi}_{xb}^{sm_s}(\underset{\sim}{r})\, d\underset{\sim}{r} \tag{4.45}$$

$$\underset{\sim}{Q} = \underset{\sim}{k}_a - \underset{\sim}{k}_b - \underset{\sim}{k}_x \tag{4.46}$$

$$\underset{\sim}{P} = \underset{\sim}{k}_b - \delta \underset{\sim}{k}_a . \tag{4.47}$$

In the Berkeley experiment [Ge77] the momentum transfer $\underset{\sim}{P}$ was small and hence the reaction selects the low momentum components of $\bar{\chi}_{xb}$. Thus in the fragmentation of ^{16}O, the reaction selects s = O and hence the fragments are left preferentially in their ground states. Neglecting any contribution from the D-state of the deuteron, the same is true for fragmentation of ^6Li. The matrix element of the interaction with the target can also be written as

$$I_1(\underset{\sim}{Q}) = \langle \underset{\sim}{k}_x | W_{aA} | \underset{\sim}{k}_a - \underset{\sim}{k}_b \rangle. \tag{4.48}$$

This is an off-shell matrix element but, for not too large values of $\underset{\sim}{k}_x$, $|I_1|^2$ may be replaced by the free-cross-section for elastic or inelastic scattering from nucleus A. For small values of $|\underset{\sim}{k}_a - \underset{\sim}{k}_b|$, we have $Q \simeq \sqrt{E_x}$ and hence $I = I_1(E_x)\, I_2(\underset{\sim}{P})$. This factorization is of similar form to that assumed by Udagawa *et al* [Ud79].

An improved approximation may be obtained by using equation (4.19) for the distorted waves (with $\gamma = 1$ for a heavy target). This gives

$$I_{sm_s}^{J_A J_A' LM} = I_1^{J_A J_A' LM}(\underset{\sim}{Q})\, I_2^{sm_s}(\underset{\sim}{P}') \tag{4.49}$$

where I_2 is as defined by equation (4.45) but with

$$\underset{\sim}{P}' = \underset{\sim}{q}_b - \delta \underset{\sim}{q}_a = \underset{\sim}{q}_b - (m_b/m_a)\, \underset{\sim}{q}_a , \tag{4.50}$$

and

$$I_1^{J_A J_A' LM}(\underset{\sim}{Q}) = \int \chi_x^{-*}(\underset{\sim}{k}_x, \underset{\sim}{R})\, \chi_b^{-*}(\underset{\sim}{k}_b, \underset{\sim}{R})\, W_{aA}^{J_A J_A' LM}(\underset{\sim}{R})\, \chi_a^{+}(\underset{\sim}{k}_a, \underset{\sim}{R})\, d\underset{\sim}{R}. \tag{4.51}$$

The requirement for factorization of the integral is thus that recoil effects are small or capable of being treated through equation (4.19).

Using either equation (4.43) or (4.49), the cross-section can be factorized into the form

$$\frac{d^3\sigma}{d\Omega_x dE_x d\Omega_b} = \frac{2\pi}{\hbar v_a} \rho_f(E) \sum_{LM} \frac{1}{2L+1} |I_1^{J_A J_A' LM}|^2 \sum_{sm_s} \frac{1}{2s+1} \left| C(s\, s_b\, s_a)\, I_2^{sm_s} \right|^2 \qquad (4.52)$$

which can be written symbolically as

$$\sigma(a+A \to b+x+A) = \delta(a, A)\, \Gamma(a \to b+x). \qquad (4.53)$$

Thus the direct reaction approach yields the factorization (4.24) observed experimentally [Ge 76] and gives the following predictions.

(i) If A is left in the ground state and compound nucleus formation is not important, $\delta(a, A)$ depends primarily on the size of the target A and not on its detailed structure. This is in agreement with the observations on high energy break-up of ^{16}O [Gr75, Li75] and on ^{6}Li break-up [Ca 78].

(ii) Cross-sections taken on different targets should show the behaviour

$$\frac{\sigma(a+A \to b+x+A)}{\sigma(a+D \to b+x+D)} = \frac{\delta(a, A)}{\delta(a, D)} \qquad (4.54)$$

which is roughly constant for small Q. If the projectile breaks up into different pairs of fragments the ratio of cross-sections for different targets is unchanged, i.e.

$$\frac{\sigma(a+A \to e+f+A)}{\sigma(a+D \to e+f+D)} = \frac{\delta(a, A)}{\delta(a, D)}. \qquad (4.55)$$

Thus the ratio is independent of the nature of the observed particle as noted in experiments with ^{16}O ions [Ge 76].

(iii) For excitation of a definite final state of the target we have

$$\sigma(a+A \to b+x+A^*) = \delta(a, A^*)\, \Gamma(a \to b+x) \qquad (4.56)$$

and

$$\frac{\sigma(a+A \to b+x+A^*)}{\sigma(a+A \to b+x+A)} = \frac{\delta(a, A^*)}{\delta(a, A)} \qquad (4.57)$$

which is essentially the ratio of the inelastic to elastic cross-sections for scattering from A.

In order to describe the light ion transfer process (4.28) we need the cluster expansion for the system B = A + x,

$$\Phi_B^{J_B M_B} = \sum_{jm_j J_A' M_A'} C(j\, J_A'\, J_B)\, (jm_j J_A' M_A' | J_B M_B)\, \Phi_A^{J_A' M_A'}(\eta)\, \psi_{xA}^{jm_j}(\underset{\sim}{R}, \xi) \qquad (4.58)$$

$$\psi_{xA}^{jm_j}(\underset{\sim}{R}, \xi) = \sum_{kqs_x' m_x'} (kq\, s_x' m_x' | jm_j)\, \chi_{xA}^{-kq}(\underset{\sim}{R})\, \Phi_x^{s_x' m_x'}(\xi) \qquad (4.59)$$

where χ is the relative cluster function. The matrix element in the prior representation of DWBA becomes

$$T_{if}(J_A s_a \to J_B s_b') = \Sigma\, C(j\, J_A' J_B)\, C(s\, s_b s_a)\, (s m_s\, s_b m_b | s_a m_a)\, (\ell m\, s_x m_x | s m_s) \times$$

$$\times\, (j m_j\, J_A' M_A' | J_B M_B)\, (k q s_x' m_x' | j m_j) \times$$

$$\times < \chi_b^-(\underset{\sim}{k}_b, \underset{\sim}{R} - \underset{\sim}{r})\, \Phi_b^{s_b' m_b'}(\lambda)\, \Phi_A^{J_A' M_A'}(\eta)\, \bar{\chi}_{xA}^{kq}(\underset{\sim}{R})\, \Phi_x^{s_x' m_x'}(\xi) | \times$$

$$\times\, | \Delta V(\xi, \eta, \lambda, \underset{\sim}{R}) | \chi_a^+(\underset{\sim}{k}_a, \underset{\sim}{R} - \delta \underset{\sim}{r})\, \Phi_b^{s_b m_b}(\eta)\, \Phi_A^{J_A M_A}(\eta)\, \bar{\chi}_{xb}^{\ell m}(\underset{\sim}{r})\, \Phi_x^{s_x m_x}(\xi) >.$$

and, assuming no excitation of particles b or x, this reduces to

$$T_{if}(J_A s_a \to J_B s_b') = \Sigma\, C(j\, J_A' J_B)\, C(s\, s_b s_a)\, (s m_s\, s_b m_b | s_a m_a)\, (\ell m\, s_x m_x | s m_s) \times$$

$$\times\, (j\, m_j\, J_A' M_A' | J_B M_B)\, (k q\, s_x m_x | j m_j)\, (J_A M_A\, L M | J_A' M_A') \times$$

$$\times \int \chi_b^{-*}(\underset{\sim}{k}_b, \underset{\sim}{R} - \underset{\sim}{r})\, \bar{\chi}_{xA}^{kq*}(\underset{\sim}{R})\, W_{aA}^{J_A J_A' LM}(\underset{\sim}{R})\, \bar{\chi}_{xb}^{\ell m}(\underset{\sim}{r})\, \chi_a^+(\underset{\sim}{k}_a, \underset{\sim}{R} - \delta \underset{\sim}{r})\, d\underset{\sim}{r}\, d\underset{\sim}{R} \tag{4.60}$$

where $W_{aA}^{J_A J_A' LM}$ is defined by equation (4.40b)

The cross-section for the transfer reaction is given by

$$\frac{d\sigma}{d\Omega} = \frac{\mu_{aA}\mu_{aB}}{(2\pi\hbar^2)^2} \frac{k_b}{k_a} \frac{1}{2J_A + 1} \frac{1}{2s_a + 1} \sum_{M_A M_B m_a m_b} |T_{if}|^2 .$$

In order to factorize this expression with the matrix element (4.60) we must make some fairly substantial approximations – we neglect the spin of particle x, neglect core excitation so that $J_A' = J_A$, and consider only $L = 0$. Then, using the factorized distorted waves defined in equation (4.19), we obtain

$$\frac{d\sigma}{d\Omega} = \frac{\mu_{aA}\mu_{bB}}{(2\pi\hbar^2)^2} \frac{k_b}{k_a} \frac{2J_B + 1}{2J_A + 1} \sum_{s m_s} \frac{1}{2s + 1} \left| C(s\, s_b s_a)\, I_2^{s m_s}(\underset{\sim}{P}') \right|^2 \times$$

$$\times \sum_{j m_j} \frac{1}{2j + 1} \left| C(j\, J_A J_B)\, I_3^{J_A j m}(\underset{\sim}{Q}') \right|^2 \tag{4.61}$$

where $\underset{\sim}{Q}' = \underset{\sim}{k}_a - \underset{\sim}{k}_b$ in the plane wave limit and

$$I_3^{J_A j m_j}(\underset{\sim}{Q}') = \int \chi_b^{-*}(\underset{\sim}{k}_b, \underset{\sim}{R})\, \bar{\chi}_{xA}^{j m_j^*}(\underset{\sim}{R})\, W_{aA}^{J_A J_A O}(\underset{\sim}{R})\, \chi_a^+(\underset{\sim}{k}_a, \underset{\sim}{R})\, d\underset{\sim}{R}. \tag{4.62}$$

Thus $I_3(\underset{\sim}{Q}')$, like $I_1(\underset{\sim}{Q})$ is an off-shell matrix element of the interaction of the projectile with the target.

The cross-section for the massive transfer process (4.31) can be derived is exactly the same way and has the form

$$\frac{d\sigma}{d\Omega} = \frac{\mu_{aA}\mu_{xC}}{(2\pi\hbar^2)^2}\frac{k_x}{k_a}\frac{2J_C+1}{2JA+1}\sum_{sm_s}\frac{1}{2s+1}\left|C(s\,s_b\,s_a)\,I_2^{sm_s}(\underset{\sim}{P}'')\right|^2 \times$$

$$\times \sum_{jm_j}\frac{1}{2j+1}\left|C(j\,J_A\,J_B)\,I_2^{J_A jm_j}(\underset{\sim}{Q}'')\right|^2 \tag{4.63}$$

where

$$\underset{\sim}{P}'' = \underset{\sim}{q}_b - (1-\delta)\,q_a = \underset{\sim}{q}_b - (m_x/m_a)\,\underset{\sim}{q}_a \tag{4.64}$$

and

$$I_3^{J_A jm_j}(\underset{\sim}{Q}'') = \int \chi_x^{-*}(\underset{\sim}{k}_x,\underset{\sim}{t})\,\chi_{bA}^{-jm_j^*}(\underset{\sim}{t})\,W_{aA}^{J_AJ_A O}(\underset{\sim}{t})\,\chi_a^{+}(\underset{\sim}{k}_a,\underset{\sim}{t})\,d\underset{\sim}{t} \tag{4.65}$$

with $\underset{\sim}{t} = \underset{\sim}{R} - \underset{\sim}{r}$, and $\underset{\sim}{Q}'' = \underset{\sim}{k}_a - \underset{\sim}{k}_x$ in the plane wave limit.

Equations (4.61) and (4.63) can be written in symbolic form as

$$\sigma_{LT}(a+A\to b+B) = \Gamma(a\to b+x)\,\Delta(A+x\to B) \tag{4.66}$$

$$\sigma_{MT}(a+A\to x+C) = \Gamma(a\to b+x)\,\Delta(A+b\to C). \tag{4.67}$$

The ratios of these cross-sections on different targets are given by

$$\frac{\sigma_{LT}(a+A\to b+B)}{\sigma_{LT}(a+D\to b+E)} = \frac{\Delta(A+x\to B)}{\Delta(D+x\to E)} \tag{4.68a}$$

$$\frac{\sigma_{MT}(a+A\to x+C)}{\sigma_{MT}(a+D\to x+F)} = \frac{\Delta(A+b\to C)}{\Delta(D+b\to F)}. \tag{4.68b}$$

The conditions for factorization of the cross-section for knock-out reactions have been examined in section 2.2 (see discussion following equation 2.65) and equation (2.67) can be re-expressed in the form

$$\sigma(a+B\to a'+x+A) = \Sigma(a+x\to a'+x)\,\Delta(B\to A+x). \tag{4.69}$$

Assuming that off-shell effects are small, the ratio of cross-sections on different targets is given by

$$\frac{\sigma(a+B\to a'+x+A)}{\sigma(a+E\to a'+x+D)} = \frac{\Delta(B\to A+x)}{\Delta(E\to D+x)} \tag{4.70}$$

which is identical with the ratio (4.68a).

The ratios (4.68a) (4.68b) and (4.70) are clearly not independent of the nature of the transferred particle, although these ratios may not be as sensitive to detailed nuclear structure as in nucleon-induced transfer and knock-out reactions. For L = 0, the quantity

$W_{aA}(\underset{\sim}{R})$ depends essentially on the nuclear matter distribution of nucleus A and for a peripheral collision the tails of the functions $\bar{\chi}^{jm_j}$ may all look rather similar in shape for low-lying excitations. This may explain why several different mechanisms are consistent with the experimental observations. However, it is clear that when the target nucleus or one of the projectile fragments is excited, the cross-section definitely will not factorize. This may explain the change in the observed behaviour of the cross-sections between 10 and 20 MeV/nucleon.

These results suggest that our understanding of direct cluster reactions induced by light ions may be used to interpret the "fast" stages of peripheral heavy ion collisions well above the Coulomb barrier.

5 Outstanding Problems

In this concluding review of the outstanding problems associated with the study of direct cluster reactions we try to distinguish between fundamental problems connected with the theoretical formalism and practical difficulties arising from computational complexities or lack of background information. This emphasis on points of difficulty should not obscure the substantial success already achieved in the development of a consistent description of scattering of composite particles and of cluster reactions.

5.1 Distortion and Localization

Surprisingly, the treatment of distortion is in general not completely satisfactory and, in consequence, deductions about the localization of certain reactions are not wholly reliable. The unified reaction theory gives a clear and definite prescription for construction of the generalized optical potential; this shows what modifications to the first order folding model are to be expected for various projectiles. This theory has been explored mainly for the real part of the potential and, with certain exceptions [Br 79c] the more difficult problem of constructing a realistic imaginary potential has not been studied. This throws much greater emphasis on the need for extensive data over a range of energies in order to establish the behaviour of a phenomenological optical potential. The absence of adequate data for this purpose and the consequent lack of a reliable general expression for the energy dependence of the imaginary part of the potential is certainly one of the main practical problems preventing successful description of a wide range of direct cluster reactions.

5.2 Definition of Spectroscopic Factors and Construction of Overlap Integrals

The difficulties discussed in section 1.4 still remain. In many respects the shell model provides the most complete formalism but this is not at present capable of handling relative or internal wavefunctions with the correct asymptotic behaviour. Other errors of order $1/A$ are present due to the absence of translational invariance. Even with the Glasgow shell model techniques [Wh 72] it is still necessary to invoke oscillator radial functions in order to construct the overlap integral [Wa 79a]. Thus the problem of developing the shell model to give thc information required for direct reaction calculations appears to be of a fundamental nature rather than a computational problem.

The cluster model, on the other hand, provides a formalism which is capable of treating the relative and internal wavefunctions correctly. Because cluster model wavefunctions have been used most frequently as variational wavefunctions a limitation to a single channel approximation or to a very small number of channels has been common. The representation of each channel function in terms of a parentage amplitude multiplying a normalized relative function is in contrast not the common procedure. However, there does not appear to be any obstacle of a fundamental nature which would prevent the calculation of spectroscopic factors within the framework of the cluster model.

5.3 Antisymmetrization

Shell model calculations of the spectroscopic factor are correctly antisymmetrized. Within the cluster formalism the procedure for correct treatment of exchange of nucleons in both the bound state and the scattering state is quite clear, and several levels of approximation have been given. The difficulties are computational ones, and are particularly severe when oscillator functions are not used. The method initiated by Rhoades-Brown [Rh 77, Ja 78] may prove useful for detailed studies but a simple ansatz is also needed for use with distorted wave calculations.

References

[Aa 61] *Aaron, R., R.D. Amado* and *B. Lee*: Phys. Rev. **121** (1961) 319

[Ab 76] *Abul-Magd, A.,* and *J. Hüfner*: Z. Physik **A 277** (1976) 379

[Ab 79] *Abdul-Jalil, I.* and *D.F. Jackson*: J. Phys. G: Nucl. Phys. **5** (1979) 1699

[Am 79] *Amado, R.*: Phys. Rev. **132** (1963) 485

[Am 67] *Amos, K.A., V.A. Madsen* and *I.E. McCarthy*: Nucl. Phys. **A 94** (1967) 103

[An 74] *Anyas-Weiss, M., J.C. Cornell, P.S. Fisher, P.N. Hudson, A. Menchaca-Rocha, D.J. Millener, A.D. Panagiotou, D.K. Scott, D. Strottmann, D.M. Brink, B. Buck, P.J. Ellis* and *T. Engeland*: Physics Reports **12 C** (1974) 201

[An 75] *Ando, T., K. Ikeda* and *Y. Suzuki*: Prog. Theor. Phys. **54** (1975) 1554

[An 77] *Ando, T., K. Ikeda* and *A. Tohsaki-Suzuki*: in Int. Pre-Symposium on Clustering Phenomena in Nuclei (Contributed Papers) (1977) 72

[An 77a] *Anantaraman, N., H.E. Gove, J.P. Trentelman, J.P. Draayer* and *F.C. Jundt*: Nucl. Phys. **A 276** (1977) 119

[An 79] *Anantaraman, N., H.E. Grove, R.A. Lindgren, J. Toke, J.P. Trentelman, J.P. Draayer, F.C. Jundt* and *G. Guillaume*: Nucl. Phys. **A 313** (1979) 445

[Ar 63] *Arvieu, R., E. Baranger, M. Veneroni, M. Baranger* and *V. Gillet*: Phys. Lett. **4** (1963) 119

[Ar 71a] *Artukh, A.G., V.V. Avdeichikov, J. Erö, G.F. Gridnev, V.L. Mikheev, V.V. Volkov* and *J. Wilczynski*: Nucl. Phys. **A 160** (1971) 511

[Ar 71b] *Artukh, A.G., V.V. Avdeichikov, J. Erö, G.C. Gridnev, V.L. Mikheev, V.V. Volkov* and *J. Wilczynski*: Nucl. Phys. **A 168** (1971) 321

[Ar 73] *Arima, A.*: in Proc. Int. Conf. on Nucl. Phys. (eds J. de Boer and H.J. Mang) (North-Holland, Amsterdam) (1973) 183

[Ar 73a] *Artukh, A.G., G.F. Gridnev, V.L. Mikheev, V.V. Volkov* and *J. Wilczynski*: Nucl. Phys. **A 211** (1973) 299

[Ar 74] *Arima, A.* and *S. Yoshida*: Nucl. Phys. **A 219** (1974) 475

[Ar 78] *Arima, A.*: in Clustering Aspects of Nuclear Structure and Nuclear Reactions (eds W.H.T. von Oers et al) AIP Conf. Proceedings No. 47 (1979) 1

[Au 70] *Austern, N.*: Direct Reaction Theories (Wiley, New York) (1970)

[Au 78] *Austern, N., C.M. Vincent* and *J.P. Farrell*: Ann. Phys. **114** (1978) 93

[Ba 65] *Balashov, V.V., A.N. Boyarkina* and *I. Rotter*: Nucl. Phys. **59** (1965) 517

[Ba 71] *Batty, C.* and *E. Friedman*: Phys. Lett. **34 B** (1971) 1

[Ba 73] *Bachelier, D., M. Bernas, O.M. Bilaniuk, J.L. Boyard, J.C. Jourdain* and *P. Radvanyi*: Phys. Rev. **C 7** (1973) 165

[Ba 74] *Barnett, A.R.* and *J.S. Lilley*: Phys. Rev. **C 9** (1974) 2010

[Ba 76] *Bachelier, D., J.L. Boyard, T. Henning, H.D. Holmgren, J.C. Jourdain, P. Radvanyi, P.G. Roos* and *M. Roy-Stephan*: Nucl. Phys. **A 268** (1976) 488

[Ba 76a] *Bang, J., C.H. Dasso, F.A. Gareev, M. Igarishi* and *B.S. Nilsson*: Nucl. Phys. **A 264** (1976) 157

[Ba 77] *Barrett, R.C.* and *D.F. Jackson*: Nuclear Sizes and Structure (Oxford University Press) (1978) Chapter 7, and references therein

[Ba 77a] *Baz, A.I., V.Z. Goldberg, N.Z. Darwisch, K.A. Gridnev, V.M. Semjonov* and *E.F. Hefter*: Lett. al Nuo. Cim. **18** (1977) 227

[Ba 77b] *Baz, A.I., V.Z. Goldberg, K.A. Gridnev, V.M. Semjonov* and *E.F. Hefter*: Z. Physik **A 280** (1977) 171

[Ba 78] *Ball, J.B., C.B. Fulmer, M.L. Mallory* and *R.L. Robinson*: Phys. Rev. Lett. **40** (1978) 1698

[Ba 78a] *Bang, J., R.H. Ibarra* and *J.S. Vaagen*: Phys. Scripta **18** (1978) 33

[Ba 78b] *Badawy, I., B. Berthier, P. Charles, M. Dost, B. Fernandez, J. Gastebois* and *S.M. Lee*: Phys. Rev. **C 17** (1978) 978

[Ba 79a] *Ball, J.B., F.E. Bertrand, R.L. Ferguson, C.B. Fulmer* and *R.L. Robinson*: Physics Division Annual Report, ORNL (1979)

[Be 62] *Berggren, T., G.E. Brown* and *G. Jacob*: Phys. Lett. **1** (1962) 88

[Be 65] *Beregi, P., N.S. Zelenskaya, V.G. Neudatchin* and *Yu F. Smirnov*: Nucl. Phys. **66** (1965) 513

[Be 65a] *Berggren, T.*: Nucl. Phys. **72** (1965) 337

[Be 75] *Becchetti, F.D., L.T. Chua, J. Jänecke* and *A.M. Van der Moeln*: Phys. Rev. Lett. **34** (1975) 225

[Be 78] *Becchetti, F.D.*: in Clustering Aspects of Nuclear Structure and Nuclear Reactions (eds W.H.T. van Oers et al) AIP Conf. Proceedings No. 47 (1978) 308

[Be 78a] *Berg, G.P.A., B. Berthier, J.P. Fouan, J. Gastebois, J.P. Le Fevre* and *M-C Lemaire*: Phys. Rev. **C 18** (1978) 2204

[Be 75b] *Betigeri, M.G., W. Chung, A. Djaloeis, C. Mayer-Böricke, W. Oelert* and *P. Turek*: In Clustering Aspects of Nuclear Structure and Nuclear Reactions (eds W.H.T. van Oers et al) AIP Conf. Proceedings No. 47 (1978) 706

[Be 79] *Becchetti, F.D., J. Jänecke, D. Overway, J.D. Coissart* and *R.L. Spross*: Phys. Rev. **C 19** (1979) 1775

[Be 81] *Betigeri, M., W. Chung, A. Djaloeis, W. Oelert* and *P. Turek*: Nucl. Phys. **A 363** (1981) 35

[Bh 78] *Bhowmik, R.K., E.C. Pollacco, N.E. Sanderson, J.B.A. England* and *G.C. Morrison*: Phys. Lett. **80 B** (1978) 41

[Bh 79] *Bhowmik, R.K., E.C. Pollacco, N.E. Sanderson, J.B.A. England* and *G.C. Morrison*: Phys. Rev. Lett. **43** (1979) 619

[Bl 74] *Blair, J.S., R.M. De Vries, K.G. Nair, A.J. Baltz* and *W. Reisdorf*: Phys. Rev. **C 10** (1974) 1856

[Bo 71] *Bondorf, J.P., F. Dickman, D.H.E. Gross* and *P.J. Siemans*: J. de Physique 32 (1971) C 6

[Bo 77] *Bodmer, A.R.* and *C.N. Panos*: Phys. Rev. **C 15** (1977) 1342

[Bo 77a] *Bonbright, D.I., J.S.C. McKee* and *J.W. Watson*: J. Phys. G: Nucl. Phys. **3** (1977) 1359

[Bo 80] *Bondorf, J.P., J.N. De, G. Fai, A.O.T. Karvinen, B. Jakobsson* and *J. Randrup*: Nucl. Phys. **A 333** (1980) 285

[Br 61a] *Britt, H.C.* and *A.R. Quinton*: Phys. Rev. **124** (1961) 877

[Br 61] *Brody, T.* and *M. Moshinsky*: Tables of Transformation Brackets for Nuclear Shell Model Calculations, Universidad Nacional Autonoma de Medico (1961)

[Br 66] *Brown, G.E.* and *A.M. Green*: Nucl. Phys. **75** (1966) 401

[Br 68] *Brown, G.E.* and *A.M. Green*: Nucl. Phys. **85** (1968) 87

[Br 70] *Brink, D.M.*: Phys. Lett. **40 B** (1970) 37

[Br 72] *Brueckner, K.A., H. Meldner* and *J.D. Perez*: Phys. Rev. C 6 (1972) 773

[Br 74a] *Braun-Munzinger, P.* and *H.L. Harney*: Nucl. Phys. **A 223** (1974) 381

[Br 74b] *Braun-Munzinger, P., H.L. Harney* and *S. Wenneis*: Nucl. Phys. **A 235** (1974) 190

[Br 75] *Brink, D.M.* and *F. Stancu*: Nucl. Phys. **A 243** (1975) 175

[Br 79a] *Bradlow, H.S., W.D.M. Rae, P.S. Fisher, N.S. Godwin, G. Proudfoot* and *D. Sinclair*: Nucl. Phys. **A 314** (1979) 171

[Br 79b] *Bradlow, H.S., W.D.M. Rae, P.S. Fisher, N.S. Godwin, G. Proudfoot* and *D. Sinclair*: Nucl. Phys. **A 314** (1979) 207

[Br 79] *Brenner, D.J., Ph.D. Thesis*: University of Surrey (1979), published as Rutherford Laboratory Report RL-79-032

[Br 79c] *Brink, D.M.*: in Microscopic Optical Potentials (ed H.V. von Geramb), Lecture Notes in Physiçs No. 89 (Springer-Verlag, Berlin) (1979) 340

[Bu 66] *Buttle, P.J.A.* and *L.J.B. Goldfarb*: Nucl. Phys. **78** (1966) 409

[Bu 68] *Buttle, P.J.A.* and *L.J.B. Goldfarb*: Nucl. Phys. **A 115** (1968) 461

[Bu 75] *Buck, B., C.B. Dover* and *J.P. Vary*: Phys. Rev. C 11 (1975) 1803

[Bu 76] *Buenerd, M., C.K. Gelbke, B.G. Harvey, D.L. Hendrie, J. Mahoney, A. Menchaca-Rocha, C. Olmer* and *D.K. Scott*: Phys. Rev. Lett. **37** (1976) 1191

[Bu 77a] *Buck, B., H. Fredrich* and *C. Wheatley*: Nucl. Phys. **A 275** (1977) 246

[Bu 77b] *Buck, B.* and *A.A. Pilt*: Nucl. Phys. **A 280** (1977) 133

[Ca 78] *Castenada, C.M., H.A. Smith, P.P. Singh, J. Jastrzebski* and *H. Karkowski*: Phys. Lett. **77 b** (1978) 371

[Ch 75] *Chant, N.S.* and *P.G. Roos* in Proc. Second. Int. Conf. on Clustering Phenomena in Nuclei (eds D.A. Goldberg et al), ERDA Technical Information Centre, Oak Ridge, Tennessee

[Ch 76] *Chirapatpimol, N., J.C. Fong, M.M. Gazzaly, G. Igo, A.D. Liberman, R.J. Ridge, S.L. Verbeck, C.A. Whitten, D.G. Kovar, V. Perez-Mendez, N.S. Chant* and *P.G. Roos*: Nucl. Phys. **A 264** (1976) 379

[Ch 77] *Chant, N.S.* and *P.G. Roos*: Phys. Rev. **C 15** (1977) 57

[Ch 78] *Chant, N.S., P.G. Roos* and *C.W. Wang*: Phys. Rev. **C 17** (1978) 8

[Ch 78a] *Chung, W., J. van Hienen, B.H. Wildenthal* and *C.L. Bennett*: Phys. Lett. **79 B** (1978) 381

[Ch 79] *Chant, N.S., P. Kitching, P.G. Roos* and *L. Antonuk*: Phys. Rev. Lett. **43** (1979) 495

[Cl 68] *Clement, D.M.* and *E.U. Baranger*: Nucl. Phys. **A 120** (1968) 25

[Co 73] *Coelho, H.T.*: Phys. Rev. **C 7** (1973) 2340

[Co 76] *Cossairt, J.D., R.D. Bent, A.S. Broad, F.D. Becchetti* and *J. Jänecke*: Nucl. Phys **A 261** (1976) 373

[Cr 76] *Cramer, J.G., R.M. DeVries, D.A. Goldberg, M.S. Zisman* and *C.F. Maguire, Phys. Rev.* **C 14** (1976) 2158

[Da 76] *Davies, W.G., R.M. DeVries, G.C. Ball, J.S. Forster, W. McLatchie, D. Shapira, J. Toke* and *R.E. Warner*: Nucl. Phys. **A 269** (1976) 477

[Da 77] *Darwisch, N.Z., K.A. Gridnev, E.F. Hefter* and *V.M. Semjonov*: Nuo. Cim. **42 A** (1977) 303

[De 68] *Deutchman, P.A.* and *I.E. McCarthy*: Nucl. Phys. **A 112** (1968) 399

[De 69] *Detraz, C., H.H. Duhm, H. Hafner* and *M. Yoshida*: Proceedings Int. Conf. on Nuclear Reactions Induced by Heavy Ions (eds Dost and W. Hering) North-Holland (1969) 319

[DeV 73] *DeVries, R.M.*: Phys. Rev. **C 8** (1973) 951

[DeV73a] *DeVries, R.M.*: Nucl. Phys. **A 212** (1973) 207

[DeV75] *DeVries, R.M.*: Phys. Rev. **C 11** (1975) 2105

[DeV76] *DeVries, R.M., J.S. Lilley* and *M.A. Franey*: Phys. Rev. Lett. **37** (1976) 481

[De 78] *Devaux, A., G. Landaud, J. Yonnet, P. Delpierre* and *J. Kanane* in Clustering Aspects of Nuclear Structure and Nuclear Reactions (eds W.H.T. van Oers et al) AIP Conf. Proceedings No. 47 (1978) 634

[Do 65] *Dodd, L.R.* and *K.R. Greider*: Phys. Rev. Lett. **14** (1965) 671

[Do 66] *Dodd, L.R.* and *K.R. Greider*: Phys. Rev. **146** (1966) 675

[Do 69] *Dodd, L.R.* and *K.R. Greider*: Phys. Rev. **180** (1969) 1187

[Do 79] *Dollhopf, W.E., C.F. Perdrisat, P. Kitching* and *W.C. Olsen*: Nucl. Phys. **A 316** (1979) 350

[Dr 66] *Drisko, R.M.* and *F. Rybicki*: Phys. Rev. Lett. **16** (1966) 275

[Dr 75] *Draayer, J.P.*: Nucl. Phys. **A 237** (1975) 157

[Dy 79] *Dyer, P., T.C. Awes, C.K. Gelbke, B.B. Back, A. Mignerey, K.L. Wolf, H. Breuer, V.E. Viola* and *W.G. Meyer*: Phys. Rev. Lett. **42** (1979) 560

[Ei 77] *Eisen, Y.* and *B. Day*: Phys. Lett. **63 B** (1977) 253

[El 55] *Elliott, J.P.* and *T.H.R. Skyrme*: Proc. Roy. Soc. **A 232** (1955) 561

[El 67] *Elton, L.R.B.* and *A. Swift*: Nucl. Phys. **A 94** (1967) 52

[Es 79] *Eswaran, M.A., H.E. Gove, R. Cook* and *B. Sikora*: Nucl. Phys. **A 325** (1979) 269

[Fa 61] *Fadeev, L.D.*: Sov. Phys. (JETP) **12** (1961) 1014

[Fe 58] *Feshbach, H.*: Ann. Phys. **5** (1958) 357

[Fe 66] *Fieldeldey, H.*: Nucl. Phys. **77** (1966) 149

[Fe 73] *Feng, D.H., R.H. Ibarra* and *M. Vallieres*: Phys. Lett. **46 B** (1973) 37

[Fe 73a] *Feshbach, H.*: in Reaction Dynamics (Gordon and Breach, New York) (1973) 171

[Fe 73b] *Feshbach, H.* and *K. Huang*: Phys. Lett. **47 B** (1973) 300

[Fl 75] *Fliessbach, T.*: Z. Physik **A 272** (1975) 39

[Fl 76] *Fliessbach, T.*: J. Phys. G: Nucl. Phys. **2** (1976) 531

[Fl 76a] *Fliessbach, T.* and *H.J. Mang*: Nucl. Phys. **A 263** (1976) 75

[Fl 78] *Fliessbach, T.*: Z. Physik **A 288** (1978) 211

[Fl 79] *Fliessbach, T.*: Nucl. Phys. **A 315** (1979) 109

[Fl 79a] *Flowers, A.G., D. Branford, J.C. McGeorge, A.C. Shotter, P. Thorley, C.H. Zimmerman, R.O. Owens* and *J.S. Pringle*: Phys. Rev. Lett. **43** (1979) 323

[Fr 67] *Frahn, W.E.*: in Fundamentals in Nuclear Theory (IAEA, Vienna) (1967) 3

[Fr 75] *Friedrich, H.* and *K. Langanke*: Nucl. Phys. **A 252** (1975) 47

[Ga 28] *Gamow, G.*: Zeits Physik **51** (1928) 204; ibid **52** (1928) 510

[Ga 70] *Gaillard, P., M. Chevalier, J.Y. Grossiord, A. Guichard, M. Gusakow* and *J.R. Pizzi*: Phys. Rev. Lett. **25** (1970) 593

[Ge 76] *Gelbke, C.K., M. Buenerd, D.L. Hendrie, J. Mahoney, M.C. Mermaz, C. Olmer* and *D.K. Scott*: Phys. Lett. **65** (1976) 227

[Ge 77] *Gelbke, C.K., M.K. Scott, M. Bini, D.L. Hendrie, J-L Laville, J. Mahoney, M.C. Mermaz* and *C. Olmer*: Phys. Lett. **70 B** (1977) 415

[Ge 78] *Gelbke, C.K., C. Olmer, M. Buenerd, D.L. Hendrie, J. Mahoney, M.C. Mermaz* and *D.K. Scott*: Phys. Reports **42** (1978) 311

[Gi 79] *Gils, H.J.*: in What Do We Know about the Radial Shape of Nuclei in the Ca-Region (eds H. Rebel et al) (Kernforschungszentrum Karlsruhe) (1979) 124

[Gl 59] *Glauber, R.J.*: in Lectures in Theoretical Physics, Vol. I (Interscience, London & New York) (1959) 315

[Gl 65] *Glendenning, N.* and *K. Harada*: Nucl. Phys. **72** (1965) 481

[Go 58] *Gottfried, K.*: Nucl. Phys. **5** (1958) 557

[Go 64] *Goldberger, M.L.* and *K.M. Watson*: Collision Theory, Wiley New York (1964), Chap. 8

[Go 70] *Goldring, G., M. Samuel, B.A. Watson, M.C. Bertin* and *S.L. Tabor*: Phys. Lett. **32 B** (1970) 465

[Go 74] *Goldberg, D.A., S.M. Smith* and *G.F. Burdzik*: Phys. Rev. **C 10** (1974) 1362

[Go 74a] *Goldhaber, A.S.*: Phys. Lett. **53 B** (1974) 306

[Go 76] *Golin, M., F. Petrovich* and *D. Robson*: Phys. Lett. **64 B** (1976) 253

[Go 79] *Gonthier, P., H. Ho, M.N. Namboodiri, J.B. Natowitz, L. Adler, O. Hartin, P. Kasiraj, A. Khodai, S. Simon* and *K. Hagel*: Progress Report 1978/79, Cyclotron Laboratory, Texas A & M University (1979) 7

[Gr 66] *Greider, K.* and *L.R. Dodd*: Phys. Rev. **146** (1966) 671

[Gr 75] *Greiner, D.E., P.J. Lindstrom, H.H. Heckman, B. Cork* and *F.S. Beiser*: Phys. Rev. Lett. **35** (1975) 152

[Gr 77] *Grossiard, J.Y., M. Bedjidian, A. Guichard, M. Gusakow, J.R. Pizzi, T. Delbar, G. Gregoire* and *J. Lega*: Phys. Rev. **C 15** (1977) 843

[Gu 71] *Guichard, A., M. Chevallier, P. Gaillard, J.Y. Groosiord, M. Gusakow, J.P. Pizzi* and *C. Ruhla*: Phys. Rev. **C 4** (1971) 700

[Gu 71a] *Gutbrod, H.H., H. Hoshida* and *R. Bock*: Nucl. Phys. **A 165** (1971) 240

[Gu 79] *Gubler, H.P., G.R. Plattner* and *I. Sick*: in What Do We Know About Radial Shapes of Nuclei in the Ca Region (eds H. Rebel et al) (Kernforschungszentrum, Karlsruhe) (1979) 210

[Ha 68] *Harada, K.* and *E.A. Rauscher*: Phys. Rev. **169** (1968) 818

[Ha 77] *Harris, J.W., T.M. Cormier, D.F. Geesamen, L.L. Lee, R.L. McGrath* and *J.P. Wurm*: Phys. Rev. Lett. **38** (1977) 1460

[Ha 78] *Hakansson, H-B, T. Berggren* and *R. Bengtsson*: Nucl. Phys. **A 306** (1978) 406

[Ha 78a] *Hanson, D.L., N. Stein, J.W. Sunier, C.W. Woods* and *R.R. Betts*: in Clustering Aspects of Nuclear Structure and Nuclear Reactions (eds W.H.T. van Oers et al) AIP Conf. Proceedings No. **47** (1978) 716

[Ha 79] *Hanson, D.L., N. Stein, J.W. Sunier, C.W. Woods* and *O. Hansen*: Nucl. Phys. **A 321** (1979) 471

[He 72] *Heckman, H.H., D.E. Greiner, P.J. Lindstrom* and *F.S. Beiser*: Phys. Rev. Lett. **28** (1972) 926

[He 75] *Hecht, K.T.* and *D. Braunschweig*: Nucl. Phys. **A 244** (1975) 365

[He 79] *Hecht, K.* and *W. Zahn*: Nucl. Phys. **A 318** (1979) 1

[Ho 77] *Ho, H., R. Albrecht, W. Dünnweber, G. Graw, S.G. Steadman* and *J.P. Wurm*: Z. Physik **A 283** (1977) 235

[Ia 75] *Iano, P.J.* and *W.T. Pinkston*: Nucl. Phys. **A 237** (1975) 189

[Ib 70] *Ibarra, R.H.* and *B. Bayman*: Phys. Rev. **C 1** (1970) 1786

[Ib 75] *Ibarra, R.H., M. Vallieres* and *D.H. Feng*: Nucl. Phys. **A 241** (1975) 386

[Ic 73] *Ichimura, M., A. Arima, E.C. Halbert* and *T. Terasawa*: Nucl. Phys. **A 204** (1973) 225

[Ig 69] *Igarishi, M., M. Kawai* and *K. Yazaki*: Prog. Theor. Phys. **42** (1969) 254

[In 77] *Inamura, T., M. Ishihara, T. Fukuda, T. Shimoda* and *H. Hirata*: Phys. Lett. **68 B** (1977) 51

[In 79] *Ingemarsson, A., O. Jonsson* and *A. Hallgren*: Nucl. Phys. **A 319** (1979) 377

[Io 76] *Ioannides, A.A.* and *R.C. Johnson*: Phys. Lett. **61 B** (1976) 4

[Io 78] *Ioannides, A.A.* and *R.C. Johnson*: Phys. Rev. **C 17** (1978) 1331

[Ja 65] *Jackson, D.F.*: Rev. Mod. Phys. **37** (1965) 393

[Ja 65a] *Jackson, D.F.* and *T. Berggren*: Nucl. Phys. **62** (1965) 353

[Ja 67a] *Jackson, D.F.*: Nuo. Cim. **51 B** (1967) 49

[Ja 69] *Jaffe, R.L.* and *W.J. Gerace*: Nucl. Phys. **A 125** (1969) 1

[Ja 69a] *Jain, A.K., N. Sarma* and *B. Banerjee*: Nuo. Cim. **62** (1969) 219

[Ja 70] *Jackson, D.F.*: Nuclear Reactions (Methuen, London) 1970

[Ja 70a] *Jain, A.K., N. Sarma* and *B. Banerjee*: Nucl. Phys. **A 142** (1970) 330

[Ja 70b] *Jain, M., P.G. Roos, H.G. Pugh* and *H.D. Holmgren*: Nucl. Phys. **A 153** (1970) 49

[Ja 73] *Jain, A.K., J.Y. Groosiord, M. Chevalier, P. Gaillard, A. Guichard, M. Gusakow* and *J.R. Pizzi*: Nucl. Phys. **A 216** (1973) 519

[Ja 74] *Jackson, D.F.* and *R.C. Johnson*: Phys. Lett. **49 B** (1974) 249

[Ja 74a] *Janus, R.T.* and *I.E. McCarthy*: Phys. Rev. **C 10** (1974) 1041

[Ja 76] *Jackson, D.F.* and *M. Rhoades-Brown*: Nucl. Phys. **A 266** (1976) 61

[Ja 76a] *Jackson, D.F.*: Nucl. Phys. **A 257** (1976) 221

[Ja 77a] *Jackson, D.F.*: Phys. Lett. **71 B** (1977) 57

[Ja 77b] *Jackson, D.F.* and *M. Rhoades-Brown*: Nucl. Phys. **A 286** (1977) 354

[Ja 77c] *Jackson, D.F.* and *M. Rhoades-Brown*: Annals of Phys. **105** (1977) 151

[Ja 77d] *Jackson, D.F.* and *M. Rhoades-Brown*: Nature **267** (1977) 593

[Ja 78] *Jackson, D.F.* and *M. Rhoades-Brown*: J. Phys. G: Nucl. Phys. **4** (1978) 1441

[Ja 78a] *Jackson, D.F.*: J. Phys. G: Nucl. Phys. **4** (1978) 1287

[Ja 81] *Jackson, D.F.* and *D.J. Brenner*: in Progress in Particle and Nuclear Physics (ed D.H. Wilkinson) (Pergamon) **5** (1981) 143

[Ja 79a] *Jänecke, J.*: in Clustering Aspects of Nuclear Structure and Nuclear Reaktions (eds W.H.T. van Oers et al) AIP Conf. Proceedings No. **47** (1979) 324

[Ja 79b] *Jänecke, J., F.D. Becchetti* and *C.E. Thorn*: Nucl. Phys. **A 325** (1979) 337

[Ja 79c] *Jackson, D.F.* and *K. Ahmad*: in Proceedings of the 8th Int. Conf. on High Energy Physics and Nuclear Structure, contributed paper

[Ja 79d] *Jain, B.K.*: Nucl. Phys. **A 314** (1979) 51

[Ja 79e] *Jain, A.K.* and *N. Sarma*: Nucl. Phys. **A 321** (1979) 429

[Ja 79f] *Jackson, D.F., E.J. Wolstenholme, L.S. Julien* and *C.J. Batty*: Nucl. Phys. **A 316** (1979) 1

[Ja 79g] *Jackson, D.F.* and *E.J. Wolstenholme*: in Clustering Aspects of Nuclear Structure and Nuclear Reactions (eds W.H.T. van Oers et al) AIP Conf. Proceedings No. **47** (1979) 530

[Ja 80] *Jänecke, J., F.D. Bechetti* and *D. Overway*: Nucl. Phys., **A 343** (1980) 161

[Jo 70] *Johnson, R.C.* and *P.J.R. Soper*: Phys. Rev. **C 1** (1970) 976

[Jo 72] *Johnson, R.C.* and *P.J.R. Soper*: Nucl. Phys. **A 182** (1972) 619

[Ka 69] *Kawai, M.* and *K. Yazaki*: Prog. Theor. Phys. **37** (1967) 638

[Ka 70] *Kadmenskii, S.G.* and *V.E. Kalechits*: Yad. Fiz. **12** (1970) 70

[Ka 70a] *Kazaks, P.A.* and *R.D. Koshel*: Phys. Rev. **C 1** (1970) 1906

[Ka 76] *Karban, O., A.K. Basak, J.A.R. Griffiths, S. Roman* and *G. Tungate, Nucl. Phys.* **A 266** (1976) 413

[Kn 75] *Knutsen, L.D.* and *W. Haeberli*: Phys. Rev. **C 12** (1975) 1469

[Ko 66] *Kolybasev, V.M.*: Sov. J. Nucl. Phys. **3** (1966) 535; ibid **3** (1966) 704

[Ko 74] *Koepke, J.A., R.E. Brown, Y.C. Tang* and *D.R. Thompson*: Phys. Rev. **C 9** (1974) 823

[Ko 79] *Koontz, R.W., C.C. Chang, H.D. Homgren* and *J.R. Wu*: Phys. Rev. Lett. **43** (1979) 1862

[Ku 72] *Kubo, K.I.* and *M. Hirata*: Nucl. Phys. **A 187** (1972) 186

[Ku 73] *Kurath, D.*: Phys. Rev. **C 7** (1973) 1390

[Ku 74] *Kurath, D.* and *I.S. Towner*: Nucl. Phys. **A 222** (1974) 1

[Ku 75] *Kukulin, V.I., V.G. Neudatchin* and *Yu F. Smirnov*: Nucl. Phys. **A 245** (1975) 429

[La 58] *Lane, A.M.* and *R.G. Thomas*: Rev. Mod. Phys. **30** (1958) 257

[Le 51] *Levinger, J.S.*: Phys. Rev. **84** (1951) 43

[Le 60] *Levinger, J.S.*: Nuclear Photo-disintegration (Oxford: Clarendon Press)

[Le 66] *Levin, F.S.*: in Reaction Dynamics (Gordon and Breach, New York (1973) 3

[Le 77] *Le Mere, M., R.E. Brown, Y.C. Tang* and *D.R. Thompson*: Phys. Rev. **C 15** (1977) 1191

[Le 78] *Leung, S.W-L.* and *H. Sherif*: Can. J. Phys. **56** (1978) 1116

[Le 79a] *Le Mere, M., E.J. Kanellopoulos, W. Sünkel* and *Y.C. Tang*: Phys. Lett. **87 B** (1979) 311

[Le 79b] *Le Mere, M., D.J. Stubeda, H. Horiuchi* and *Y.C. Tang*: Nucl. Phys. **A 320** (1979) 449

[Li 64] *Lim, K.* and *I.E. McCarthy*: Phys. Rev. **133** (1964) B 1006

[Li 66] *Lim, K.* and *I.E. McCarthy*: Nucl. Phys. **88** (1966) 433

[Li 73] *Liebert, T.B., K.H. Purser* and *R.L. Burman*: Nucl. Phys. **A 216** (1973) 335

[Li 75] *Lindstrom, P.J., D.E. Greiner, H.H. Heckman, B. Cork* and *F.S. Beiser*: LBL-3650 (1975)

[Lo 64] *Lovelace, C.*: Phys. Rev. **135** (1964) B 1225

[Lo 79] *Love, W.G.*: in Microscopic Optical Potentials (ed H.V. von Geramb), Lecture Notes in Physics No. 89 (Springer-Verlag, Berlin) (1979) 350

[Lu 75] *Lukyanov, V.K.* and *A.I. Titov*: Phys. Lett. **57 B** (1975) 10

[Ma 64] *Mang, H.J.*: Ann. Rev. Nucl. Sci. **14** (1964) 1

[Ma 68] *Mathews, J.L., W. Bertozzi, S. Kowalski, C.P. Sargent* and *W. Turchinetz*: Nucl. Phys. **A 112** (1968) 654

[Ma 72b] *Marchese, C.J., N.M. Clarke* and *R.J. Griffiths*: Phys. Rev. Lett. **29** (1972) 660

[Ma 72a] *Malmin, R.E.*: Argonne National Laboratory Report PHY-1972F (1972)

[Ma 74] *MacDonald, W.M.*: Nucl. Phys. **54** (1964) 393

[Ma 78] *Macfarlane, M.*: in Nuclear Structure Physics (eds S.J. Hall and J.M. Irvine), Scottish Universities Summer School in Physics (1978) 1

[Ma 78a] *Mallet-Lemaire, M-C.*: in Clustering Aspects of Nuclear Structure and Nuclear Reactions (eds W.H.T. van Oers et al) AIP Conf. Proceedings No. **47** (1978) 271

[Ma 78b] *Majka, Z.*: Phys. Lett. **76 B** (1978) 161

[Ma 78c] *Majka, Z., J.J. Gils* and *H. Rebel*: Z. Physik **A 288** (1978) 139

[Ma 78d] *Mallett, P.R.*: M. Phil. Thesis, University of Surrey (1978) unpublished

[McC61] *McCarthy, I.E.* and *D.L. Pursey*: Phys. Rev. **122** (1961) 578

[McC68] *McCarthy, I.E.*: Introduction to Nuclear Theory (Wiley, New York) (1968)

[McC69] *McCarthy, I.E.* and *A.W. Thomas*: Nucl. Phys. **A 135** (1969) 463

[Me 69] *Meboniya, J.V.*: Phys. Lett. **30 B** (1969) 153

[Me 71] *Meldner, H.* and *J.D. Perez*: Phys. Rev. **A 4** (1971) 1388

[Mi 71] *Miller, H.G., W. Buss* and *J.A. Rawlins*: Nucl. Phys. **A 163** (1971) 637

[Mi 77] *Michel, F.* and *R. Vanderpoorten*: Phys. Rev. **C 15** (1977) 142

[Mi 77a] *Milder, F.L., J. Jänecke* and *F.D. Becchetti*: Nucl. Phys. **A 276** (1977) 72

[Mo 78] *Moszkowski, S.A.*: Nucl. Phys. **A 309** (1978) 273

[Mo 79] *Motobayashi, T., I. Kohno, T. Ooi* and *S. Namajima*: Nucl. Phys. **A 331** (1979) 193

[Na 81] *Nadesen, A., P. Schwandt, P.P. Singh, A.D. Bacher, P.T. Debevec, W.W. Jacobs, M.D. Kaitchuk* and *J.T. Meek*: Phys. Rev. **C 23** (1981) 1023

[Ne 71] *Neudatchin, V.G., V.I. Kukulin, V.L. Korotkikh* and *V.P. Korennoy*: Phys. Lett. **34 B** (1971) 581

[Ne 72] *Negele, J.* and *D. Vautherin*: Phys. Rev. **C 5** (1972) 1472

[Ng 75] *Ngo, C., B. Tamain, M. Beiner, R.J. Lombard, D. Mas* and *H.H. Deubler*: Nucl. Phys. **A 252** (1975) 237

[No 78] *Noble, J. V.* and *H.T. Coelho*: Phys. Rev. **C 3** (1971) 1840

[No 78] *Nomura, T., H. Utsunomiya, T. Motobayashi, T. Inamura* and *M. Yanokura*: Phys. Rev. Lett. **40** (1978) 694

[Oe 70] *von Oertzen, W.*: Nucl. Phys. **A 148** (1970) 529

[Oe 78] *Oelert, W., A. Djaloeis, C. Mayer-Böricke, P. Turek* and *S. Wiktor*: Nucl. Phys. **A 306** (1978) 1

[Oe 78a] *Oelert, W., A. Djaloeis, C. Mayer-Böricke, P. Turek* and *S. Wiktor*: in Clustering Aspects of Nuclear Structure and Nuclear Reactions (eds W.H.T. van Oers et al) AIP Conf. Proceedings No. **47** (1978) 708

[Oe 79] *Oelert, W., M. Betigeri, W. Chung, A. Djaloeis, C. Mayer-Böricke* and *P. Turek*: Nucl. Phys. **A 329** (1979) 192

[Oe 79a] *Oelert, W., W. Chung, M. Betigeri, A. Djaloeis, C. Mayer-Böricke* and *P. Turek*: Phys. Rev. **C 20** (1979) 459

[Pa 72] *Park, J. Y., W. Scheid* and *W. Greiner*: Phys. Rev. **C 6** (1972) 1565

[Pa 75] *Partridge, R.A., R.E. Brown, Y.C. Tang* and *D.R. Thompson*: in Clustering Phenomena in Nuclei II (ed *D.A. Goldberg, J.B. Marion* and *S.J. Wallace*) ORO-4856-25 Nat. Tech. Inf. Service, US Department of Commerce (1975) 225

[Pa 76] *Partridge, R.A., Y.C. Tang, D.E. Thompson* and *R.E. Brown*: Nucl. Phys. **A 273** (1976) 341

[Pe 62] *Perey, F.G.* and *B. Buck*: Nucl. Phys. **32** (1962) 353

[Pe 63] *Perey, F.G.*: in Direct Interactions and Nuclear Reaction Mechanisms (ed E. Clementel and C. Villi) (Gordon and Breach, New York) (1963) 125

[Pe 67] *Perey, F.G.* and *G.R. Satchler*: Nucl. Phys. **A 97** (1967) 515

[Pe 75] *Perkin, D.G., A.M. Kobos* and *J.R. Rook*: Nucl. Phys. **A 245** (1975) 343

[Pe 76] *Peng, J.C., J.V. Maher, W. Delert, D.A. Sink, C.M. Cheng* and *H.S. Song*: Nucl. Phys. **A 264** (1976) 312

[Ph 68] *Philpott, R.J., W.T. Pinkston* and *G.R. Satchler*: Nucl. Phys. **A 119** (1968) 241

[Pi 65] *Pinkston, W.T.* and *G.R. Satchler*: Nucl. Phys. **72** (1965) 641

[Pi 69] *Pizzi, J.R., M. Gaillard, P. Gaillard, A. Guichard, M. Gusakow, G. Reboulet* and *C. Ruhla*: Nucl. Phys. **A 136** (1969) 496

[Pi 76] *Pinkston, W.T.*: Nucl. Phys. **A 269** (1976) 281

[Pi 77] *Pinkston, W.T.*: Nucl. Phys. **A 291** (1977) 342

[Pi 79] *Pilt, A.A.*: in Clustering Aspects of Nuclear Structure and Nuclear Reactions (ed W.H.T. van Oers et al) AIP Conf. Proceedings No. **47** (1979) 173

[Pi 79a] *Pinkston, W.T.* and *P.J. Iano*: Nucl. Phys. **A 330** (1979) 91

[Pr 69] *Prakash, A.* and *N. Austern*: Ann. of Phys. **51** (1969) 418

[Pu 69] *Pugh, H.G., J.W. Watson, D.A. Goldberg, P.G. Roos, D.I. Bonbright* and *R.A.J. Riddle*: Phys. Rev. Lett. **22** (1969) 408

[Pu 79] *Put, L.W.*: in Microscopic Optical Potentials (ed H.V. von Geramb), Lecture Notes in Physics No. 89 (Springer-Verlag, Berlin) (1979) 302

[Ra 63] *Rasmussen, J.O.*: Nucl. Phys. **44** (1963) 93

[Ra 65] *Rasmussen, J.O.*: in Alpha-, Beta- and Gamma-Ray Spectroscopy (ed K. Siegbahn) (North-Holland, Amsterdam) (1965) 701

[Ra 67] *Rauscher, E.A., J.O. Rasmussen* and *K. Harada*: Nucl. Phys. **A 94** (1967) 33

[Ra 68] *Rawlins, J.A., C. Glavina, S.H. Ku* and *Y.M. Shin*: Nucl. Phys. **A 122** (1968) 128

[Ra 75] *Rawitscher, G.*: Phys. Rev. **C 11** (1975) 1152

[Re 70] *Redish, E.F., G.J. Stephenson* and *G.M. Lerner*: Phys. Rev. **C 2** (1970) 1665

[Re 74] *Redish, E.F.*: Nucl. Phys. **A 235** (1974) 82

[Re 77] *Redish, E.F.*: in Momentum Wave Functions – 1976 (ed D.W. Devins) AIP Conf. Proceedings No. **36** (1977) 111

[Rh 77] *Rhoades-Brown, M., Ph.D. Thesis*: University of Surrey (1977) unpublished

[Ro 68] *Rotter, I.*: Nucl. Phys. **A 122** (1968) 567

[Ro 69] *Rotter, I.*: Nucl. Phys. **A 135** (1969) 378

[Ro 69a] *Roos, P.G., H. Kim, M. Jain* and *H.D. Holmgren*: Phys. Rev. Lett. **22** (1969) 242

[Ro 76] *Roos, P.G., D.A. Goldberg, N.S. Chant, R. Woody* and *W. Reichert*: Nucl. Phys. **A 257** (1976) 317

[Ro 77] *Roos, P.G., N.S. Chant, A.A. Cowley, D.A. Goldberg, H.D. Holmgren* and *R. Woody*: Phys. Rev. **C 15** (1977) 69

[Ro 79] *Rotter, I.*: preprint

[Ru 78] *Rust, N.J.A., M.R. Clover, R.M. DeVries, R. Ost, R.N. Cherry* and *H.E. Gove*: in Clustering Aspects of Nuclear Structure and Nucelar Reactions (eds W.H.T. van Oers et al) AIP Conf. Proceedings No. **47** (1978) 718

[Sa 69] *Saito, S.*: Prog. Theor. Phys. **41** (1969) 705

[Sa 73 a] *Saito, S., S. Okai, R. Tamagaki* and *M. Yasuno*: Prog. Theor. Phys. **50** (1973) 1561

[Sa 73] *Santos, F.D.*: Nucl. Phys. **A 212** (1973) 341

[Sa 74] *Santos, F.D.*: Phys. Lett. **48 B** (1974) 193

[Sa 75] *Satchler, G.R.*: Phys. Lett. **58 B** (1975) 4

[Sa 76] *Satchler, G.R.* and *W.G. Love*: Phys. Lett. **65 B** (1976) 415

[Sa 78] *Sandulescu, A., I. Silisteanu* and *R. Wünsch*: Nucl. Phys. **A 305** (1978) 205

[Sc 69] *Schaeffer, R.*: Nucl. Phys. **A 132** (1969) 186

[Sc 74] *Schier, H.* and *B. Schoch*: Nucl. Phys. **A 229** (1974) 93

[Sc 78] *Scott, D.K.*: in Nuclear Structure Physics (eds S.J. Hall and J.M. Irvine), Scottish Universities Summer School in Physics (1978) 557

[Sc 78a] *Scott, D.K., M. Bini, P. Doll, C.K. Gelbke, D.L. Hendrie, J-L Laville, J. Mahoney, A. Menchaca-Rocha, M.C. Mermaz, C. Olmer, T.J.M. Symons, Y.P. Viyogi, K. Van Bibber, H. Wieman* and *P.J. Siemens*: LBL-7729 (1978)

[Sh 65] *Shapiro, I.S., C.M. Kolybasov* and *J.P. Augst*: Nucl. Phys. **61** (1965) 353

[Sh 68] *Scherk, L.* and *E.W. Vogt*: Canad. J. Phys. **46** (1968) 1119

[Sh 79] *Sherman, J.D., D.L. Hendrie* and *M.S. Zisman*: Phys. Rev. **C 13** (1976) 20

[Si 72] *Siemssen, R.H., H.F. Fortune, A. Richter* and *J.W. Tippie*: Phys. Rev. **C 5** (1972) 1839

[Si 79] *Sinha, B.*: in Microscopic Optical Potentials (ed H.V. von Geramb), Lecture Notes in Physics No. **89** (Springer-Verlag, Berlin) (1979) 372

[Si 79a] *Siwek-Wilczynska, K., E.H. du Marchie van Voorthuysen, J. van Popta, R.H. Siemssen* and *J. Wilczynski*: Nucl. Phys. A. **A 330** (1979) 150

[Sm 61] *Smirnov, Yu F.*: Nucl. Phys. **27** (1961) 177

[Sm 62] *Smirnov, Yu F.*: Nucl. Phys. **39** (1962) 346

[Sp 75] *Sprung, D., M. Vallieres, X. Campi* and *Ko Che-Ming*: Nucl. Phys. **A 253** (1975) 1

[St 78] *Stubeda, D.J., M. Le Mere* and *Y.C. Tang*: Phys. Rev. **C 17** (1978) 447

[Su 77] *Suzuki, Y.*: in Int. Pre-Symposium on Clustering Phenomena in Nuclei (contributed papers) (1977) 6

[Su 79] *Sünkel, W.* and *Y.C. Tang*: Nucl. Phys. **A 329** (1979) 10

[Su 79a] *Sugihawa, T.T.*: Progress Report 1978/79, Cyclotron Laboratory, Texas A & M University (1979) 32

[Sw 67] *Swan, P.*: Phys. Rev. Lett. **19** (1967) 245

[Sy 78] *Symons, T.J.M., P. Doll, M. Bini, D.L. Hendrie, J. Mahoney, G. Mantzouris, D.K. Scott, K. Van Bibber, Y.P. Viyogi, H.H. Wieman* and *C.K. Gelbke*: LBL-8379 (1978)

[Ta 52] *Talmi, I.*: Helv. Phys. Acta **25** (1952) 185

[Ta 62] *Tang, Y.C., K. Wildermuth* and *L.D. Pearlstein*: Nucl. Phys. **32** (1962) 504

[Ta 74] *Tamura, T.* and *K.S. Low*, Computer Phys. Comm. **8** (1974) 349

[Ta 78b] *Takimoto, K., R. Wada, E. Takada, M. Fukada, T. Yamaya, K. Umeda, H. Endo, T. Suehiro, J. Schimizu* and *Y. Ohkuma*: in Clustering Aspects of Nuclear Structure and Nuclear Reactions (eds W.H.T. van Oers et al) AIP Conf. Proceedings No. **47** (1978) 710

[Ta 78] *Tang, Y.C.* and *L. Le Mere*: Phys. Reports **47** (1978) 167

[Ta 78a] *Tanimura, O.*: Nucl. Phys. **A 309** (1978) 233

[Ta 79] *Tang, Y.C.*: in Microscopic Optical Potentials (ed H.V. von Geramb), Lecture Notes in Physics No. **89** (Springer-Verlag, Berlin) (1979) 322

[Ta 79a] *Tanimura, O.* and *T. Tazawa*: Nucl. Phys. **A 321** (1979) 490

[Te 52] *Teichmann, T.* and *E.P. Wigner*: Phys. Rev. **87** (1952) 123

[Th 54] *Thomas, R.G.*: Prog. Theor. Phys. **12** (1954) 253

[To 61] *Tobocman, W.*: Theory of Direct Nuclear Reaction (Oxford University Press (1961)

[To 76] *Tobocman, W.*: Phys. Rev. **13** (1976) 790

[To 77] *Tohsaki-Suzuki, A., K. Naito, T. Ando* and *K. Ikeda*: in Int. Pre-Symposium on Clustering Phenomena in Nuclei (Contributed Papers) (1977) 52

[To 77a] *Towner, I.S.*: A Shell Model Description of Light Nuclei (Oxford University Press) (1977)

[To 78] *Tobocman, W.*: Phys. Rev. **17** (1978) 2205

[To 79] *Tonozuka, I.* and *A. Arima*: Nucl. Phys. **A 323** (1979) 45

[Ud 79] *Udagawa, T., T. Tamura, T. Shimoda, H. Frölich, M. Ishihara* and *K. Nagatani*: Phys. Rev. **C 20** (1979) 1949

[Va 76] *Vallieres, M., D.H. Feng* and *R.H. Ibarra*: Nucl. Phys. **A 256** (1976) 21

[Va 79] *Vaagen, J.S., B.S. Nilsson, J. Bang* and *R.H. Ibarra*: Nucl. Phys. **A 319** (1979) 143

[Vi 73] *Vincent, C.M.*: Phys. Rev. **C 8** (1973) 929

[Vi 77] *Videback, F., R.B. Goldstein, L. Grodzins, S.G. Steadman, T.A. Belote* and *J.D. Garrett*: Phys. Rev. **C 15** (1977) 954

[Vo 74] *Volta, L.G., P.G. Roos, N.S. Chant* and *R. Woody*: Phys. Rev. **C 10** (1974) 520

[Vo 76] *Volkov, V.V.*: Sov. J. Nucl. Phys. **6** (1976) 240

[Wa 57] *Watson, K.M.*: Phys. Rev. **105** (1957) 388

[Wa 58] *Watson, K.M.*: Rev. Mod. Phys. **30** (1958) 565

[Wa 58a] *Watanabe, S.*: Nucl. Phys. **8** (1958) 484

[Wa 71a] *Watson, B.A., C.C. Chang* and *S.L. Tabor*: Particles and Nuclei **2** (1971) 376

[Wa 71] *Watson, J.W., H.G. Pugh, P.G. Roos, D.A. Goldberg, R.A.J. Riddle* and *D.I. Bonbright*: Nucl. Phys. **A 173** (1971) 513

[Wa 72] *Watson, J.W.*: Nucl. Phys. **A 198** (1972) 129

[Wa 78] *Walter, H.K.*: in Clustering Aspects of Nuclear Structure and Nuclear Reactions (eds W.H.T. van Oers et al) AIP Conf. Proceedings No. **47** (1978) 444

[Wa 80] *Wang, C.W., N.S. Chant, P.G. Roos, A. Nadesen* and *T.A. Carey*: Phys. Rev. **C 21** (1980) 1705

[Wa 80a] *Watt, A., D. Kelvin* and *R.R. Whitehead*: T. Phys. G: Nucl. Phys.. **6** (1980) 31

[We 64] *Weinberg, S.*: Phys. Rev. **133** (1964) B 232

[Wh 71] *Whitehead, R.R.* and *A. Watt*: Phys. Lett. **35 B** (1971) 189

[Wh 72] *Whitehead, R.R.*: Nucl. Phys. **A 182** (1972) 290

[Wi 47] *Wigner, P.P.* and *L. Eisenbud*: Phys. Rev. **72** (1947) 29

[Wi 58] *Wildermuth, K.* and *Th. Kanellopoulos*: Nucl. Phys. **7** (1958) 150

[Wi 62] *Wildermuth, K.*: Nucl. Phys. **31** (1962) 478

[Wi 66] *Wildermuth, K.* and *W. McClure*: Cluster Representation of Nuclei (Springer-Verlag, Berlin) (1966)

[Wi 77] *Wildermuth, K.* and *Y.C. Tang*: A. Unified Theory of the Nucleus (Vieweg, Braunschweig) (1977)

[Wi 79] *Wildermuth, K., F. Fernandez, E.J. Kanellopoulos* and *W. Sünkel*: J. Phys. G: Nucl. Phys. **6** (1980) 603

[Wi 79a] *Wilczynski, J., R. Kamermans, J. van Popta, R.H. Siemssen, K. Siwek-Wilcynska* and *S.Y. van der Werf*: Phys. Lett. **88 B** (1979) 65

[Wo 76] *Wozniak, G.J., D.P. Stahel, J. Cerny* and *N.A. Jelley*: Phys. Rev. **C 14** (1976) 815

[Wu 78] *Wu, J.R., C.C. Chang* and *H.D. Holmgren*: Phys. Rev. Lett. **40** (1978) 1013

[Wu 79] *Wu, J.R., C.C. Chang, H.D. Holmgren* and *R.W. Koontz*: Phys. Rev. **20** (1979) 1284

[Ya 79] *Yamada, H., D.R. Zolnowski, S.E. Cala, A.C. Kahler, J. Pierce* and *T.T. Sugihara*: Phys. Rev. Lett. **43** (1978) 605

[Yo 62] *Yoshida, S.*: Nucl. Phys. **33** (1962) 685

[Yo 73] *Yoshida, H.*: Phys. Lett. **47 B** (1973) 411

[Yo 76] *Yoshida, H.*: Nucl. Phys. **A 257** (1976) 348

[Zo 78] *Zolnowski, D.R., H. Yamada, S.E. Cala, A.C. Kahler* and *T.T. Sugihara*: Phys. Rev. Lett. **41** (1978) 92

[Zo 79] *Zolnowski, D.R., S.E. Cala, A.C. Kahler, H. Yamada, J.M. Pierce* and *T.T. Sugihara*: Progress Report 1978/79, Cyclotron Laboratory, Texas A & M University (1979) 33

Index

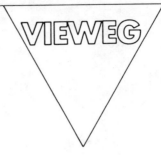

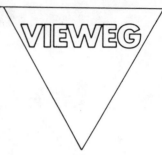